PROPERTIES
OF
MATERIALS

P. F. Kelly

CRC Press
Taylor & Francis Group
Boca Raton London New York

CRC Press is an imprint of the
Taylor & Francis Group, an **informa** business

CRC Press
Taylor & Francis Group
6000 Broken Sound Parkway NW, Suite 300
Boca Raton, FL 33487-2742

First issued in paperback 2020

© 2015 by Taylor & Francis Group, LLC
CRC Press is an imprint of Taylor & Francis Group, an Informa business

No claim to original U.S. Government works

ISBN-13: 978-1-4822-0622-7 (hbk)
ISBN-13: 978-0-367-78368-6 (pbk)

Visit the Taylor & Francis Web site at
http://www.taylorandfrancis.com

and the CRC Press Web site at
http://www.crcpress.com

Contents

List of Examples

Preface

My original motivation for writing this series of books, *Elements of Mechanics*, *Properties of Materials*, and *Electricity and Magnetism*, was to provide students attending my classes with advance copies of notes for each lecture. Successive accretions of explanatory material [much of it generated in response to student questions] transformed this project into something more substantial, which could be read as

- A sole source for a sequence of introductory physics classes

- A student supplement to a standard textbook

- A review in preparation for graduate/professional/comprehensive examinations

The provenance of these volumes as notes for lecture is manifest throughout in style and content. Physics is, in this telling, an engaging endeavour rather than a spectator sport. Also, attempts are made to ensure that the reasons for studying physics:

LEARN	to acquire knowledge about nature,
CONTROL	to exercise dominion over nature,
CREATE	to experience the joys of invention and artistry,

are not lost entirely under the weight of detailed investigation of specific models for particular phenomena. Finally, while occasional mention is made of *au courant* topics, our primary concern is with the foundations of classical physics. Readers who wish to more fully appreciate the "modern" developments in physics are encouraged to work through these notes first.

Very early on, while each book in the trilogy was scarcely more than an inkling, they were dubbed $\mathcal{MAPS}$, $\mathcal{SPAM}$, and $\mathcal{AMPS}$. These names derived fuller significance by interpretation as acronyms obtained from permutations of the four Latin words

Scientia	*Physica*	*Ars*	*Mechanica*.

This artifice is a nod toward the deeper mathematical structures hinted at in these notes. The words themselves have import too.

$\mathcal{S\,P}$ The *Science of Physics* is the ordering of knowledge[1] of the physical world. This ordering culminates in the discernment of fundamental symmetries and the formulation of conservation laws and theorems which characterise [some would dare to say "govern"] the behaviour of physical systems.

---

[1]This aspect of the courses hews closely to traditional notions of liberal education.

xiv

$\mathcal{A M}$ The *Art of Mechanics* conveys the notion that the practice of physics is essentially creative. In another riff on this pair of words, the *Artes Mechanicae*, as traditionally understood, are practical skills[2] such as those employed by artisans in the production of useful goods and decorative artwork. Throughout these notes, we shall craft many mathematical models providing serviceable approximations to [pertinent aspects of] physical systems.

The three volumes in the series are partially sequenced: $\mathcal{MAPS}$ is propaedeutic, while $\mathcal{SPAM}$ and $\mathcal{AMPS}$ may be subsequently read in either order. There is also a gradual shift in writing style from a more discursive tone [with text and mathematical syntactic redundancy] in $\mathcal{MAPS}$ to a somewhat terser one in $\mathcal{SPAM}$ and $\mathcal{AMPS}$. An aim is to be welcoming to those new to the study of physics. Our intent is that upon completing this series students be prepared for specialised-subject upper-level undergraduate physics courses and textbooks.

A burden of thanks is owed to all of the students whom I have taught, and especially to those who greatly enjoyed it.

[This came as a surprise to some!]

Special credit is owed to those who have assisted me in the preparation of figures, especially Andy Geyer, and Carl and Jaspar von Buelow.

---

[2]In the past, *Artes Mechanicae* suffered invidious comparison with *Artes Liberales* or "Liberal Arts."

Preface for Volume II

Although the acronym for this volume, SPAM, is the string reversal of MAPS, it is certainly not our intention to reverse our progress and un-learn the subjects covered there. Rather, we shall have a close-eyed[3] look at [some of the] properties of matter as found in the real[4] world.

Our perspective is somewhat inverted, insofar as we go from being principles driven in MAPS to being empirically inspired in SPAM.

> ASIDE: To say that this sounds like Plato *vs.* Aristotle is to posit a false dichotomy. In MAPS we were [ultimately] attempting to reproduce (to some specified degree of accuracy and precision) aspects of particular physical systems. In SPAM, model descriptions of phenomena are developed in light of, and justified by appeal to, physical principles.

All of the material objects that we encountered in MAPS were taken to be solid and perfectly rigid. These bodies did not deform in response to external forces. Applying a force to any part was tantamount to applying it to the whole. These assumptions allowed us to make manifest the Newtonian laws of motion, conservation laws, and physical principles.

An impossible-to-paper-over difficulty is that applied forces cause real objects to dimple and dent and squish and stretch. Objects STRAIN under the application of [external] STRESS. We shall develop linear models of elasticity to correlate and quantify the changes in an object's shape which occur in response to the application of a constant force.

Giving up rigidity is the first step toward abandoning solidity, too. We shall study ideal fluids—liquid and gas—both when quiescent and when flowing. The empirical gas laws shall play a prominent rôle in our later presentation of thermodynamics. In developments we touch upon in this introductory volume, it turns out that these gas laws may be derived from the simple particle mechanics [*à la* MAPS] of vast numbers of [barely interacting] point-like molecules!

Nature provides a great many examples of physical systems experiencing [quasi-]oscillation, or [nearly] repetitive behaviour. The simplest model consists of a block moving on a horizontal frictionless plane under the influence of an ideal Hookian spring. A richer version incorporates the effects of a linear drag force. A specialisation of this augmented model describes the incremental, but inevitable, slowing of a[n ideal] pendulum. Another, less obvious, specialisation describes the settling of a pickup truck when a load is placed in its box, or one steps on the bumper to clamber aboard. We will also study the behaviour of oscillating systems subjected to time-dependent external applied forces.

An object's dimpling or denting or squishing or stretching is seldom static and isolated to one point or small region. Rather, effects "ripple" out to other, possibly distant, parts of the body as waves. We shall consider two extreme cases: isolated pulses and infinitely extended repetitive structures. LINEAR SUPERPOSITION is invoked to analyse the combined effects of two or more waves simultaneously present in the same material medium, including standing waves, beating, interference, and diffraction, to name but a few.

The waves which we most directly perceive are sound and light.

> ASIDE: Ironically, while these are sensed by our ears and eyes, that they are waves is not so directly apparent.

---

[3] *Pace* our snarky or extremely tired friends, we shall not be closed-eyed.

[4] As was suggested in the Preface for VOLUME I, reality is a heavily freighted concept!

We shall briefly consider acoustics and optics, with some mention of human factors. The wave model provides a [local] means by which the phenomenological rules of classical geometric optics may be derived. In an interesting twist, the same empirical rules can also be derived via the [non-local, variational] MINIMAL TIME PRINCIPLE.[5]

A brief introduction to thermodynamics is taken up in the final parts of this volume. After motivating discussions of temperature, heat, and thermometry, the Laws of Thermodynamics are introduced and applied to ideal gas systems. Thermodynamic cycles will be shown to have oscillatory aspects. Some of the deeper and more formal aspects of thermodynamic systems are mentioned in the last few chapters.

Threaded throughout our investigations of particular phenomena is the question of how one might go about modelling the behaviour of composites assembled from simple "pure" constituents. By crafting combinatorics for composites we obviate the need to measure phenomenological parameters[6] in all prospective cases. Knowing the pure parameters is sufficient to infer the composite's effective values. When there are only two constituents, they are [almost] constrained to combine in one of two possible manners. The evocative terms that we use to describe these are series and parallel.

SERIES If the constituents in the composite are affected one-after-the-other (in a generalised sense), they are deemed to be connected in series.

PARALLEL If the elements are affected all-at-once, they are arranged in parallel.

With three or more constituents, MIXED configurations are possible.

MIXED The elements forming the composite may possess a non-trivial [global, topological] structure. We shall treat such cases[7] by iterated analysis of their SERIES and PARALLEL sub-structures.

EXAMPLE [*Bertal's Mixed (series–parallel) Trajectory through Introductory Physics*]

All of Bertal's physics confrères began their studies with MAPS. Most proceeded to SPAM the following semester, and then to AMPS. [The others chose to do AMPS first, then SPAM.] Not Bertal. He took both SPAM and AMPS in the semester after having had MAPS.

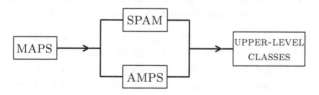

FIGURE 1 Bertal Took SPAM and AMPS Together in Parallel, in Series with MAPS

---

[5] A signal propagates from one point to another via the path which takes the MINIMAL amount of time. Extensions of this idea have proven to be enormously fruitful in modern physics.

[6] These cannot, in practice or in principle, be predicted or derived using extant models, and must perforce be measured.

[7] Perhaps not the most general cases, but certainly many of those which are classified as "planar."

Part I

Materials

Chapter 1

'Tidings' of Rigidity's Breakdown

The goal of this chapter is to demonstrate that the assumption of **rigidity**, implicit throughout the entirety of VOLUME I, cannot be maintained in all circumstances.

Q: Couldn't we just say, "Let's relax the assumption of rigidity?"

A: Yes. Instead we shall demonstrate the necessity of its relaxation.

Let's recollect a few features of Newtonian gravity, further generalising the analysis to extended bodies, before giving over the remainder of the chapter to a discussion of terrestrial tides produced by the Moon and the Sun.

Newton's Law of Universal Gravitation, NUG,[1] posits that the interaction force between two point masses [*a.k.a.* gravitational charges], M_1 and M_2, separated by distance r_{12}, is directed along the line joining the points and has magnitude dictated by the product of a universal coupling constant, the masses, and the inverse-square of the separation distance. According to this Law, the force exerted **by** the point source, 1, **on** the point object, 2, is

$$\vec{F}_{\text{G},12} = -\frac{G\,M_1\,M_2}{r_{12}^2}\,\hat{r}_{12}\,.$$

Action-at-a-distance concerns militate for the introduction of gravitational fields permeating all of space. The field at a particular location is defined by determining the gravitational force on a [point-like] "test" mass situated there, dividing by the magnitude of the test mass, m_0, and taking the limit in which m_0 approaches zero:

$$\vec{g}_1 = \lim_{m_0 \to 0} \frac{\vec{F}_{\text{G},10}}{m_0}\,.$$

The gravitational field produced by a point source with mass M_1 must then admit the following explicit form:

$$\vec{g}_1(\vec{r}_{10}) = \lim_{m_0 \to 0} \frac{-\frac{G\,M_1\,m_0}{r_{10}^2}}{m_0}\,\hat{r}_{10} = \lim_{m_0 \to 0} -\frac{G\,M_1}{r_{10}^2}\,\hat{r}_{10} = -\frac{G\,M_1}{r_{10}^2}\,\hat{r}_{10}\,.$$

The hard fact is that not all gravitating sources and/or objects can be approximated as point particles. Four possible combinations present themselves:

$$\Big(\text{point, extended}\Big) \otimes \Big(\text{source, object}\Big) = \begin{cases} \text{point source} & \text{point object} \\ \text{extended source} & \text{with} \quad \text{point object} \\ \text{point source} & \text{extended object} \\ \text{extended source} & \text{extended object} \end{cases}\,.$$

The point–point case is the realm of NUG. The case of extended sources and point objects was considered in Chapter 48 in VOLUME I, by employing the PARTITION, COMPUTE, SUM [and REFINE] strategy. This work uncovered a restricted version of Gauss's Law stating that, external to a spherically symmetric extended source, the gravitational field is identical to that of a point source.

---

[1] For details, see Lecture 47 *et seq.* in VOLUME I.

Consider the effects of a point-like gravitational field acting on an extended object. For concreteness, we investigate a very primitive model in which the extended object is comprised of four nearby point particles. In what follows, we designate these constituents as "feet," "head," "left hand," and "right hand" respectively. The present situation is illustrated in Figure 1.1.

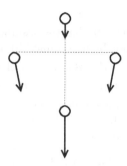

FIGURE 1.1 An Extended Object Comprised of Four Point Particles Responding to the Gravitational Field of a Point-Like Source

As the four constituent particles do not all reside at precisely the same point in space, the local accelerations due to gravity that they experience differ.

$$\text{The} \begin{bmatrix} \text{feet} \\ \text{and} \\ \text{head} \end{bmatrix} \text{have accelerations with} \begin{Bmatrix} \text{common direction} \\ \text{but} \\ \text{different magnitude} \end{Bmatrix}.$$

$$\text{The accelerations of the} \begin{bmatrix} \text{left} & \text{and} & \text{right} \end{bmatrix} \text{hands have} \begin{Bmatrix} \text{common magnitude} \\ \text{but} \\ \text{different direction} \end{Bmatrix}.$$

The geometry of the four-particle system will become distorted as the system evolves in response to the variable local gravitational field, as suggested by the local acceleration vectors in Figure 1.1. This is a manifestation of **gravitational tidal effects**.

- Tides arise from the gradient of the gravitational field, *i.e.,* differences between neighbouring points, and not the strength, *per se*, of the field itself.

  Hence, it is always possible to imagine situations in which tidal effects may be large [on any scale], and as a consequence any object that one might consider, however rigid, cannot always avoid tidal deformation.

- Tides may be extremely strong in the vicinity of black holes. It is an ironical fact that one may remain blissfully unaware whilst falling through the event horizon[2] of a distant super-massive black hole, and yet be severely "bent out of shape" by a nearby light-weight mini black hole even while evading its clutches.

- The familiar ocean tides, encountered at the seaside, are produced by the superposition of Moon–Earth and Sun–Earth tidal interactions.

_____
[2]From within the event horizon of a black hole, not even light can escape.

TIDES ON THE EARTH DUE TO THE MOON AND SUN

FIGURE 1.2   Tides on the Earth Arising from the Influence of the Moon

The greatest contributor to the Earth's ocean tides is the gravitational field of the Moon. It may seem curious that high tide bulges appear on both the side of the Earth facing the Moon and that opposite the Moon.

[This is consistent with the observed pattern of tides flowing in and out twice daily.]

The reason for the high tide on the far side of the Earth is that the gradient of the lunar field causes the Earth's core/mantle/crust to accelerate away from its surface/ocean. This slight falling away is manifest, to us, as an increase in the relative height of the land[3]/sea.

The Sun is the third major actor on the tidal stage. As the Earth–Moon system revolves around the Sun, it rotates, thus allowing for a variety of geometries giving rise to [maximally reinforcing] SPRING tides, [partially cancelling] NEAP tides, and everything in between.

FIGURE 1.3   Spring Tides in the Sun–Earth–Moon System (not to scale)

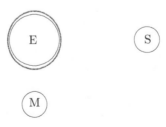

FIGURE 1.4   Neap Tides in the Sun–Earth–Moon System (also not to scale)

ASIDE: There are galactic tidal effects too, but these are incredibly small.

---

[3]Land tides are rather small owing to the greater rigidity of rock, soil, and ice *vis-à-vis* liquid water.

Two additional bits close out this chapter.

- Tidal friction is dissipative. Ocean tides crash onto shorelines as the Earth rotates under the bulging waters. Concomitant production of heat is associated with bending and flexing of the crust.

    [A similar effect occurs in an underinflated tire on a moving automobile.]

    The most spectacular nearby instance of this is the vulcanism observed on Io, one of the Galilean moons of Jupiter.

- The energy dissipated by tidal friction is usually drawn from the rotational kinetic energy[4] of the system. Drawing down the rotational kinetic energy is accompanied by a diminution of the angular frequency [assuming that the moment of inertia remains effectively constant], and thus a lengthening of the period of rotation.

    ASIDE: Our days on Earth are getting longer! Every so often, the world's high-precision laboratories will, in concert, insert a *leap second* into their synchronised chronometers.

Eventually, after the passage of eons, the rotational period of the Earth will match the orbital period of the Moon. That is, the length of each day will match the duration of a month, and thenceforth the Moon shall only be visible from one-half of the Earth's surface. This phenomenon is called TIDAL LOCKING.

- – The Moon is already tidally locked with respect to the Earth. Only one side/face of the Moon is visible[5] to us on Earth because the lunar orbital and rotational periods are equal.

- – When the Earth becomes tidally locked with the Moon, real estate agents will tout the property value enhancement afforded by a Moon view.

---

[4] Another possible source is the gravitational potential energy of parts of the system.

[5] Not until the 1960s, when spacecraft orbited the Moon and sent back photographic images of the far side, were notions of "little green lunar denizens" finally debunked.

Chapter 2

Elastic Properties of Solids

Until the previous chapter, it was implicitly assumed that extended objects were rigid. These objects underwent no changes in shape, *i.e.*, their internal, or relative, geometry remained constant. In fact, however, whenever a force is applied to a solid object, a STRESS is produced within the object. The physical effect/manifestation of this stress is that the object experiences some form and amount of deformation, a STRAIN.

ELASTIC MODULUS   An elastic modulus is the ratio of a particular type of physical stress and its resultant strain, *i.e.*,

$$\text{Elastic Modulus} \; = \; \frac{\text{stress}}{\text{strain}}.$$

Bodies with greater rigidity have correspondingly larger values of elastic moduli. "True," or "complete," rigidity arises in the limit in which the elastic moduli all tend toward infinity.   [Typical values of elastic moduli for metallic solids are on the order of 10^{10}.]

That there are three elastic moduli derives from the spatial tri-dimensionality of the world.

Young's Modulus	resistance to changes in length
Shear Modulus	resistance to transverse shift
Bulk Modulus	resistance to compression/expansion

Each of these moduli is examined, in turn, below.

YOUNG'S MODULUS

Let's consider a uniform bar fabricated of some specific homogeneous material. **Uniformity** strongly constrains the shape [geometry] of the bar to have constant cross-sectional area A and length L_0. **Homogeneity** ensures that the material properties of the bar are constant throughout its extent.

[By virtue of its identification as a "bar," the unnecessary assumption $L_0 \gg \sqrt{A}$ is implicit.]

FIGURE 2.1   Young's Modulus

One end of the bar is firmly affixed to an immobile anchor. IF an applied force, of magnitude F_A and direction parallel to the bar, acts on the free end, THEN the bar will s-t-r-e-t-c-h.

TENSILE STRESS   The tensile stress endured by the bar is

$$\text{tensile stress} = \frac{\text{applied force}}{\text{cross-sectional area}} = \frac{F_A}{A}.$$

The units of tensile stress are newtons per square metre, $[\text{N}/\text{m}^2]$. This expression for the stress has the correct *intuitive*[1] form, being directly proportional to the applied force and inversely proportional to the cross-sectional area.

TENSILE STRAIN   The tensile strain characterises the response of the bar to the applied stress:

$$\text{tensile strain} = \frac{\text{change in length}}{\text{original length}} = \frac{\Delta L}{L_0} = \frac{L_A - L_0}{L_0}.$$

In the final equality, L_A is the equilibrium length of the stretched bar, once the end of the bar has come to rest.

[The applied force might act for some considerable time before the bar attains rest.]

Tensile strain is a ratio of lengths, and thus it is a dimensionless quantity. The relative[2] amount of stretching is what matters.

YOUNG'S MODULUS   Young's modulus is the ratio of tensile stress to tensile strain, *i.e.*,

$$Y = \frac{\text{tensile stress}}{\text{tensile strain}} = \frac{F_A/A}{\Delta L/L_0}.$$

ASIDE: In consideration of Young's modulus, one deliberately neglects the fact that the cross-sectional area of the bar must change [ever so slightly] as the bar is stretched or compressed.

Gedanken Experiment: Elastic Regimes

If one were to take a uniform and homogeneous bar, affix one end to an anchor, and pull on the other with a steadily increasing applied force, then a graph of applied stress *vs.*[3] resultant strain would evince the three regimes[4] portrayed in Figure 2.2.

Elastic   The curve intersects the axes at $(0, 0)$, since an absence of strain occurs only in the absence of stress. For small amounts of stress, the stress/strain relationship is linear, with slope equal to Young's modulus for the material.

Throughout this linear regime it is the case that, IF the applied force is reduced to zero, THEN the bar returns to its original length. Generalising, one realises that the strain at a given stress is unique and independent of the manner in which the stress has previously varied.

----

[1] It is quite natural that stress is proportional to force. Also, it is consistent with one's experiences with rods and ropes that the force may be deemed to be distributed over the cross-sectional face and hence the stress should vary inversely with A. If this latter argument seems a bit of a stretch, then think about how breaks in ropes and rods tend to occur at thin spots or at cracks.

[2] A longer rope or rod has more *give* than a shorter one (of the same material and thickness).

[3] Although this seems *the wrong way around*, it is conventional.

[4] Some authors would say four regimes, as will be explained in an upcoming ASIDE.

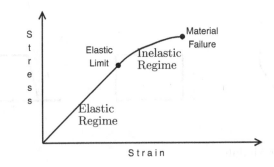

FIGURE 2.2 Stress *vs.* Strain

[The elastic regime is reminiscent of Hooke's Law, *eh?*]

Inelastic   Eventually, with increasing stress, the linear relationship breaks down at what is termed the **elastic limit**. If the system is stressed beyond its elastic limit, the material undergoes *plastic deformation*.[5] Once the material has entered into this regime, it does not return to its original length even when the applied force is reduced to zero.

> ASIDE: The boundary between the elastic and inelastic regimes is not really a critical point as portrayed above. Just beyond the onset of non-linearity [at the "proportional limit"], the system is able to avoid irreversible plastic deformation and return to its original length. When the elastic limit is broached, the shape [or composition] of the substance endures irreversible change. Our simple model ignores this narrow [fourth] regime by identifying the end of linearity with the elastic limit.

Rupture   At some value of stress, above and beyond the elastic limit, the bar will fracture into two or more pieces. This *catastrophic failure* is quite irreversible.

SHEAR MODULUS

A uniform and homogeneous rectangular parallelepiped [henceforth often abbreviated "∥-piped"] has face area A and height h. One face of the ∥-piped is firmly affixed to a plane surface. If an applied force, of magnitude F_A and direction parallel to the plane surface, acts on the open face, then the ∥-piped will **shear**, *i.e.*, the open face will shift, becoming offset from the fixed face, while a pair of opposite rectangular sides will become parallelograms.

SHEAR STRESS   The shear stress experienced by the block of material is

$$\text{shear stress} = \frac{\text{applied force}}{\text{face area}} = \frac{F_A}{A}.$$

The units of shear stress are N/m^2. This expression for the stress has the correct intuitive[6] form, being directly and inversely proportional to the applied force and the face area, respectively.

---

[5]The geometry/arrangement of substructures in the material is permanently changed. This may take the form of relative motion of crystal grains, or the rearrangement of [H-] bonds in extended hydrocarbons, or the tearing of fibres, or what have you.

[6]The scaling with transverse force is eminently sensible. The inverse scaling with face area is consistent with the notion that to bolster an object against shear one places another object alongside, effectively making the original object wider or longer.

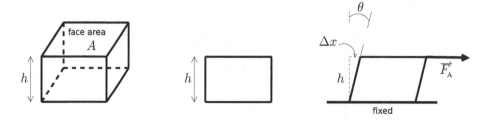

FIGURE 2.3   Shear Modulus

SHEAR STRAIN   The shear strain characterises the parallelepiped's response to the applied stress:

$$\text{shear strain} = \frac{\text{transverse shift}}{\text{height}} = \frac{\Delta x}{h} = \tan(\theta).$$

Shear strain is dimensionless. In the final equality, the strain has been re-expressed in terms of the **shear angle**, θ. In many circumstances it is experimentally advantageous to measure θ. Physical intuition[7] suggests that the relative amount by which the top face shears is more significant than the absolute amount, corroborating the definition.

SHEAR MODULUS   The shear modulus is equal to the shear stress divided by the shear strain, *i.e.*,

$$M_S = \frac{\text{shear stress}}{\text{shear strain}} = \frac{F_A/A}{\Delta x/h} = \frac{F_A/A}{\tan(\theta)}.$$

> ASIDE: That the height of the block must change [ever so slightly] in conjunction with the shearing of the opposite faces is not incorporated into the shear modulus.

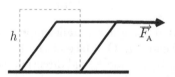

FIGURE 2.4   Exaggerated Shear with Concomitant Diminution of Height

Shear also exhibits the three [four] regimes indicated in Figure 2.2.

Bulk Modulus

A homogeneous and uniform cube has total face area A and volume V_0. If a collection of forces, each of magnitude F_A, act inward on all six of the plane surfaces comprising the boundary of the cube, then the cube will experience some amount of compression.

---

[7] A tall block has more *sway* than a shorter one (comprised of the same material and having the same face area). CAVEAT: If the faces do not remain parallel, Young's Modulus effects may dominate.

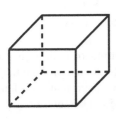
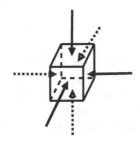

FIGURE 2.5   Bulk Modulus

BULK STRESS   The bulk stress suffered by the cube is

$$\text{bulk stress} = \frac{\text{applied force on each face}}{\text{face area}} = \frac{F_{\text{A}}}{A}.$$

The units of bulk stress are N/m^2. The stress is directly and inversely proportional to the applied force and the face area, respectively, as might be expected.[8]

BULK STRAIN   The bulk strain characterises the response of the cube to the applied stress:

$$\text{bulk strain} = \frac{\text{change in volume}}{\text{original volume}} = \frac{\Delta V}{V_0}.$$

The change in volume under compression is [almost[9]] always negative, and thus the bulk strain is [typically] negative in this case, as well as dimensionless, as usual.[10]

BULK MODULUS   The bulk modulus is equal to the ratio of the bulk stress to the magnitude of the bulk strain:

$$B = -\frac{\text{bulk stress}}{\text{bulk strain}} = \frac{F_A/A}{\left| \Delta V/V_0 \right|}.$$

The bulk modulus is a positive quantity [in almost all circumstances].

ASIDE: That the face area of the volume of substance must decrease [ever so slightly] as it is compressed is neglected in this model.

The regime behaviour of the bulk modulus is similar to that shown in Figure 2.2. However, instead of "material failure" in the sense of fracture, the substance usually undergoes a change of **state** or a change of **phase** [*e.g.*, gas → liquid or liquid → solid].

[Thermodynamical aspects of phase changes will be discussed in Chapter 38.]

Another sort of phase change is a configurational change affecting the chemical nature and properties of the substance [*e.g.*, carbon → diamond].

---

[8]One may prevent the implosion of an object by reducing the compressive force or distributing the force over a larger area.

[9]There are a few exotic materials making this counter-intuitive qualifier necessary.

[10]A larger cube *squishes* by a greater amount than does a small one subjected to the same compression.

The three elastic moduli represent linear, planar, and volume effects.

Two essential CAVEATS about elastic moduli are listed below.

Linear   The introduction of elastic moduli amounts to a linearisation, or tangent-line approximation, to more complicated non-linear behaviour. [A more fully explicated instance of this occurs in the designation of electrical resistivity for a material, found in Chapter 19 of VOLUME III.] Predictions based on elastic considerations become progressively less accurate as the magnitude of the effect grows large.

Multi-Parameter   The moduli have multi-parametric dependence on the manner and circumstances in which stresses are applied.

For example, implicit in the present discussion of elastic moduli is the notion that there was no flow of heat into or out of the system while[11] the stress was applied. These would then be more properly described as the **adiabatic** elastic moduli.

Conversely, the stress may be applied slowly enough to a body in contact with a large thermal reservoir that the system's temperature remains constant. The elastic modulus in this case is dubbed **isothermal**.

We shall leave these considerations for now, and simply refer to elastic moduli as though they were unique. The distinction between adiabatic and isothermal will be sharply delineated when we discuss the Laws of Thermodynamics in Chapter 42 *et seq.*

---

[11]Perhaps the time scale on which the force(s) act is short compared to the time scale(s) on which heat flows.

Chapter 3

Elastic Solids in Series and Parallel

For grins, let's compute effective values of the elastic moduli for particular **series** and **parallel** combinations of blocks or bars of homogeneous[1] material.

EXAMPLE [*Series Composition and Young's Modulus*]

Two uniform bars with common cross-sectional area A are firmly[2] attached together, and one end of the composite is affixed to an anchor point as shown in Figure 3.1. The two bars are made of different materials, with Young's moduli Y_1 and Y_2, and possess lengths L_1 and L_2, respectively. The length of the composite bar[3] is

$$L_{(12)} = L_1 + L_2 \,,$$

where we have neglected [or subtracted away] any increase in length ascribable to the means of fastening the constituent bars.

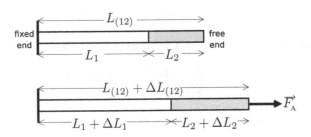

FIGURE 3.1   Composite Young's Modulus: Series Case

When a constant force, F_A, is applied to the free end of the composite bar, it stretches by an amount[4]

$$\Delta L_{(12)} = \Delta(L_1 + L_2) = \Delta L_1 + \Delta L_2 \,.$$

The tension force exerted by the clamp or glue on each of the homogeneous bars [where they are joined] is equal in magnitude to the applied force [when the bar is in equilibrium].

---

[1]**Q:** And what does one do if one's building blocks are not homogeneous? **A:** Step 1: Partition the block into subunits which are [approximately] homogeneous and are arranged in a manner which can be recast in series and parallel terms. Step 2: Compute an effective value of the modulus for the block employing the methods developed in this chapter and in Chapter 14.

[2]Two assumptions are made about the manner in which the bars are attached to each other and to fixed anchors. One is that the clamp or glue employed is perfectly rigid, *i.e.*, does not stretch or bend at all under the influence of an external force. The other assumption is that the thickness of the attaching mechanism or substance is either self-consistently subtracted away or neglected in the analysis.

[3]Quantities pertinent to the series combination of bars 1 and 2 are labelled by the subscript "(12)."

[4]It may require some elapse of time for the endpoints of the bar to come to relative rest.

ASIDE: This follows from N2, once the bar has stabilised and remains at rest.

Since each constituent suffers the same stress, the Young's moduli are:

$$Y_1 = \frac{F_A/A}{\Delta L_1/L_1}, \qquad Y_2 = \frac{F_A/A}{\Delta L_2/L_2}, \qquad \text{and} \qquad Y_{(12)} = \frac{F_A/A}{\Delta L_{(12)}/L_{(12)}}.$$

The respective strains are:

$$\frac{\Delta L_1}{L_1} = \frac{F_A/A}{Y_1}, \qquad \frac{\Delta L_2}{L_2} = \frac{F_A/A}{Y_2}, \qquad \text{and} \qquad \frac{\Delta L_{(12)}}{L_{(12)}} = \frac{F_A/A}{Y_{(12)}}.$$

Folding these results into $\Delta L_{(12)} = \Delta L_1 + \Delta L_2$ yields

$$\frac{F_A}{A} \frac{L_{(12)}}{Y_{(12)}} = \frac{\Delta L_{(12)}}{L_{(12)}} L_{(12)} = \Delta L_{(12)} = \Delta L_1 + \Delta L_2 = \frac{\Delta L_1}{L_1} L_1 + \frac{\Delta L_2}{L_2} L_2 = \frac{F_A}{A} \left[\frac{L_1}{Y_1} + \frac{L_2}{Y_2} \right].$$

Thus,

$$\frac{L_{(12)}}{Y_{(12)}} = \frac{L_1}{Y_1} + \frac{L_2}{Y_2}$$

expresses the Young's modulus for a series combination of two homogeneous materials.

Suppose that the entire bar is truly homogeneous with Young's modulus $\widetilde{Y}$, and hence $Y_1 = \widetilde{Y} = Y_2$. In this case, the partition into two constituents is formal and arbitrary.

LOGIC   The *faux*-composite bar must have $Y_{(12)} = \widetilde{Y}$.

SERIES
EXPR'N   Application of the compact formula derived above yields

$$\frac{L_{(12)}}{Y_{(12)}} = \frac{L_1}{Y_1} + \frac{L_2}{Y_2} = \frac{L_1 + L_2}{\widetilde{Y}} = \frac{L_{(12)}}{\widetilde{Y}},$$

confirming that the model produces the expected behaviour.

EXAMPLE [*Parallel Composition and Young's Modulus*]

Two uniform bars with common length L have Young's moduli Y_1 and Y_2 and cross-sectional areas A_1 and A_2 respectively. We shall restrict ourselves to symmetric circumstances not unlike those illustrated in Figure 3.2.

[Such symmetry avoids twisting, bending, or shear, which might complicate the analysis.]

If one end of this composite[5] bar is clamped, and a constant force, F_A, is applied at the other end, then its length increases by an amount ΔL.

The force applied to the bar is assumed to be distributed over its entire face area in such a way that each constituent bar is stretched by the same amount. Hence,

$$F_A = F_1 + F_2 \qquad \text{and} \qquad A_{[12]} = A_1 + A_2.$$

---

[5]Quantities pertinent to the parallel combination of bars 1 and 2 are labelled by the subscript "[12]."

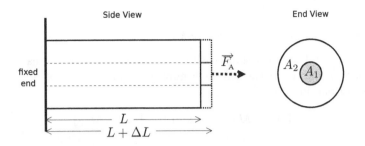

FIGURE 3.2 Composite Young's Modulus: Parallel Case

As each constituent endures the same strain, the Young's moduli are

$$Y_1 = \frac{F_1/A_1}{\Delta L/L}, \qquad Y_2 = \frac{F_2/A_2}{\Delta L/L}, \qquad \text{and} \qquad Y_{[12]} = \frac{F_A/A_{[12]}}{\Delta L/L}.$$

Thus, the respective forces must be

$$F_1 = A_1 Y_1 \frac{L}{\Delta L}, \qquad F_2 = A_2 Y_2 \frac{L}{\Delta L}, \qquad \text{and} \qquad F_A = A_{[12]} Y_{[12]} \frac{L}{\Delta L}.$$

Inserting these relations into the total force/area ratio, $F_A/A_{[12]} = (F_1 + F_2)/(A_1 + A_2)$, yields

$$Y_{[12]} \frac{\Delta L}{L} = \frac{F_A}{A_{[12]}} = \frac{F_1 + F_2}{A_1 + A_2} = \frac{[A_1 Y_1 + A_2 Y_2]}{A_1 + A_2} \frac{\Delta L}{L}.$$

Hence, the Young's modulus of a parallel combination of two homogeneous bars is the **area-weighted average of the Young's moduli of the constituents**,

$$Y_{[12]} = \frac{A_1 Y_1 + A_2 Y_2}{A_1 + A_2}.$$

Suppose that the entire bar is homogeneous, with Young's modulus $\widetilde{Y}$, and the [formal] partition becomes arbitrary. Once again, then, $Y_1 = \widetilde{Y} = Y_2$.

LOGIC   The *faux*-composite bar must have $Y_{[12]} = \widetilde{Y}$.

PARALLEL   Computation of the area-weighted average Young's modulus,
EXPR'N

$$Y_{[12]} = \frac{A_1 Y_1 + A_2 Y_2}{A_1 + A_2} = \frac{A_1 \widetilde{Y} + A_2 \widetilde{Y}}{A_1 + A_2} = \widetilde{Y},$$

gives rise to the expected result.

There are other equivalent ways of expressing the effective Young's modulus for series and parallel combinations of materials. One's taste dictates one's preferred representation. The important aspects to be gleaned from these two examples are that:

- meaningful effective values of Young's modulus for series and parallel composites of materials exist, and

- these effective values are computable in terms of the Young's moduli and geometric parameters of the pure materials.

EXAMPLE [*Series Composition and Shear Modulus*]

Two homogeneous and uniform slabs with common cross-sectional area A are fused[6] into a vertical stack as shown on the left in Figure 3.3. The two slabs are characterised by respective sets of parameters:

$$\{A, h_1, M_{S1}\} \quad \text{and} \quad \{A, h_2, M_{S2}\}.$$

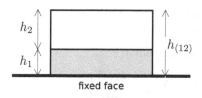 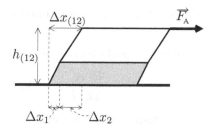

FIGURE 3.3   Composite Shear Modulus: Series Case

If one face is firmly anchored and the other is subjected to a constant transverse applied force, F_A, as is illustrated on the right in Figure 3.3, then the composite slab of height

$$h_{(12)} = h_1 + h_2$$

shears by an amount

$$\Delta x_{(12)} = \Delta x_1 + \Delta x_2.$$

The [assumedly] static nature of the situation demands that the force exerted by the adhesive or clamp on each of the slabs be equal in magnitude to the applied force. This observation is the key to the analysis which follows.

Each of the slabs and the stack suffer the same stress, and thus the shear moduli are

$$M_{S1} = \frac{F_A/A}{\Delta x_1/h_1}, \qquad M_{S2} = \frac{F_A/A}{\Delta x_2/h_2}, \qquad \text{and} \qquad M_{S(12)} = \frac{F_A/A}{\Delta x_{(12)}/h_{(12)}}.$$

The respective amounts of strain in each slab are

$$\frac{\Delta x_1}{h_1} = \frac{F_A/A}{M_{S1}}, \qquad \frac{\Delta x_2}{h_2} = \frac{F_A/A}{M_{S2}}, \qquad \text{and} \qquad \frac{\Delta x_{(12)}}{h_{(12)}} = \frac{F_A/A}{M_{S(12)}}.$$

Combining these relations into the constraint on the strains, $\Delta x_{(12)} = \Delta x_1 + \Delta x_2$, yields

$$\frac{F_A}{A}\frac{h_{(12)}}{M_{S(12)}} = \frac{\Delta x_{(12)}}{h_{(12)}}\,h_{(12)} = \Delta x_{(12)} = \Delta x_1 + \Delta x_2 = \frac{\Delta x_1}{h_1}\,h_1 + \frac{\Delta x_2}{h_2}\,h_2 = \frac{F_A}{A}\left\{\frac{h_1}{M_{S1}} + \frac{h_2}{M_{S2}}\right\}.$$

Thus, the shear modulus of two homogeneous materials combined in series is determined by the relation

$$\frac{h_{(12)}}{M_{S(12)}} = \frac{h_1}{M_{S1}} + \frac{h_2}{M_{S2}}.$$

---

[6]Employing glue or a rigid clamp, of course.

In the event that the entire slab is homogeneous, with shear modulus $\widetilde{M}_S$, the partition into two slabs becomes artificial, and $M_{S1} = \widetilde{M}_S = M_{S2}$.

LOGIC   The *faux*-composite slab must have $M_{S(12)} = \widetilde{M}_S$.

SERIES EXPR'N   Explicit computation using the above formula yields

$$\frac{h_{(12)}}{M_{S(12)}} = \frac{h_1}{M_{S1}} + \frac{h_2}{M_{S2}} = \frac{h_1 + h_2}{\widetilde{M}_S} = \frac{h_{(12)}}{\widetilde{M}_S},$$

illustrating the consistency of this approach to modelling elastic moduli.

EXAMPLE [*Parallel Composition and Shear Modulus*]

Two uniform slabs with characteristic parameters

$$\left\{A_1, h, M_{S1}\right\} \quad \text{and} \quad \left\{A_2, h, M_{S2}\right\}$$

are laminated into a single block as shown in Figure 3.4.

> ASIDE: It is implicitly assumed that there is no twisting, only shear, of the laminate upon application of the transverse external force.

If one face of this laminate slab is held fixed while the other is acted upon by a transverse force, F_A, then the slab shears by an amount $\Delta x_{[12]}$.

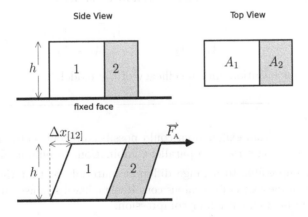

FIGURE 3.4 Composite Shear Modulus: Parallel Case

The force exerted on the laminate is split among the slabs in such a way that each shears by the same amount, $\Delta x = \Delta x_{[12]} = \Delta x_1 = \Delta x_2$, and the force is assumed to be distributed over the entire top face area. Hence,

$$F_A = F_1 + F_2 \quad \text{and} \quad A_{[12]} = A_1 + A_2.$$

As each of the slabs experiences the same strain, $\Delta x/h$, the shear moduli are

$$M_{S1} = \frac{F_1/A_1}{\Delta x/h}, \qquad M_{S2} = \frac{F_2/A_2}{\Delta x/h}, \qquad \text{and} \qquad M_{S[12]} = \frac{F_A/A_{[12]}}{\Delta x/h}.$$

Therefore, the transverse forces applied to the slabs are

$$F_1 = A_1 M_{S1} \frac{\Delta x}{h}, \qquad F_2 = A_2 M_{S2} \frac{\Delta x}{h}, \qquad \text{and} \qquad F_A = A_{[12]} M_{S[12]} \frac{\Delta x}{h}.$$

Insertion of these relations into the ratio of the total force to total area, $F_A/A_{[12]} = (F_1 + F_2)/(A_1 + A_2)$, yields

$$M_{S[12]} \frac{\Delta x}{h} = \frac{F_A}{A_{[12]}} = \frac{F_1 + F_2}{A_1 + A_2} = \frac{[A_1 M_{S1} + A_2 M_{S2}]}{A_1 + A_2} \frac{\Delta x}{h}.$$

Thus, the shear modulus for a laminate slab comprised of two slabs in parallel is the **area-weighted average value of the shear moduli** of the constituents:

$$M_{S[12]} = \frac{A_1 M_{S1} + A_2 M_{S2}}{A_1 + A_2}.$$

Corroboration of this result occurs in the trivial case in which the partition into two horizontal slabs is merely formal and arbitrary.

LOGIC   IF the two slabs have the same shear modulus, $M_{S1} = \widetilde{M}_S = M_{S2}$, THEN the *faux*-laminate slab must have $M_{S[12]} = \widetilde{M}_S$.

PARALLEL
EXPR'N   Application of the combinatorial formula derived above yields

$$M_{S[12]} = \frac{A_1 M_{S1} + A_2 M_{S2}}{A_1 + A_2} = \frac{A_1 \widetilde{M}_S + A_2 \widetilde{M}_S}{A_1 + A_2} = \widetilde{M}_S,$$

again confirming our intuition and the efficacy of the model.

Well! That was fun!

Q: What factors, aside from exhaustion, could possibly prevent us from *turning the crank* once more and considering series and parallel combinations of bulk moduli?

A: It is well-nigh impossible to arrange different materials so that they do not undergo internal shears [or, in the case of filamentous constituents, lineal strains] while the system as a whole is undergoing bulk expansion or compression.

In *reality*™, materials science gets complicated fast!

Chapter 4

Fluid Statics

In Chapter 1, ocean tides were invoked to justify relaxation of the assumption of rigidity which underlies all of VOLUME I. Here, and in the next few chapters, we shall abandon rigidity altogether and briefly touch upon the subject of FLUID MECHANICS. First, we'll investigate FLUID STATICS, by which we mean fluid systems which are at rest and quiescent [*i.e.*, the bulk characteristics of the fluid do not exhibit explicit time dependence].

FLUID   A fluid is a substance whose shape is dictated by the geometry of its container. Both liquids and gases fall into this category. More technically: **a fluid substance is unable to withstand shear**, *i.e.*, $M_S \to 0$. IF a fluid is subjected to shear stress, THEN the shear strain is [practically] unbounded.

[A complicating factor is that the deformation/flow of the fluid may occur very slowly.[1]]

AVERAGE MASS DENSITY   The average mass density within a particular region of space with volume ΔV is

$$\rho_{\text{av}} = \frac{\Delta M}{\Delta V} ,$$

where ΔM is the total mass of all substances found within ΔV. The SI units of density are kg/m^3. So-called *c-g-s* units, g/(cm)^3, are also in common usage.

MASS DENSITY   The local mass density[2] at a point in space, $\mathcal{P}$, is the limiting value of the average mass density throughout volumes which envelop $\mathcal{P}$ as these volumes tend to zero:

$$\rho = \lim_{\Delta V \to 0} \frac{\Delta M}{\Delta V} .$$

In common usage, the adjectives *local* and *mass* are elided.

AVERAGE PRESSURE   The average pressure found in a fluid is the force per unit area acting perpendicular to an imaginary planar surface element with area ΔA. The unit vector, $\hat{n}$, extends perpendicular to the plane containing the surface element. The magnitude of an increment of average pressure produced by a force,[3] $\Delta \vec{F}$, is

$$\Delta P_{\text{av}} = \frac{\Delta \vec{F} \cdot \hat{n}}{\Delta A} = \frac{\Delta F_{\perp}}{\Delta A} .$$

---

[1]When you next slake your thirst with a glass of your favourite beverage, do pause to consider that glass is, in many respects, a fluid. On normal time scales, the flow of glass in response to shear stress is imperceptible.

[2]Local mass density was the factor employed in VOLUME I to effect integrations over mass elements proper to an extended [rigid] body.

[3]The Δ adorns the force to indicate both that it is constitutive (not necessarily net) and that it is deemed to be distributed across the surface element.

The numerator in the final equality, $\Delta F_\perp$, is the component of the force acting perpendicular to the area element. The total average pressure, P_{av}, is the sum of all of the pressure increments arising from various forces. The SI unit of pressure is the pascal [Pa]:

$$1\,Pa = 1\,N/m^2.$$

Other pressure units include: torr [torr], *a.k.a.* mm of mercury [mmHg]; atmospheres [atm]; and pounds per square inch [psi]. The elastic moduli have the same dimension as pressure.

PRESSURE   The local pressure at a particular point in the fluid, $\mathcal{P}$, is the limit of the average pressure on surfaces passing through $\mathcal{P}$, as the areas of the surfaces tend to zero. Isotropy is concomitant with locality in almost all situations, thus rendering the directional aspects moot. Hence, local pressure is quoted as

$$P = \lim_{\Delta A \to 0} \frac{\Delta F}{\Delta A}.$$

This expression holds for constituent pressures as well as for the net pressure.

The essential empirical fact about pressure in fluids is that:

In a column of fluid in static equilibrium, the pressure increases with depth.

This is incorporated into PASCAL'S PRINCIPLE and expressed in PASCAL'S FORMULA.

PASCAL'S PRINCIPLE   Pressure, P_A, applied to a fluid in static equilibrium, is transmitted undiminished to every point within the fluid and to the walls of the container in which the fluid resides.

PASCAL'S FORMULA   The pressure at a specified point in the fluid is

$$P = P_0 + \rho\,g\,h.$$

Here, P_0 is the pressure at a particular reference point in [or on the boundary of] the fluid, ρ is the density of the fluid [assumed to be constant], g is the magnitude of the acceleration due to gravity [also assumed to be constant], and h denotes the relative depth of the specified point with respect to the reference point.

PROOF of PASCAL'S PRINCIPLE

The assumptions which undergird Pascal's Formula and Principle are now laid bare.

- The fluid fills a vertical column[4] with cross-sectional area A and height h.

- The column of fluid resides in a constant local gravitational field[5] with magnitude g, acting downward along the vertical axis.

---

[4] In truth, the column need not be vertical and its bounding walls may be comprised of fluid.
[5] This assumption places restrictions on the height and width of the column.

- The fluid is assumed to be homogeneous[6] with uniform mass density[7] ρ.

- Homogeneity and uniformity together imply **incompressibility** of the fluid.

- An applied force with magnitude F_0 is exerted directly downward on the fluid at the top of the column.

The volume, V, mass, M, and weight, W, of the fluid in the column are

$$V = A h, \qquad M = \rho V = \rho A h, \qquad \text{and} \qquad W = M g = \rho A h g.$$

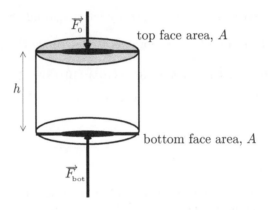

FIGURE 4.1 A Column of Fluid in Static Equilibrium under an External Force

For the fluid to remain in static equilibrium, the net external force on the column must vanish. Therefore, a force, $\vec{F}_{\text{bot}}$, acts upward on the base of the column. The magnitude of this force must be

$$F_{\text{bot}} = F_0 + W = F_0 + \rho A h g,$$

so as to cancel the force applied downward at the top of the column and to support the weight of the fluid in the column. Hence, the fluid pressure at the base of the column is

$$P_{\text{bot}} = \frac{F_{\text{bot}}}{A} = \frac{F_0 + \rho A h g}{A} = \frac{F_0}{A} + \rho g h = P_0 + \rho g h.$$

GENERALISED PASCAL FORMULA   The local version of Pascal's formula is

$$dP = \rho g \, dh,$$

allowing the density and g to vary with position. Pressure differences between points in the fluid are obtained by integration.

---

[6] For instance, it does not consist of layers or regions of different, **immiscible** fluids. Such cases, discussed in Chapter 8, are handled by a fluid regimes approach.

[7] Uniform density is more apposite to liquid than gas, since liquids, typically, have much larger bulk moduli than do gases. In Chapter 10, linear variation of density with pressure will be considered.

EXAMPLE [*In the Swim of Things*]

When diving into deep water, one might attain a depth of almost twice one's height.

Q: What is the water pressure in a freshwater lake at a depth of 3 metres?

A: A crude estimate of the pressure at this depth is provided by Pascal's Formula. Approximate values of the relevant parameters are:

P_0 Atmospheric pressure, $P_0 \simeq 10^5$ Pa, is exerted everywhere on the surface.

ρ The density of water is $\rho_{\text{H2O}} \simeq 10^3$ kg/m^3.

g The acceleration due to gravity has magnitude $g \simeq 10$ m/s^2.

Substituting these into Pascal's formula to determine the pressure at a depth of $h = 3$ m yields

$$P_3 = P_0 + \rho_{\text{H2O}}\, g\, h \simeq 10^5 + \left(10^3\right)(10)(3) = 1.3 \times 10^5 \text{ Pa} \simeq 1.3 \text{ atm}.$$

The pressure that a diver can expect to encounter at three metres' depth is approximately 1.3 atmospheres.

EXAMPLE [*A Pressure Gauge*]

The pressure gauge described below and pictured in Figure 4.2 affords a reliable determination of the [assumed to be constant[8]] fluid pressure within an enclosed chamber.

This device consists of a U-shaped tube, open at both ends, a sufficient volume of [relatively inert and incompressible] fluid, and the plumbing needed to connect one open end to the chamber.

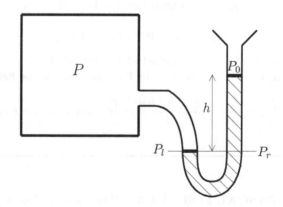

FIGURE 4.2 A Pressure Gauge in Operation

Various significant points are marked and their associated pressures are noted in the figure. The difference in the heights of the fluid across the arms of the U-shaped tube is denoted by h. A gas–fluid boundary is termed a **meniscus**. It is slightly curved owing to **surface tension** effects across its area and at the edge where the fluid meets the interior surface of the tube. Most commonly, the centre of the meniscus is lower than its edge.

---

[8]The chamber must be small, and the gas of low density, for the pressure to be effectively constant.

ASIDE: Convention dictates that the height assigned to an end of a fluid column be at the lowest part of the meniscus [*i.e.*, where the fluid first completely fills the perpendicular cross-section of the tube].

P   The pressure in the chamber is the sought-after quantity.

P_l The pressure at the top of the fluid column on the left side of the gauge [open to the chamber] is labelled P_l. For low-density gases and small height differences, $P_l \simeq P$.

P_r According to Pascal, the pressure in the right column at elevation directly across from the fluid meniscus on the left side is

$$P_r = P_l + \rho_{\text{fluid}}\, g\, h_{lr}\,,$$

where h_{lr} is the height difference between the two locations. As $h_{lr} = 0$ by assumption, it follows that $P_r = P_l$.

P_0 The right arm of the tube is open to the air. Thus, the fluid/air boundary on the right experiences [local] atmospheric pressure across its surface.

Pascal's formula (applied to the right column) determines the pressure P_r to be

$$P_r = P_0 + \rho_{\text{fluid}}\, g\, h\,.$$

The relations among the pressures combine to yield

$$P \simeq P_l = P_r = P_0 + \rho_{\text{fluid}}\, g\, h\,.$$

Thus, knowledge of the atmospheric pressure, P_0, fluid density, ρ_{fluid}, and acceleration due to gravity, g, along with a measurement of the difference in height between the right and left columns, h, is sufficient to determine the gas pressure within the chamber.

This, in essence, is how a pressure gauge works. A question remains.

Q: How might P_0 be ascertained?

A1: If one's interest is confined to pressure increments, then it is possible to ignore P_0.

A2: A **barometer**[9] is typically employed to measure P_0, as we shall see in the next example.

EXAMPLE [*A Barometer*]

A barometer is also a U-shaped tube partially filled with liquid. The chief difference is that the left side in Figure 4.3 is capped rather than connected to a chamber. The enclosed volume above the fluid level on the capped side is as near a vacuum as is practicable, meaning that whatever gas is enclosed in the bulb is at negligible pressure, $P \simeq 0$.

Adopting the notation employed in the pressure gauge example,

$$0 \simeq P\,, \qquad P_l = P_r\,, \qquad P_r = P_0\,, \qquad \text{and} \qquad h = -H\,.$$

Combining these relations and applying Pascal's formula,

$$0 \simeq P_0 + \rho_{\text{fluid}}\, g\, (-H) \qquad \Longrightarrow \qquad P_0 = \rho_{\text{fluid}}\, g\, H\,.$$

The SI standard value [at sea level and at 20 C] for atmospheric pressure is $P_0 = 101.3\,\text{kPa}$. This is equivalent to the oft-quoted value of 760 mmHg or 760 **torr**.

---

[9]Credit for the invention of the barometer is ascribed to Evangelista Torricelli (1608–1647), whom we shall encounter again in Chapter 7. A unit of pressure, the **torr**, is so-named in his honour.

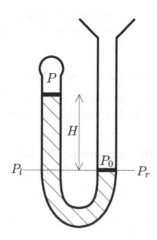

FIGURE 4.3   A Barometer

THE PRINCIPLE OF HYDRAULICS

Suppose that fluid completely fills a reservoir whose shape may be changed by manipulating the positions of two pistons sliding frictionlessly within cylinders. Several simplifying assumptions are made.

- The fluid is incompressible, *i.e.*, its volume is fixed.

- Dissipative forces are negligible. [The fluid flows without internal friction, *i.e.*, viscosity is absent. The pistons slide in their respective cylinders without friction.]

- Any $\rho_{\text{fluid}}\, g\, \Delta h$ contributions to the local pressure, from Pascal's formula, are negligible on the scale set by the external pressures.

- The pistons have cross-sectional areas A [LARGE] and a [SMALL].

IF an external force of magnitude f is applied directly downward on the small piston, THEN a force of magnitude F is exerted upward on the large piston. This situation is illustrated in Figure 4.4.

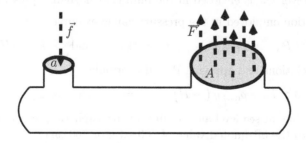

FIGURE 4.4   A Hydraulic System

Pascal's Principle applies and, in conjunction with the idealisations and approximations made above, it follows that the pressure is the same throughout the reservoir. Hence, at the pistons,

$$P_A = P_a \, .$$

It is also the case that $P_A = \frac{F}{A}$, while $P_a = \frac{f}{a}$. Thus,

$$\frac{F}{A} = \frac{f}{a} \qquad \Longrightarrow \qquad F = \frac{A}{a} \, f \, .$$

This simple hydraulic device provides "force multiplication." The **mechanical advantage**, A/a, is set by the ratio of the sizes of the respective pistons.

ASIDE: Similar force multiplication effects occur with levers and systems of pulleys. In the former, the mechanical advantage is equal to the ratio of arm lengths, and the torque is the invariant quantity. For the latter, the mechanical advantage is determined by the number of paired pulleys, while the invariant quantity is the tension in the rope.

EXAMPLE [*Lifting an Automobile with One Finger*]

PK carries a hydraulic jack in the boot of his Triumph Spitfire. The jack consists of a large cylinder with a piston capped by a cleat on which the object to be lifted rests, a small piston activated by a lever, a one-way-flow valve, and a pressure release and fluid recovery switch. In operation, the small piston is used to pump [nearly incompressible] fluid into the reservoir, increasing its volume and thereby driving the larger piston, along with the cleat and its load, upward.

Estimates of the total weight of the car and that supported by the jack are $M \sim 10000 \, \text{N}$ and $M/5 \sim 2000 \, \text{N}$, respectively. The hydraulic part of the system is comprised of pistons with areas $a = 8 \, (\text{mm})^2$ and $A = 80 \, (\text{mm})^2$, while the lever lengths for the handle and the small piston driver are $L = 40 \, \text{cm}$ and $l = 5 \, \text{cm}$.

The mechanical advantage gleaned from the lever is [*cf.* the discussions of torque in VOLUME I] given by the ratio $L/l = 8$, while that obtained from the hydraulic system is $A/a = 10$. The combined mechanical advantage is the product of the two separate mechanical advantages, $8 \times 10 = 80$, and is the factor relating the force exerted on the handle, f, to that exerted by the large piston, F. Supporting the load on the jack, $\sim 2000 \, \text{N}$, requires a force on the handle of

$$f \sim \frac{2000 \, \text{N}}{80} = 25 \, \text{N} \, .$$

Thus, the applied force necessary to operate the jack which lifts the car is roughly 25 N. This corresponds to the weight of a 2.5 kg [5 lbs, in the imperial system] bag of flour or sugar.

THERE IS NO FREE [ENERGY] LUNCH

While mechanical advantage produces force multiplication, it does not convey an energy advantage. In fact, mechanical work is conserved for the idealised hydraulic system.

[Inclusion of dissipative effects entails a net loss of mechanical energy. *C'est la vie!*]

Mechanical work is input to the reservoir system by the external force f acting downward on the piston with area a through a displacement d:

$$W_{\text{input}} = f\,d\,.$$

The hydraulic system does mechanical work via the exertion of an upward force, F, by the piston with area A through the displacement D:

$$W_{\text{output}} = F\,D\,.$$

Incompressibility entails that the volumes of fluid displaced by the respective piston strokes are equal:

$$a\,d \equiv A\,D \qquad \Longrightarrow \qquad D = \frac{a}{A}\,d\,.$$

Hence,

$$W_{\text{output}} = F\,D = \left(\frac{A}{a}\,f\right)\left(\frac{a}{A}\,d\right) = f\,d = W_{\text{input}}\,!$$

While hydraulic systems may effect force multiplication [to a degree established by each one's mechanical advantage], the energetic situation is, at best, neutral.

Chapter 5

Eureka!

Archimedes of Syracuse (*circa* 287–212 BC) has been called the greatest mathematician of antiquity by none other than Gauss.[1] Archimedes was a *Renaissance Man*[2] with abiding interests in mathematics, physics, engineering, and philosophy. In addition, he wrote a set of lecture notes for University Physics entitled *Mechanica Scientia*.[3] Syracuse's leader at the time, King Hieron II, requested [read "demanded"] that Archimedes devise a non-destructive method of determining whether a crown that he had commissioned consisted of pure gold or an adulterated mixture of gold and other metals. The inspiration for Archimedes' solution occurred to him while he was partaking of his bath. As legend has it, upon realising the import of his discovery, he exclaimed "*ευρηκα!*" and ran—naked—through the Syracusan streets back to his office.

> ASIDE: While most scholars of ancient Greek agree that "*Eureka!*" should be translated as "*I am in a state of having found it!*," an alternative interpretation was presented in an episode of Doctor Who [an authoritative source in our estimation] in which The Doctor declares that, in the Greek dialect of that locale and era, "*Eureka!*" means "*This bath is too hot!*"

In this chapter, we shall first examine Archimedes' Principle and the buoyant force that it defines. Afterward, we will illustrate Archimedes' technique for determining the specific gravity of an irregular solid object.

ARCHIMEDES' PRINCIPLE   A body which is wholly or partially submerged in a fluid experiences a BUOYANT FORCE equal in magnitude and opposed in direction to the weight of the displaced fluid.

PROOF of ARCHIMEDES' PRINCIPLE

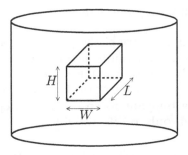

FIGURE 5.1   A Parallelepiped of Fluid within a Column in Static Equilibrium

Suppose that the assumptions rendering Pascal's Principle operative are all in force.

---

[1] Carl Friedrich Gauss (1777–1855) knew a thing or two about what it takes to be a great mathematician.
[2] Pardon the anachronism!
[3] Later authors have shamelessly attempted to ride on his coattails by adopting similar working titles.

[*I.e.*, the system is in static equilibrium, the fluid has uniform density, and g is constant.]

A rectangular parallelepiped of fluid, with dimensions

$$\text{length} = L, \quad \text{width} = W, \quad \text{and} \quad \text{height} = H,$$

is immersed in a larger volume of fluid, as shown in Figure 5.1. The fluid pressures at the top and bottom surfaces of the parallelepiped, P_{top} and P_{bot}, are related by

$$P_{\text{bot}} = P_{\text{top}} + \rho_{\text{fluid}}\, g\, H\,,$$

according to Pascal's formula. Multiplying both sides of this expression by the common area of the bottom and top faces, $A = L\,W$, yields

$$\text{LHS} = P_{\text{bot}} \times A = F_{\text{bot}}\,, \qquad \text{RHS} = \left(P_{\text{top}} + \rho_{\text{fluid}}\, g\, H \right) \times A = F_{\text{top}} + \rho_{\text{fluid}}\, g\, H\, L\, W\,.$$

The volume of the parallelepiped is $V = L\,W\,H$ and the mass of fluid that it contains is $M_{\text{fluid}} = \rho_{\text{fluid}} \times V$. Thus, $\text{RHS} = F_{\text{top}} + M_{\text{fluid}}\, g$. Setting the LHS equal to the RHS and rearranging, one obtains

$$F_{\text{bot}} - F_{\text{top}} = M_{\text{fluid}}\, g = F_{\text{B}}\,.$$

The magnitude of the net buoyant force acting upon the wholly submerged parallelepiped, F_{B}, is equal to the weight of the fluid contained in the parallelepiped.

Let's tidy this up with a few definitions and expand the scope of the above proof.

DISPLACED FLUID   The displaced fluid is that amount of fluid which was occupying the volume now occupied by the parallelepiped. It is the fluid that was *shoved aside* [*a.k.a.* displaced] to accommodate the [fluid-filled] parallelepiped at its location in the larger reservoir.

SUB-VOLUME CONTENTS   Careful reflection reveals that the above argument relied solely on properties [*viz.*, pressure and density] of the fluid medium adjacent to the sub-volume. The actual content of the parallelepiped, fluid or otherwise, does not alter the buoyant force that it experiences. **The buoyant force is always equal in magnitude and opposite in direction to the weight of displaced fluid irrespective of the composition of the interior of the displacing parallelepiped.**

SUB-VOLUME SHAPE   The regular geometry of the parallelepiped served only to simplify the analysis. Any sub-volume geometry can be accommodated by means of the PARTITION, COMPUTE, and SUM method. That is, PARTITION the volume into a collection of [approximately] regular geometric shapes, COMPUTE the buoyant force on each, and SUM to obtain the net buoyant force. **The buoyant force does not depend on the shape of the volume of displaced fluid, only on its weight.**

BUOYANT FORCE   Objects which are wholly or partially immersed in a fluid medium experience a buoyant force equal to minus the weight of the fluid displaced by the object, *i.e.*,

$$\vec{F}_{\text{B}} = -\vec{W}_{\text{disp}} = -M_{\text{disp}}\, \vec{g}\,.$$

The immersed object may be a sub-volume of the fluid itself, in which case its own weight is cancelled by the buoyant force. **The buoyant force allows fluids to exist in static equilibrium.** In this respect, the buoyant force acts somewhat like a normal force [except that the magnitude of the buoyant force is fixed].

ASIDE: For compressible fluids, the buoyant force is variable [generally within a narrow range], and this property gives rise to interesting dynamical behaviour. A playful example is found in a water toy which floats somewhat below the surface of the body of water in which it is immersed.

↕ If one pushes the toy further under and lets go, then the increased buoyant force produced by the denser surrounding water causes the toy to rise.

↕ If one raises the toy closer to the surface and releases it, then the diminished buoyant force is unable to prevent the toy from sinking.

SPECIFIC GRAVITY   The specific gravity of a material is experimentally defined as the ratio of the weight of a sample in air to the weight of an equivalent volume of pure water [H_2O].

$$\text{Specific Gravity} = \frac{\text{Weight in air}}{\text{Weight of equiv. vol. of } H_2O} = \frac{\rho_{\text{material}} \times \text{Volume} \times g}{\rho_{H2O} \times \text{Volume} \times g} = \frac{\rho_{\text{material}}}{\rho_{H2O}}.$$

[In the second equality above, the buoyant force of the displaced air is neglected.]

Thus, the specific gravity is also the relative density of the material with respect to that of water. Specific gravity depends only on the substance and not at all on the geometric properties [size and shape] of the sample.

EXPERIMENTAL METHOD FOR DETERMINATION OF SPECIFIC GRAVITY

STEP ONE   Determine the weight of the sample in air, W_{air}, using a device of the sort illustrated on the left in Figure 5.2.

FIGURE 5.2   Weighing a Sample Immersed in Air, Then in Water

STEP TWO   Determine the weight of the sample while it is wholly submerged in water, as in the right panel in Figure 5.2, thus obtaining W_{H2O}.

STEP THREE   The difference $W_{\text{air}} - W_{H2O}$ is equal to the weight of an equivalent volume of water. Thus, the specific gravity of the material comprising the sample is

$$\text{Specific Gravity} = \frac{W_{\text{air}}}{W_{\text{air}} - W_{H2O}}.$$

By employing this technique, Archimedes was able to demonstrate that the density of the material comprising Hieron II's crown differed significantly from that of gold.

Q: Instead of bothering with specific gravity, why don't we simply determine densities directly using

$$\text{density} = \frac{\text{mass}}{\text{volume}} ?$$

A: More accurate and precise estimates of the densities of substances are obtained via determination of their specific gravities, because, in practice, it proves surprisingly difficult to measure volumes.

Chapter 6

Fluid Dynamics: Flux

Fluids are not always quiescent, as is assumed to be the case in Fluid Statics. In this chapter we shall begin to investigate fluid flow.

IDEAL FLUID FLOW The following properties, taken together, constitute the defining characteristics of IDEAL FLUID FLOW.

- **Laminar** The velocity of the fluid at all locations remains constant in time. The fluid flows steadily, and with no **turbulence**.

  [*Real fluids swirl and whirl, forming eddies and whorls.*]

- **Non-Viscous** Viscosity [*a.k.a.* internal friction] is absent.

  [*For real fluids, neighbouring bits in relative motion rub together, dissipating energy.*]

- **Incompressible** The fluid is incompressible, *i.e.*, $B_{fluid} \to \infty$.

  [*Real fluids compress and rarefy in response to pressure changes.*]

- **Uniform** An often-made assumption is that the velocity of the fluid is the same at all points throughout a particular region. This assumption ignores boundary layer effects.

  [*In streambeds, real fluid flows more speedily in the centre than near the edges.*]

A non-viscous incompressible fluid is often dubbed "perfect."

CONSERVATION IDEAS

A perfect fluid flows through a compound pipe consisting of two straight sections with constant cross-sectional areas A_1 and A_2, respectively, joined by a segment with variable cross-section, as illustrated in Figure 6.1. Three assumptions are made: the walls of the pipe sections are rigid, the compound pipe is symmetric about the axis running along its mid-line, and the fluid moves uniformly, with speeds v_1 and v_2, in the two regions with constant cross-section. The flow is non-uniform in the interpolating segment.

EQUATION of CONTINUITY An equation of continuity is the mathematical expression of a conservation law. In the case of incompressible fluids flowing ideally, both the volume and the mass of fluid are conserved.

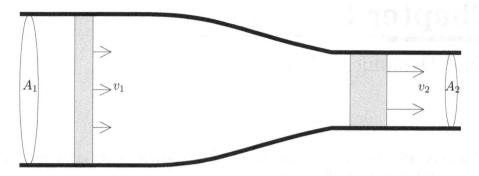

FIGURE 6.1   Fluid Flow through a Pipe with Variable Cross-Section

IF the fluid is perfect AND contained within a rigid-walled[1] pipe, THEN equal volumes of fluid must flow **into** and **out of** the compound pipe throughout **any** time interval Δt.

$$A_1 \left(v_1 \, \Delta t \right) = A_2 \left(v_2 \, \Delta t \right) \qquad \Longrightarrow \qquad v_1 \, A_1 = v_2 \, A_2 = \text{a constant.}$$

Conservation [continuity] implies that no fluid is lost or gained, except at SINKS and SOURCES.

$$A \quad \left\{ \begin{array}{c} \text{SOURCE} \\ \text{SINK} \end{array} \right\} \quad \text{is a localised region} \quad \left\{ \begin{array}{c} \text{OUT OF} \\ \text{INTO} \end{array} \right\} \quad \text{which fluid flows.}$$

The notions of source and sink are rather general. For instance, in Figure 6.1, the open end of the pipe to the left acts as a source, while the open end on the right acts as a sink.

STREAMLINES   Streamlines correspond to the unique trajectories of particular [infinitesimal] fluid elements.

- Under laminar flow, the pattern of streamlines is unchanging in time.

- Streamlines may not cross.
  [If two streamlines were to intersect at a point in space, then the trajectory of a fluid element residing at the intersection point at time t is ill-defined afterward. Thus, by *reductio ad absurdum*, neither could be an actual streamline for the fluid.]

- Sheaves of streamlines form surfaces and volumes of flow.
  [Adjacent and contiguous streamlines remain so except in the presence of obstacles, junctions, sources, or sinks. Thus, 2- and 3-d bundled structures of streamlines inherit their laminar properties.]

- No fluid may cross surface boundaries consisting of sheaves of streamlines.
  [This property follows directly from the ones above.]

---

[1]The minimal condition is that the cross-sectional area of the pipe remain constant. Rigidity of the walls is sufficient, although not necessary.

MASS FLUX The mass flux, $\Phi_{m,S}$, is equal to the time rate at which the mass associated with the flowing matter passes through an imaginary surface, S. The mass flux is computed via

$$\Phi_{m,S} = \int_S \rho\,\vec{v}\cdot d\vec{A}\,.$$

On the RHS, ρ is the local density, and $\vec{v}$ the velocity field of the streaming fluid elements. The integral is performed over the imaginary surface S. A more detailed parsing of this expression is undertaken below.

$\Phi_{m,S}$ The LHS represents the mass flux through a particular [imaginary] surface, S. The SI units of $\Phi_{m,S}$ are $\mathrm{kg/s}$, *i.e.*,

$$[\Phi_{m,S}] = \left[\frac{\mathrm{kg}}{\mathrm{m}^3} \times \frac{\mathrm{m}}{\mathrm{s}} \times \mathrm{m}^2\right] = \left[\frac{\mathrm{kg}}{\mathrm{s}}\right],$$

as befits the time-rate-of-flow of mass.

$\vec{v}$ The velocity field is a vector-valued function of position describing the speed and direction of the flow of fluid elements.

- Knowledge of the velocity field restricted to the surface is sufficient to determine the flux.

- The velocity field vectors at neighbouring [up- and down-stream] points consistently join together [*a.k.a.* integrate] to form the streamlines.

$d\vec{A}$ The vector area element at a particular location on the imaginary surface has as its magnitude the infinitesimal size of a patch of area[2] on the surface. Its direction is perpendicular [*a.k.a.* normal] to the imaginary surface. Barring pathologies [kinks, conical points, discontinuities, or what have you], the direction of $d\vec{A}$ is determined up to ORIENTATION.[3] The mathematical expression of $d\vec{A}$ depends on the coordinates used to parameterise the surface. In Cartesian coordinates,

$$d\vec{A} = A_x\,dy\,dz\,\hat{\imath} + A_y\,dz\,dx\,\hat{\jmath} + A_z\,dx\,dy\,\hat{k} = \left(A_x\,dy\,dz\,,\ A_y\,dz\,dx\,,\ A_z\,dx\,dy\right).$$

$\vec{v}\cdot d\vec{A}$ The DOT PRODUCT of the local velocity field with the infinitesimal area element provides the volume amount of fluid passing through the surface at a point in space per unit time interval.

ρ The mass density, ρ, *weights* the integrand, producing the mass flux.

ASIDE: Setting the weight factor equal to 1 yields the volume flux,

$$\Phi_{V,S} = \int_S \vec{v}\cdot d\vec{A}\,,$$

with SI units of cubic metres per second.

$\int_S$ The net flux is the sum of the local fluxes through all points on the surface. It is most often the case that the surface is bounded and of finite extent, but this isn't always necessary for the flux to be defined.

---

[2]Tiny planar patches of area are often called "plaquettes."

[3]Although the line normal to a plaquette is unambiguously defined, the vector area element may be directed in one of two ways. Common terms employed to distinguish these are: UP/DOWN, LEFT/RIGHT, FRONT/BACK, TOP/BOTTOM.

EXAMPLE [*Not Flummoxed by Flux*]

An ideal fluid, with constant density ρ_0, flows uniformly with $\vec{v} = v\,\hat{j}$, as in the three instances shown in Figure 6.2. The mass flux through each of the planar shaded surfaces shown shall be determined.

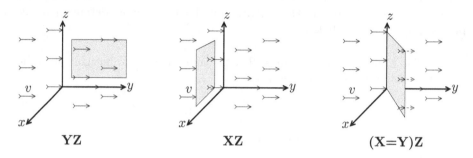

FIGURE 6.2   Mass Fluxes through Various Planar Surfaces

YZ   The surface S_{yz}, over which the flux is to be computed, lies within the yz-plane. Therefore,
$$\rho_0\,\vec{v} \cdot d\vec{A} = 0$$
everywhere on this surface. Thus, $\Phi_{m,S_{yz}} = 0$. Two comments are:

- The flux is zero, because none of the fluid passes through S_{yz}.

- The direction of the surface area vector is ambiguous, $\pm\hat{\imath}$. However, in either case the integrand vanishes.

XZ   The surface S_{xz} lies in a portion of the xz-plane. Therefore,
$$\rho_0\,\vec{v} \cdot d\vec{A} = \rho_0\,v\,dA\,\cos(0) = \rho_0\,v\,dA$$
everywhere on this surface. The orientation ambiguity has been resolved, by *fiat*, through our choosing the integrand to be positive. As the density and speed are both assumed to be constant across the surface, the mass flux is
$$\Phi_{m,S_{xz}} = \rho_0\,v\,A_{xz}\,,$$
where A_{xz} is the area of the planar surface S_{xz}.

(X=Y)Z   The surface $S_{(x=y)z}$ lies in a plane which incorporates both the z-axis and the line $y = x$. Again exercising our freedom to render the flux positive,
$$\rho_0\,\vec{v} \cdot d\vec{A} = \rho_0\,v\,dA\,\cos(\pi/4) = \frac{1}{\sqrt{2}}\,\rho_0\,v\,dA$$
everywhere on the surface of interest. Thus,
$$\Phi_{m,S_{(x=y)z}} = \frac{1}{\sqrt{2}}\,\rho_0\,v\,A_{(x=y)z}\,,$$
where $A_{(x=y)z}$ is the area of the planar surface.

Upon reflection, it is clear that the magnitude of the flux through S_{xz} is maximal, while through S_{yz} it is minimal. The essential point is that the flux through planar surfaces depends strongly on the relative angle between the velocity field and the surface normal.

Conservation Laws pertinent to volume and matter take the following forms.

CONSERVATION of VOLUME   A closed non-self-intersecting[4] surface, S, bounds a region of space through which incompressible fluid is flowing. IF the influx from sources located within the bounded region is precisely cancelled by a matching outflow from sinks within the same region, THEN the net volume flux into the region is exactly zero, *i.e.,*

$$\Phi_{V,S} = \oint_S \vec{v} \cdot d\vec{A} = 0.$$

> ASIDE: Closed non-self-intersecting surfaces divide space into two disjoint regions: **inside** and **outside**. This enables resolution of the orientation ambiguity. By convention, the areal normal points from the inside to the outside.

CONSERVATION of MATTER   A closed non-self-intersecting surface, S, bounds a region of space through which fluid is flowing. IF there is no net flux from sources/sinks within the surface, THEN the net mass flux into the region is exactly zero, *i.e.,*

$$\Phi_{m,S} = \oint_S \rho\,\vec{v} \cdot d\vec{A} = 0.$$

These two laws are summed up nicely by the phrase:

What goes in, must come back out.

These two laws are distinct,[5] because situations [compressible fluids, changes of phase, chemical reactions] do arise in which the matter content and the volume do not remain commensurate.

---

[4] Closed self-intersecting surfaces are unsuited for our present purposes.
[5] The material mode is more fundamental than the volume version.

Chapter 7

Bernoulli's Equation

Bernoulli's[1] equation encompasses a great many particular dynamical circumstances.

BERNOULLI'S EQUATION   For an ideal fluid, the quantity

$$P + \frac{1}{2}\rho v^2 + \rho g y$$

assumes the same constant value[2] throughout the fluid.

PROOF and EXPLANATION of BERNOULLI'S EQUATION

An ideal fluid is confined to a pipe in which there is both a constriction and a change in elevation. At an initial time, t, two equal-volume chunks of fluid are located in separate regions of pipe, as illustrated by the darker-shaded bands in Figure 7.1.

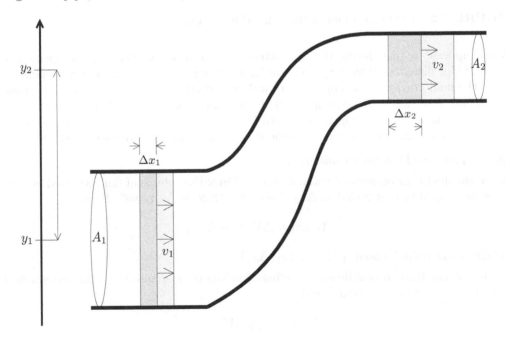

FIGURE 7.1   Fluid in a Rising and Narrowing Pipe at Time t

---

[1]Daniel Bernoulli (1700–1782) was a scion of an illustrious family of mathematicians and scientists.

[2]The important fact is that the quantity is constant; its actual value is incidental. The same is true of the total mechanical energy, as was noted in Chapter 27 of VOLUME I.

A brief moment later, at time $t + \Delta t$, the fluid chunks have progressed[3] along the pipe to regions immediately adjacent to their locations at time t, as indicated by the lighter-shaded bands in Figure 7.1 and the darker-shaded zones in Figure 7.2.

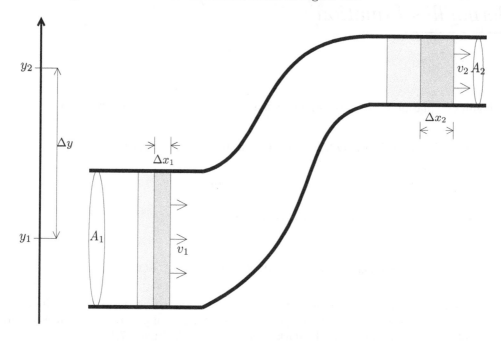

FIGURE 7.2 The Same Fluid in the Same Pipe at Time $t + \Delta t$

What happens[4] is that during the time interval of duration Δt, the fluid occupying the region of pipe demarcated by Δx_1 in Figure 7.1 moves forward [remaining within the pipe] by Δx_1. The fluid previously occupying that locale was itself displaced forward, and iteration of this argument under the assumption that the fluid is incompressible requires that the fluid in the Δx_2 region be displaced forward by distance Δx_2. The net effect is a transfer of fluid **from** the Δx_1 [initial] region **to** the Δx_2 [final] region, as indicated in Figure 7.3.

Q: How much fluid has been transferred?

Av: IF the fluid is incompressible [as is assumed], THEN the volume of fluid removed at Δx_1 is precisely equal to that added at Δx_2 [CONSERVATION OF VOLUME]. Hence,

$$A_1\, \Delta x_1 = \Delta V = A_2\, \Delta x_2\,,$$

yielding a constraint[5] among $\{A_1, A_2, \Delta x_1, \Delta x_2\}$.

Am: IF the fluid has constant density ρ_0 [often a corollary of incompressibility and homogeneity], THEN the amount of mass transferred is

$$\Delta M = \rho_0\, \Delta V\,,$$

for the constrained ΔV expressed directly above.

---

[3]Here the inessential supposition of uniformity assists in visualising the flow.

[4]*Reality*™ is rather more complicated. These analyses are confined to ideal fluids.

[5]That this constraint is equivalent to an equation of continuity is abundantly clear upon dividing both sides by the time interval, Δt.

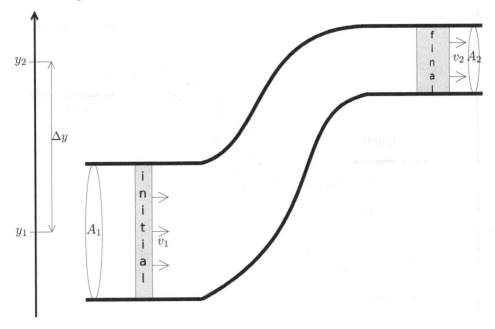

FIGURE 7.3 The Net Motion of Fluid in the Pipe

ASIDE: In circumstances where the density is variable, it is [almost[6] always] the case that the mass is conserved even when the volume is not.

It cannot have escaped notice that, in the course of being transferred from region 1 to region 2, the fluid experienced a change in elevation. Considered in bulk, the transferred fluid has risen by distance

$$\Delta y = y_2 - y_1 \, ,$$

where y_1 and y_2 are the heights of the centres of mass of the initial and final fluid chunks.

ASIDE: When the pipe has sufficient symmetry and is completely full of homogeneous fluid, the centres of mass of the fluid in regions 1 and 2 coincide with the geometric centres of the local pipe segments.

An illustration which should help make this clear appears in Figure 7.4, in which the trajectories of three fluid elements making their way from region 1 to region 2 are shown. Despite the evident fact that each of the three fluid elements shown has risen by an amount which differs from that of the others, it is possible to prove by inference, or by explicit computation, that the difference between the initial and final heights of the CofM is precisely equal to the height difference between the centres of the pipes.

In effect, the mass ΔM has been raised a vertical distance Δy, and the transferred fluid experiences a net change in its **gravitational potential energy** equal to

$$\Delta U_{\mathrm{g}}(y) = (\Delta M) \, g \, (\Delta y) = \rho \, \Delta V \, g \, (y_2 - y_1) \, .$$

---

[6]This CAVEAT is to accommodate situations in which mass–energy intraconversion occurs.

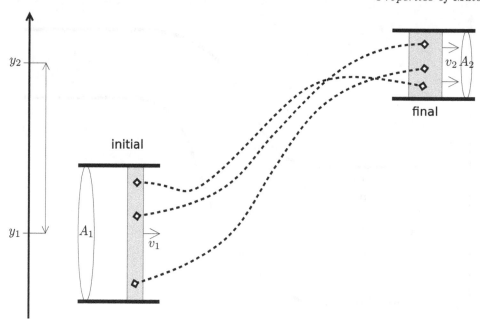

FIGURE 7.4 Particular Fluid Elements May Change Their Relative Positions

Also, upon transfer, the speed of the fluid changes, as dictated by the equation of continuity. The assumed uniformity of flow renders the computation of the kinetic energies readily tractable:

$$K_1 = \frac{1}{2}\,(\Delta M)\,v_1^2 = \frac{1}{2}\,\rho\,\Delta V\,v_1^2, \quad \text{and} \quad K_2 = \frac{1}{2}\,(\Delta M)\,v_2^2 = \frac{1}{2}\,\rho\,\Delta V\,v_2^2.$$

Thus, the change in the kinetic energy of the transferred fluid is

$$\Delta K = K_2 - K_1 = \frac{1}{2}\,\rho\,\Delta V\left(v_2^2 - v_1^2\right).$$

It is not as though the fluid elements flow in the absence of forces. The net force which pushes the fluid in region 1, $F_{\text{net},1}$, acts through Δx_1, as the fluid is displaced forward by this distance. For laminar flow, $F_{\text{net},1}$ must be constant throughout the time interval, Δt, in which the motion occurs, and thus the net work done on the fluid in region 1 is

$$W_1\left[F_{\text{net},1}\right] = F_{\text{net},1}\,\Delta x_1.$$

In similar fashion, the net force retarding the forward progress of the fluid in region 2, $F_{\text{net},2}$, acts through the displacement Δx_2, and the net work done on this fluid is

$$W_2\left[F_{\text{net},2}\right] = -F_{\text{net},2}\,\Delta x_2.$$

ASIDE: *Wait a minute!* This seems quite asymmetrical. In addition, it appears to have been forgotten that the fluid in region 2 is pushed forward while that in region 1 has its forward motion resisted. All of these concerns are true enough. However, the forces which advance 2 and retard 1 are internal to the system comprised of all of the fluid in the pipe between the two regions. Hence, these internal forces do not contribute net external work.

The propulsive and resistive forces in the fluid are non-conservative. The net work performed on the fluid by these forces is

$$W_{\text{NET, non-C}} = W_1\left[F_{\text{net,1}}\right] + W_2\left[F_{\text{net,2}}\right] = F_{\text{net,1}}\,\Delta x_1 - F_{\text{net,2}}\,\Delta x_2\,.$$

The assumption of incompressibility ensures that $A_1\,\Delta x_1 = \Delta V = A_2\,\Delta x_2$. Therefore, **the net work done on the fluid by non-conservative forces** is

$$W_{\text{NET, non-C}} = \frac{F_{\text{net,1}}}{A_1}\,A_1\,\Delta x_1 - \frac{F_{\text{net,2}}}{A_2}\,A_2\,\Delta x_2 = \frac{F_{\text{net,1}}}{A_1}\,\Delta V - \frac{F_{\text{net,2}}}{A_2}\,\Delta V = \left(P_1 - P_2\right)\Delta V\,.$$

The pressure experienced in each region is equal to the net [advancing/retarding] force experienced by the respective fluid chunks divided by the relevant cross-sectional area.

These net changes in the gravitational potential energy and the kinetic energy combine with the non-conservative work in the WORK–ENERGY THEOREM:

$$W_{\text{NET, non-C}} = \Delta E = \Delta K + \Delta U\,.$$

Elaboration, and recognition of the common factor of ΔV, yields

$$\left(P_1 - P_2\right)\Delta V = \frac{1}{2}\rho\,\Delta V\left(v_2^2 - v_1^2\right) + \rho\,\Delta V\,g\left(y_2 - y_1\right)$$

$$\implies \quad \left(P_1 - P_2\right) = \frac{1}{2}\rho\left(v_2^2 - v_1^2\right) + \rho\,g\left(y_2 - y_1\right)\,.$$

Rearranging to separate the dependencies on quantities pertinent to the distinct regions yields

$$P_1 + \frac{1}{2}\rho v_1^2 + \rho\,g\,y_1 = P_2 + \frac{1}{2}\rho v_2^2 + \rho\,g\,y_2\,.$$

As the subscripts "1" and "2" are merely labels, this relation must hold for all pairs of points lying within the [contiguous] fluid, which is only logically possible if the LHS and RHS are separately equal to the same constant. Hence, Bernoulli's equation reads

$$\text{constant} = P + \frac{1}{2}\rho v^2 + \rho\,g\,y\,.$$

EXAMPLE [*Pascal from Bernoulli*]

Pascal's formula is obtained from application of Bernoulli's equation under the condition that the body of fluid is at rest, *i.e.*, $v = 0$ everywhere.

In Figure 7.5, $\mathcal{P}_0$ is the reference point, at which the fluid pressure is P_0, while at some other point, $\mathcal{P}$, the pressure is P. Setting $y = 0$ at the reference height, the height of $\mathcal{P}$ [at relative depth h] must be $y = -h$. Applying Bernoulli to each of the points and then identifying their constant values yields the following.

$$\left.\begin{aligned}\text{Bernoulli } &= P_0 + \frac{1}{2}\rho\,(0)^2 + \rho\,g\,(0) = P_0 \\[2mm] \text{Bernoulli } &= P + \frac{1}{2}\rho\,(0)^2 + \rho\,g\,(-h) = P - \rho\,g\,h\end{aligned}\right\} \quad \implies \quad P = P_0 + \rho\,g\,h\,.$$

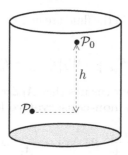

FIGURE 7.5   Bernoulli's Equation Contains Pascal's Formula

EXAMPLE [*Torricelli's Law*]

An open tank of liquid has a small hole in its side. It shall be assumed that the rate at which fluid leaves the tank [the flux] is small enough that the level of fluid drops very slowly. It is also assumed that the atmospheric pressure, P_{atm}, the density, ρ, and the local acceleration due to gravity, g, are effectively constant throughout a region of space which includes both the top surface of the fluid and the hole in the side of the tank.

Q: What is the speed at which fluid squirts out of the hole in the side of the tank?

A: Setting $y = 0$ at the top surface of the fluid, realising that the external pressure both

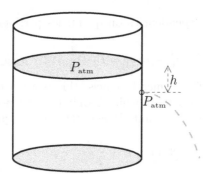

FIGURE 7.6   Torricelli's Law

at the top surface and at the hole is provided by the atmosphere, and recognising that the speed at which the top surface moves is negligible according to our assumptions about the flux enables the computation of Bernoulli's constant at both locations:

$$\text{Bernoulli at top surface} = P_{\text{atm}} + \frac{1}{2}\,\rho\,(0)^2 + \rho\,g\,(0) = P_{\text{atm}}\,,$$

$$\text{Bernoulli at hole} = P_{\text{atm}} + \frac{1}{2}\,\rho\,v_{\text{hole}}^2 + \rho\,g\,(-h)\,.$$

Thus,

$$P_{\text{atm}} = P_{\text{atm}} + \frac{1}{2}\,\rho\,v_{\text{hole}}^2 - \rho\,g\,h \qquad \Longrightarrow \qquad v_{\text{hole}} = \sqrt{2\,g\,h}\,.$$

The speed at which fluid leaves the hole depends only on the local [constant] gravitational field and the depth of the hole beneath the surface of the liquid.

Each fluid element exits the hole at precisely the speed that it would have attained, at that depth, had it been dropped from the top surface through air or vacuum, starting at rest!

This amusing little result is called **Torricelli's Law**, in honour of Evangelista Torricelli. Upon reflection, one realises that Torricelli's Law is but a straightforward consequence of the principle of energy conservation.

EXAMPLE [*Venturi Effect*]

In Figure 7.7, ideal fluid flows through a horizontal pipe possessing a constriction.

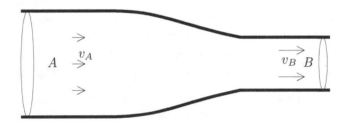

FIGURE 7.7   Venturi Effect

Q: How do the pressures in the constricted, B, and unconstricted, A, regions compare?

A: To find out, let's apply Bernoulli's equation and the equation of continuity.

$$\left. \begin{array}{l} \text{Bernoulli at } A = P_A + \dfrac{1}{2}\rho v_A^2 + \rho g y \\[2mm] \text{Bernoulli at } B = P_B + \dfrac{1}{2}\rho v_B^2 + \rho g y \end{array} \right\} \implies P_A + \frac{1}{2}\rho v_A^2 = P_B + \frac{1}{2}\rho v_B^2 .$$

By continuity, the fluid moves faster in the constricted region, $v_B > v_A$, and thus

$$P_B < P_A .$$

That the fluid pressure is lower where the fluid is moving faster is called the **Venturi Effect**, after Giovanni Venturi (1746–1822). There are many important applications and implications of the Venturi Effect. Two are quoted below. The former appears straightforward, while the latter seems counterintuitive.

○   The "venturi" is a device employed to measure the speed of a vehicle with respect to the fluid in which it is immersed. At its most basic, it consists of an open tube protruding into the fluid, connected to a pressure gauge. Fluid streams transversely by the mouth of the tube as a consequence of the motion of the vehicle. The concomitant diminishing of the pressure in the tube, below that of still fluid, is speed-dependent.

○   Constriction of the arteries supplying blood to vital organs is deleterious to one's health mainly[7] because of the associated lower arterial pressure in the neighbourhood of the constriction. [CAVEAT: Complicating factors include (1) that the arterial walls are not rigid, and may collapse when the pressure of the flowing blood is reduced, (2) that blood is far from an ideal fluid,[8] and (3) that increased flow speed can lead to turbulence.]

> **Q:** But wait. Isn't this backwards? People with arterial constriction generally exhibit high, not low, blood pressure.
>
> **A:** True enough! What happens is that one's body compensates for the localised reduction of blood pressure in the vicinity of the constriction by a systemic increase in blood pressure.

....

EXAMPLE [*Gunfight at the H_2O-K Corral*]

Bertal reinforces his SuperSoaker™ water pistol with duct tape to enable it to withstand a pressure difference, with respect to atmospheric pressure, of 450 kPa.

Q1: How fast does the water emerge from the barrel of the toy?

Q2: What is the maximum range of Bertal's water pistol [assuming no drag]?

A1: Since the water pistol is small, $\Delta y \simeq 0$, the Venturi form of Bernoulli's equation is operative. The density of water is $\rho = 1000\,\text{kg/m}^3$. With $\Delta P = 450\,\text{kPa}$ and $v_{\text{tank}} = 0\,\text{m/s}$, it follows that

$$\frac{1}{2}\,\rho\,v_{\text{stream}}^2 = 450,000\,\left[\frac{\text{N}}{\text{m}^2}\right] \qquad \Longrightarrow \qquad v_{\text{stream}} = 30\,\text{m/s}\,.$$

A2: Analysis of projectile motion in the standard case [*cf.* VOLUME I, Chapter 7] provides an expression for the [maximum] range:

$$R_{\text{max}} = \frac{v_0^2}{g}\,\sin\left(2\,\theta_0\right)\Big|_{\theta_0 = \pi/4} = \frac{v_0^2}{g}\,.$$

Setting $g = 10\,\text{m/s}^2$ and $v_0 = v_{\text{stream}}$ then determines the maximal range to be $R = 90\,\text{m}$.

Hence, neglecting air resistance, Bertal might hope to douse his target from a distance which is roughly the length of a football field!

---

[7]A secondary concern is that the cause of the constriction might become dislodged and occlude another, smaller artery.

[8]For instance, blood is rather viscous (see Chapter 9).

Chapter 8

No Confusion, It's Just Diffusion

Much of our discussion of fluids has been predicated upon the assumption that the fluid substance under consideration is homogeneous.

Q: How restrictive or realistic is this?

A: In situations involving pure substances, homogeneity seems quite reasonable on scales which are small enough that bulk modulus effects are not significant. For mixtures of substances, experience suggests two divergent possibilities [along with intermediate behaviour].

Miscible substances mix freely to yield a homogeneous composite. For example, acetic acid and water combine to form vinegar.

> ASIDE: One might cavil that this example should be regarded as dilution rather than mixing since acetic acid is most often already mixed with water. Such quibbling can drive one to drink [a mixture of ethyl alcohol, water, and various flavouring agents].

It is important to note that the approach to homogeneity occurs without any specific external intervention or effort. Of course, stirring or agitating will speed the process.

Immiscible substances do not mix. Despite one's best efforts, such fluids are incapable of blending. A common instance of this occurs when olive oil is added to the vinegar solution noted just above to make salad dressing. One can shake the salad dressing vigorously to disperse both fluids throughout the container, and yet when the agitation ceases the dressing reverts to a pool of oil floating on a bed of vinegar, as shown on the left in Figure 8.1.

For intermediate behaviour, we need look no further than the spoonful of honey added to a piping-hot cup of tea. The honey settles somewhat, producing a sweetness gradient which is discernable as the tea is sipped.

BILAYER FLUIDS

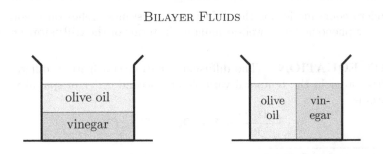

FIGURE 8.1 Physical and Unphysical Arrangements of Fluids with Different Densities

Immiscible fluids with differing densities naturally organise into layers. In a salad dressing, olive oil [with approximate density[1] $918\,\mathrm{kg/m^3}$] floats upon the table vinegar [about $1005\,\mathrm{kg/m^3}$]. Such a layered fluid is in stable equilibrium, and thus it is static. The pressure as a function of depth in the fluid may be readily determined by treating the system as a sequence of fluid density regimes.

> ASIDE: Although horizontal layering may be traditional, vertical striping is *haute couture*, and perhaps *haute cuisine*, too, this season.
>
> Convince yourself by an argument involving Pascal, Archimedes, and fluid conservation that stable equilibrium is not possible for this arrangement of fluids.
>
> While the amusing arrangement of vertical striping cannot be sustained for any length of time, it is possible to have long-lived metastable states of higher density fluids lying on top of lower density ones. The most familiar instances of this are "temperature inversions," which occur when a layer of cooler, higher density air lies upon a pocket of warmer, lower density air. Inversions happen with greater frequency in valleys ringed by mountains or hills. The ground-level air quality is often compromised when inversions occur because pollutants which are normally dispersed upward in the atmosphere become trapped.

In the case of miscible fluids, spontaneous[2] **diffusion** leads inexorably to a homogeneous final mixed state.

DIFFUSION   The spreading of substances from regions of higher concentration to those of lower concentration by essentially random local microscopic motions is called diffusion. A familiar and tasty example of diffusion is the manner in which the aroma of baking cookies gradually permeates the kitchen.

> **CONCENTRATION**   Concentration is a measure of [local] number density.[3] In practice, this is expressed in terms of molecules per cubic metre, `mole` per `L`, or what have you. Common measures of **relative concentration** are "parts per thousand/million/billion."
>
> **MOTIONS**   Thermal energy [*a.k.a.* heat] powers the random microscopic motions. Diffusion occurs more rapidly for hotter substances.
>
> **NET**   That there is spontaneous net flow from high to low concentration and not the converse may be understood by means of the statistical entropic argument to be advanced in Chapter 48.

The idea of molecular diffusion is applicable to analyses of heat conduction in thermodynamics and the spread of disease in epidemiology, as well as to other fields of study.

While the microscopic model for the diffusion of substances relies on random behaviour, the macroscopic phenomenon is well-formulated in terms of the **diffusion equation**.

DIFFUSION EQUATION   The diffusion equation is a [partial] differential equation which is first order in the temporal variable and second order in spatial variable(s). Its general form is

$$\frac{\partial c(t,\vec{r})}{\partial t} = \vec{\nabla} \cdot \left[\mathcal{D}(c,\vec{r})\, \vec{\nabla} c(t,\vec{r}) \right]\,,$$

---

[1] Temperature and other effects influence the precise value of the density.

[2] This is—later—to be appreciated in the context of the Second Law of Thermodynamics (Chapter 47).

[3] Number density is distinct from mass density.

where $c(t, \vec{r})$ denotes the [local] concentration at the position $\vec{r}$ and the instant t. The **diffusion coefficient**, $\mathcal{D}(c, \vec{r})$, may be concentration[4] and position dependent. The simplest instance, in which the diffusion coefficient is constant, $\mathcal{D}(c, \vec{r}) = D$, is dubbed the HEAT EQUATION by many authors. Linearity of the heat equation ensures that linear combinations of known solutions are themselves solutions. Below, we explicitly solve the heat equation when space is one-dimensional.

ASIDE: An alternative expression of the diffusion equation, favoured by some authors,

$$\vec{J} = -\mathcal{D}\,\vec{\nabla} c \qquad \text{and} \qquad \frac{\partial c}{\partial t} + \vec{\nabla} \cdot \vec{J} = 0 \,,$$

introduces the **current density**, $\vec{J}$, for the diffusive flow.

ISOTROPIC DIFFUSION EQUATION WITH CONSTANT COEFFICIENT IN 1-D

The diffusion equation in 1-d with a constant diffusion coefficient reduces to

$$\frac{\partial c(t, x)}{\partial t} = D\,\frac{\partial^2 c(t, x)}{\partial x^2} \,.$$

Despite this being the simplest form of the diffusion equation, obtaining its solutions is a non-trivial exercise. Below, we outline three diverse analyses.

I   Green's Function Solutions

Trial and error eventually lead us[5] to the following form:

$$c_{\mathrm{GF}}(t, x) = \frac{C_0}{\sqrt{4\pi D t}}\,\exp\left(-\frac{x^2}{4 D t}\right) \,, \qquad \bigl[\text{Notation alert:} \quad \exp(y) \equiv e^y\bigr]$$

for C_0, a constant bearing the appropriate dimensions.

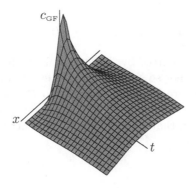

FIGURE 8.2   Green's Function Solution to the 1-D Heat Equation

Several comments are warranted.

---

[4]Such dependence produces a non-linear differential equation for the concentration. These beasties are typically very difficult to solve.

[5]We don't really do this ourselves. Instead, following Newton, we *stand on the shoulders of giants*, and appropriate this known result.

$t \to 0$ Despite the appearance of factors which diverge, this solution is [relatively] well-behaved as $t \to 0$. For $x \neq 0$, $c_{\mathrm{GF}}(t, x) \to 0$ as $t \to 0^+$. However, when $x = 0$, the limit $t \to 0^+$ of $c_{\mathrm{GF}}(t, x)$ is ∞.

These behaviours are revealed in Figure 8.2 where, in essence, a narrow peak residing at the origin at time $t = 0$ flattens and broadens as time advances. Note that the solution is interpretable only for future times, *i.e.*, $t \geq 0$.

SHIFT   Our ability to freely choose the origin of coordinates and the zero of time points to an immediate generalisation of the minimal solution. Thus,

$$c_{\mathrm{GF}}(t, x \,;\, t_0, x_0) = \frac{C_0}{\sqrt{4 \pi D (t - t_0)}} \exp\left(-\frac{(x - x_0)^2}{4 D (t - t_0)} \right)$$

represents the diffusion of a concentration spike initially localised at x_0 at the instant $t = t_0$. Again, the solution is only valid for times $t > t_0$. Some authors expressly enforce this by multiplying the Green's Function solution above by a temporal **Heaviside**[6] **step function**,

$$\theta(t_0, t) = H(t - t_0) = \begin{cases} 1 \,, & t - t_0 > 0 \\ 0 \,, & t - t_0 < 0 \end{cases}.$$

The precise value of the step function at zero argument, $t - t_0 = 0$, is a matter of convention, with the common choice being $H(0) \equiv 1/2$.

DIST'N   The Green's Function solutions ought to be interpreted in a distributional sense. Phenomenologically relevant solutions may be constructed via [continuous] linear superposition of appropriately weighted Green's Function solutions.

To wit, suppose that a source of the diffusing substance is described by the function $s(t^*, x^*)$. The source typically has both temporal and spatial dependence. The concentration at the spatial position x at time t is obtained by employing the Green's Function solution to disperse all of the substance dispensed by the [extended] source at all prior times and superposing the results. Thus, in mathematical form,[7]

$$c(t, x) = \int \left\{ c_{\mathrm{GF}}(t, x \,;\, t^*, x^*) \, H(t - t^*) \right\} s(t^*, x^*) \, dt^* \, dx^* \,,$$

where causality is enforced by the temporal step function. Provided that the source function is not so singular as to prevent convergence of the integral, the orders of integration (with respect to t^* and x^*) and differentiation (t and x) may be interchanged, in which case it is readily apprehensible, by direct substitution, that the convolved[8] concentration function satisfies the diffusion equation.

II   Fourier Solutions

A standard approach to multi-variable differential equations is to restrict one's attention to **separable** solutions. In the case at hand, these are of the form

$$c(t, x) = T(t) \, X(x) \,,$$

where $T(t)$ is a function of t alone, while $X(x)$ depends solely on x.

---

[6] Oliver Heaviside (1850–1925) was an English scientist and inventor.
[7] The technical term for this type of operation is **convolution**.
[8] Technically, it is convolved, but it may be *convoluted*, too.

ASIDE: Despite this being a draconian restriction on the form of the concentration function, the underlying linearity of the diffusion equation allows the formation of more general solutions by superposition of these separable solutions.

Assuming both separability and non-zero concentration leads to

$$\frac{1}{T}\frac{dT}{dt} = \frac{D}{X}\frac{d^2 X}{dx^2}.$$

As the LHS is exclusively a function of time while the RHS depends solely on position each must be separately equal to the same [dimensionless] constant! Identifying this constant as $-k^2 D$, for convenience,[9] we separately obtain

$$\frac{d\ln(T)}{dt} = -k^2 D, \quad \text{and} \quad \frac{d^2 X}{dx^2} = -k^2 X.$$

The solution to the temporal equation is the exponential function, $\exp\left(-k^2 D t\right)$. The spatial part is of SIMPLE HARMONIC OSCILLATOR [SHO] form,[10] with sinusoidal solutions. Assembling these results yields concentrations of the form

$$c_{\mathrm{F}}(t,x\,;\,k) = e^{-k^2 D t}\left(A_c \cos\left(k\,x\right) + A_s \sin\left(k\,x\right)\right),$$

where $\left\{A_s, A_c\right\}$ are the constants of integration necessary to accommodate initial/boundary conditions, while k is derived from the separation constant.

These solutions, too, are best understood in a distributional sense. A view of the pure $k = 3$ concentration is sketched in the left panel of Figure 8.3. The equal-weight superposition of all k from 2 to 10 appears in the right panel.

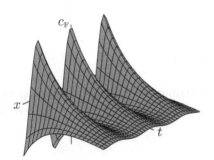
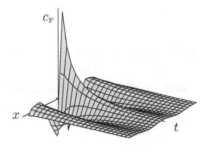

FIGURE 8.3 Fourier Solutions to the 1-D Heat Equation

The Fourier solutions are multi-modal (unless all k appear) and evince the flattening and broadening of concentration peaks.

---

[9]The important bit here is that the constant is negative definite. Positive and zero values are unphysical. The "D" appears as magnitude only [without its units].

[10]Chapters 11 through 20 are consumed with investigations of aspects of harmonic oscillation. The SHO solutions, quoted here, were discussed in VOLUME I.

ASIDE: This is a foretaste of Fourier analysis, which will be more fully investigated in Chapter 23. The Fourier solutions to the diffusion equation are
dispersive, in that the shape of the concentration function changes with
time as inhomogeneities are flattened and broadened. When we encounter
mechanical waves, in Chapter 21 *et seq.*, we shall restrict ourselves to the
dispersionless case. We'll revisit dispersion in the context of optics in
Chapter 33.

III   Random Walk Simulations of Diffusive Behaviour

At the submicroscopic level, individual molecules are best thought of as moving
randomly. In 1-d isotropic systems, the *a priori* probability of moving in one direction is equal to that of moving in the other [in a given time interval]. Formalising
this into a mathematical model, one has the so-called "random walk."

Output from a simple computer programme simulating a 1-d random walk consisting of 10 equal-sized steps is presented in Figure 8.4. In each plot, the height
of the histogram bar represents the relative frequency with which the corresponding displacement from the starting point was obtained in various-sized samples
of walks. Isotropy suggests that the distributions should be symmetric about the
origin. The asymmetries exhibited in the histograms can be ascribed to statistical
fluctuations whose effects wane as the number of trials increases.

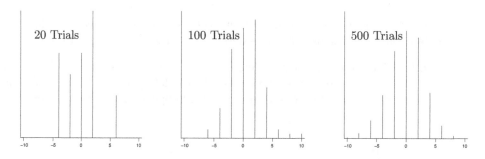

FIGURE 8.4 Random Walk Trials to Model Diffusion

As the number of walks becomes larger, the distribution of final positions appears
increasingly Gaussian[11] [albeit discrete, owing to the crudeness of the model with a
fixed stepsize].

---

[11]Chapter 50 contains an analysis of certain properties of the Gaussian distribution in the context of the
Maxwell–Boltzmann distribution of molecular velocities found when an ideal gas is in a state of thermal
equilibrium.

Chapter 9

Baby, It's Viscous Outside

Of the assumptions underlying the model of ideal fluids:

Laminar,   Non-Viscous,   Incompressible,   (Uniform),

eschewing viscosity has the gravest consequences. Viscosity is

    o   associated with the perceived "thickness" or "stiffness" of a fluid

> E.g., maple syrup is *stiffer* than blood, which is *thicker* than water,
> which in turn is more viscous than methanol (wood alcohol).

    o   often described as a form of internal friction [We did exactly this in Chapter 6.]

    o   and defined for fluids in a manner which parallels that of shear modulus for solids, studied in Chapters 2 and 3

[That there are a number of different types of viscosity further complicates matters.]

VISCOSITY   The viscosity of a fluid undergoing laminar flow relates the gradient of the fluid's velocity field to the shear stress needed to maintain the flow.

To further develop this notion, consider the situation in which homogeneous fluid completely fills a region of space lying between two [large[1]] parallel planar surfaces. A microscopic layer of fluid binds tightly to each of the surfaces. Suppose that one of the surfaces is forced into uniform "sideways" motion at speed v_0, while remaining parallel. A constant applied force, F_{A}, is required to keep the plate moving at constant speed.

> In the absence of viscosity, the moving plate would slide without resistance,
> and no applied force would be required to maintain its speed.

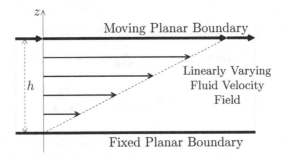

FIGURE 9.1   Couette Fluid Flow between Boundary Surfaces in Relative Motion

---

[1]For sufficiently large surfaces, the perimeter edge effects are negligible.

The layer of bound fluid is carried along with the moving surface. Nearby fluid is also pulled along, but at reduced speed. This effect persists all the way through the bulk of the fluid, until the other, immobile plane is reached. The fluid experiences **Couette flow**,[2] as illustrated in Figure 9.1, with the following set of assumptions.

○   The areas of the moving and fixed surfaces are sufficiently large.[3]

[In **Taylor–Couette flow**, the fluid resides between concentric rotating cylinders. When the cylinder diameters are both large compared to the spacing between them, the local situation is practically indistinguishable from that shown in Figure 9.1.]

○   The system is in steady state, speeds are low, and the fluid flow is laminar.[4]

In Figure 9.1, the lower planar surface is at rest, while the upper surface is sustained in motion at speed v_0 [→] by the continued application of a force, F_A. The fluid speeds adjacent to the top and bottom surfaces are v_0 and 0, respectively. The assumptions concerning the common area of the plates, A, ensure that $\sqrt{A} \gg h$.

ASIDE: The same Couette flow, as seen from the perspective of an observer moving along with the upper surface, appears in Figure 9.2. Viscous effects are [inertial] observer independent.

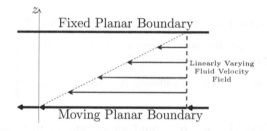

FIGURE 9.2   Couette Fluid Flow from a Different Perspective

The viscosity, η, is[5]

$$\eta = \frac{F_A/A}{\Delta v/\Delta h} \, .$$

The numerator is the shear stress, *cf.* Chapter 2, while the denominator is the average rate of change of the [laminar] flow speed with respect to the perpendicular distance from the surface judged to be at rest. Here,

$$\frac{\Delta v}{\Delta h} = \frac{v_0 - 0}{h}, \qquad \Longrightarrow \qquad \eta = \frac{h}{v_0} \frac{F_A}{A} = \frac{h}{v_0} \left[\begin{array}{c} \text{shear} \\ \text{stress} \end{array} \right] .$$

Several aspects of this quantification of viscosity are elaborated upon below.

UNITS   The SI units of viscosity are $[\eta] = \frac{m}{m/s}$ Pa = Pa · s. Expressed in fundamental SI units, this[6] becomes

$$[\eta] = \text{Pa} \cdot \text{s} = \frac{\text{kg}}{\text{m} \cdot \text{s}^2} \cdot \text{s} = \frac{\text{kg}}{\text{m} \cdot \text{s}} .$$

---

[2]Maurice Couette (1858–1943) was a French physicist who specialised in Rheology.

[3]Here, "large" means that the two surfaces remain effectively overlapping, despite their relative motion, throughout all relevant time scales.

[4]At higher fluid speeds, the system passes through stages of increasing turbulence.

[5]CAVEAT: Engineers often employ "μ" to denote viscosity.

[6]Some authors dub the Pa · s a poiseuille after Jean Louis Poiseuille (1797–1869), who developed a model of laminar flow of viscous incompressible blood through narrow cylindrical arteries and veins.

ASIDE: The CGS unit, the `poise`, may be readily converted to `Pa · s`:

$$\texttt{poise} = \frac{\text{g}}{\text{cm} \cdot \text{s}} \quad \Longleftrightarrow \quad 1 \, \text{Pa} \cdot \text{s} = 10 \, \texttt{poise} \, .$$

TYPES   In this presentation of viscosity, it has been implicitly suggested that η is constant. This need not be the case. The common varietals of viscosity are listed in the table below.

$\eta = $ constant	Newtonian Fluid
$\eta \uparrow$ as $v \uparrow$	Shear Thickening
$\eta \downarrow$ as $v \uparrow$	Shear Thinning
$\eta \uparrow$ with stirring and time	Rheopectic
$\eta \downarrow$ with stirring and time	Thixotropic
$\cdots$	$\cdots$

The Newtonian Fluid is the simplest approximate model of a viscous fluid. Some automotive transmission and differential systems employ shear thickening fluid to transfer torque to the wheels. Paint is typically shear thinning so as to stay on the applicator and its intended surface, but to flow readily from one to the other. Whipped cream and meringue are tasty instances of rheopectic fluids.

Cream and egg whites go into respective mixing bowls. After being subjected to the beaters for a time, each emerges stiff, glossy, and capable of sustaining "peaks."

Certain clays are thixotropic, in that they become softer as they are worked.

ALTERNATE   The **kinematic viscosity**, $\nu = \eta/\rho$, where ρ is the fluid density, provides another description of viscous phenomena. The SI units of kinematic viscosity are

$$[\nu] = \frac{\text{Pa} \cdot \text{s}}{\text{kg/m}^3} = \frac{\text{kg}}{\text{m} \cdot \text{s}} \frac{\text{m}^3}{\text{kg}} = \frac{\text{m}^2}{\text{s}} \, .$$

ASIDE: The CGS equivalent is called the `stokes`,[7]

$$\text{St} = \frac{\text{cm}^2}{\text{s}} = 10^{-4} \frac{\text{m}^2}{\text{s}} \, .$$

More convenient still is the `centistokes`,

$$\text{cSt} = 10^{-2} \frac{\text{cm}^2}{\text{s}} = 10^{-6} \frac{\text{m}^2}{\text{s}} \, ,$$

because it well-approximates the kinetic viscosity of liquid water.

CONVERSE   The reciprocal of viscosity is termed **fluidity**. The viscous behaviour of mixtures may be more readily expressed in terms of the fluidities of the constituents.

P, T Another complication is that viscosity is pressure and temperature dependent.

---

[7]Sir George Stokes (1819–1903), a distinguished mathematical physicist, occupied the Lucasian Chair of Mathematics at Cambridge. This is the professorship once held by Newton, Dirac, and Hawking!

EXAMPLE [*Viscosity and Taylor–Couette—1*]

Homogeneous fluid completely fills the region of space lying between concentric and completely overlapping cylindrical shells with common length L and inner and outer radii R_i and R_o (such that $h = R_o - R_i \ll R_i$). A torque with magnitude τ_0 is required to keep the outer cylinder rotating slowly and uniformly, with angular speed ω_0, about the fixed inner cylinder.

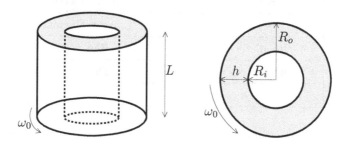

FIGURE 9.3   Concentric Cylindrical Shells with Interstitial Viscous Fluid

Q: What is the viscosity of the fluid lying between the cylinders?

A: The tangential speed of the outer cylinder is $v = \omega_0 R_o$, and thus

$$\frac{\Delta v}{\Delta h} = \frac{\omega_0 R_o}{R_o - R_i} = \frac{\omega_0 R_o}{h}.$$

To determine the shear stress, we shall ascertain the area of the outer surface from its geometry, and the force needed to keep it in uniform motion from the torque. The area of the outer cylinder is $A = 2\pi R_o L$, while the moment arm of the shear force is R_o. Therefore,

$$\tau_0 = F_{\text{A}} R_o, \qquad \text{and hence} \qquad F_{\text{A}} = \frac{\tau_0}{R_o} \qquad \Longrightarrow \qquad \frac{F_{\text{A}}}{A} = \frac{\tau_0}{2\pi R_o^2 L}.$$

> ASIDE: Note that it is too simplistic to conclude that the shear stress is inversely proportional to L and R_o^2, because the critical value of the torque is expected to depend on the geometry of the cylinders, too.

Continuing the analysis to determine the viscosity, we obtain

$$\eta = \frac{F_{\text{A}}/A}{\Delta v/\Delta h} = \frac{1}{\Delta v/\Delta h} \left[\begin{array}{c} \text{shear} \\ \text{stress} \end{array} \right] = \frac{h}{\omega_0 R_o} \frac{\tau_0}{2\pi R_o^2 L} = \frac{h\,\tau_0}{2\pi \omega_0 R_o^3 L}.$$

[Recall the implied assertion that $R_o \sim R_i$.]

EXAMPLE [*Viscosity and Taylor–Couette—2*]

Glycerine, with viscosity $\eta_G = 1.4\,\mathrm{Pa \cdot s}$, completely fills the space between two concentric overlapping cylinders with common length $L = 40\,\mathrm{cm}$, and radii $R_i = 15\,\mathrm{cm}$ and $R_o = 16\,\mathrm{cm}$, respectively.

Q: What torque is required to turn the inner cylinder so that its period of rotation is $10\,\mathrm{s}$ (while the outer cylinder remains at rest)?

A: The uniform angular speed is $\omega_0 = \frac{2\pi}{T} = \frac{\pi}{5}\,\mathrm{rad/s}$, and thus the tangential speed of the inner cylinder is $v_t = \omega_0\,R_i = 0.03\,\pi\,\mathrm{m/s}$.

The shear stress necessary to keep the cylinder in motion is

$$\left[\begin{array}{c} \text{shear} \\ \text{stress} \end{array} \right] = \eta_G\,\frac{v_0}{h} = 1.4 \times \frac{0.03\,\pi}{0.01} = 4.2\,\pi\,\mathrm{Pa}.$$

The area of the inner cylinder is $A = 2\,\pi\,R_i\,L = 0.12\,\pi\,\mathrm{m}^2$, and therefore the force distributed across the moving surface is

$$F_{\text{A}} = 4.2\,\pi \times 0.12\,\pi = 0.504\,\pi^2\,\mathrm{N}.$$

The torque associated with such a force acting at the radius of the inner cylinder is

$$\tau = F_{\text{A}} \times R_i = .504\,\pi^2 \times 0.15 \simeq 0.746\,\mathrm{N \cdot m}.$$

ASIDE: We can check this result by artfully rearranging the expression obtained in the previous example. There,

$$\eta = \frac{h\,\tau_0}{2\,\pi\,\omega_0\,R_o^3\,L}.$$

Hence, in the present case, we write,

$$\tau = \frac{2\,\pi\,\eta_G\,\omega_0\,R_i^3\,L}{h}.$$

Substituting first $\omega_0 = 2\,\pi/T$, and then the parameter values, yields

$$\tau = \frac{4\,\pi^2\,\eta_G\,R_i^3\,L}{T\,h} = \frac{4\,\pi^2 \times 1.4 \times 0.15^3 \times 0.4}{10 \times 0.01} = 0.0756\,\pi^2 \simeq 0.746\,\mathrm{N \cdot m}.$$

REYNOLDS NUMBER

All of the analyses that we have undertaken have avoided turbulance and assumed laminar flows. Rather than explicitly verifying [or merely hoping] that this assumption is borne out in particular cases, one seeks a discriminant between the laminar and turbulent regimes. The Reynolds Number plays this vital rôle.

REYNOLDS NUMBER   The Reynolds Number,[8] Re, is a ratio of relevant quantities and scales pertinent to situations in which fluid is moving with respect to a solid surface. We'll quote a simplified version pertaining to fluid flowing with mean speed[9] $\bar{v}$ in a cylindrical pipe with diameter D and cross-sectional area $\mathcal{A} = \frac{\pi}{4}\,D^2$. The volume flux is $\Phi_V = \frac{1}{4}\,\pi\,D^2\,\bar{v}$. The Reynolds number, in this context, is

$$Re = \frac{\rho\,\bar{v}\,D}{\eta} = \frac{\rho\,\Phi_V\,D}{\eta\,\mathcal{A}} = \frac{\Phi_V\,D}{\nu\,\mathcal{A}} = \frac{4\,\Phi_V}{\pi\,D\,\nu}.$$

---

[8]Osborne Reynolds (1842–1912) studied fluid flow in pipes and around ships.

[9]This is the speed at which a uniform flow across the entire available cross-section of the pipe yields the actual volume flux.

SMALL   IF the Reynolds number is less than one thousand or so, THEN the fluid flow is laminar.

INTER-   WHEN the Reynolds number is in the mid-thousands to the mid-ten thou-
MEDIATE   sands, crossover from laminar to turbulent flow occurs.

LARGE   IF the Reynolds number is greater than one hundred thousand, THEN the fluid flows in a turbulent manner.

The significance of the Reynolds number is that it compares the relative net strengths of the microscopic internal fluid forces, which attempt to guide fluid elements along streamlines, with the effects of inertia, which can lead the fluid elements to career in wild, unpredictable, **turbulent** manners.

EXAMPLE [*Reynolds Numbers*]

The kinetic viscosity of water is approximately $1\,\text{centistokes} = 10^{-6}\ \text{m}^2/\text{s}$. The Reynolds number for water completely filling a pipe with inside diameter of $80\,\text{cm}$ is

$$Re = \frac{4\,\Phi_V}{\pi\,D\,\nu} = \frac{4\,\Phi_V}{\pi \times \frac{4}{5} \times 10^{-6}} = \frac{5}{\pi} \times 10^6\ \Phi_V\,.$$

• At low flux, say $\frac{2\pi}{25} \times 10^{-3}\ \text{m}^3/\text{s} \simeq 0.251\,\text{L/s}$, for which the mean flow speed is $1/2\ \text{mm/s}$, the Reynolds number is $Re = 4 \times 10^2$. Such a sedate flow is certain to be laminar.

• At an intermediate rate of flux, say $\frac{\pi}{1250}\ \text{m}^3/\text{s} \simeq 2.51\,\text{L/s}$, the mean flow speed is $1/2\ \text{cm/s}$. The Reynolds number in this case, $Re = 4 \times 10^3$, lies in the transition region. Turbulence may arise even at such low speeds.

• For a relatively large flux, say $\frac{2\pi}{25}\ \text{m}^3/\text{s} \simeq 251\,\text{L/s}$, the mean flow speed is $1/2\ \text{m/s}$, and hence the Reynolds number is $Re = 4 \times 10^5$, indicating that the flow is definitely turbulent.

NAVIER–STOKES EQUATION

The Navier–Stokes Equation provides a general [local] expression of N2 within a fluid substance. While admitting a relatively simple and aesthetic form, the N–S equation is a ferociously complicated set of non-linear coupled second-order differential equations for the fluid velocity field. In the [simplest] case of an incompressible Newtonian fluid [constant density ρ and constant viscosity η] one obtains

$$\rho\left(\frac{\partial \vec{v}}{\partial t} + (\vec{v} \cdot \vec{\nabla})\,\vec{v}\right) = -\vec{\nabla}P + \eta\,\nabla^2\vec{v} + \vec{F}_{\text{ext'l}}$$

for the velocity field $\vec{v}(t,\vec{r})$, where $\vec{\nabla}$ is the gradient[10] and the **Laplacian** is $\nabla^2 \equiv \vec{\nabla} \cdot \vec{\nabla}$. The LHS is comprised of a local "$m\,\vec{a}$" term and a covariant flow term. The RHS receives contributions from the pressure gradient, the shear stress, and any external forces present.

Although the general N–S equation has been studied since the mid-1800s, mathematical issues of existence and uniqueness of its solutions are, as yet, still unresolved!

---

[10]The gradient was introduced in Chapter 26 of VOLUME I.

Chapter 10

Gas Gas Gas

Let's recall what has transpired thus far. We realised in Chapter 1 that the mere fact that tidal forces are able to increase without bound demands that the rigidity of objects, assumed throughout VOLUME I, perforce be relaxed. Relegation of rigidity invited the introduction of Young's, shear, and bulk elastic moduli for solids.

Shear was done away with in our study of fluids.

ASIDE: Recollect that fluids eventually conform to the shape of the container in which they reside on account of there being no lasting insurmountable impediment to rearrangement of neighbouring agglomerations of substance.

Chapters 6 and 7 were concerned with the static and dynamic properties of ideal liquids. In this chapter, the incompressibility restriction shall be relaxed, thus admitting "small" values of the bulk modulus coefficient. Fluids with small bulk moduli are typically gaseous.

EMPIRICAL GAS LAWS

Robert Boyle (1627–1691) performed various experiments on sealed canisters of gases. Using pistons to regulate the volumes, and pressure gauges [not unlike those described in Chapter 6], he discovered his eponymous law.

BOYLE'S LAW Boyle's Law,

$$P \propto \frac{1}{V},$$

holds for dilute gases maintained at constant temperature, provided that phase changes[1] do not occur.

Somewhat later, numerous eighteenth century scientists conducted many experiments with various gases exposing several empirical regularities.

CHARLES' LAW That the volume of a sample of gas [at fixed pressure] changed in proportion to its temperature was reported by Joseph Louis Gay-Lussac (1778–1850), but he credited this discovery to Jacques Alexandre César Charles (1746–1823). A quantified expression of Charles' Law is

$$V(T) = V_0 \left[1 + \frac{T - T_0}{273} \right],$$

where V_0 is the volume of the gas at the temperature T_0, expressed in Celsius. The factor appearing in the denominator, $273\,\text{C}$, is an empirical constant.[2]

---

[1] *E.g.*, condensation of gas into liquid or deposition of gas into solid.

[2] The particular value of this constant cannot fail to pique one's interest given that the ABSOLUTE ZERO of temperature, to be discussed in Chapter 37, is deemed to occur at $-273.15\,\text{C}$.

GAY-LUSSAC'S LAW   Gay-Lussac[3] determined that [under most circumstances]

$$P \propto T,$$

when the sample of gas is maintained at fixed volume.

Further investigations, by Amadeo Avogadro (1776–1856)[4] led to the introduction of AVO-GADRO'S PRINCIPLE.

AVOGADRO'S PRINCIPLE   Avogadro's Principle is that equal volumes of gas at the same pressure and temperature contain the same number of molecules.[5] A corollary to Avogadro's Principle is that cramming more gas into a fixed volume at constant temperature engenders higher pressure, *i.e.*,

$$P \propto \text{\# of gas particles.}$$

All four empirical relations, Boyle's Law, Charles' Law, Gay-Lussac's Law and Avogadro's Principle, are subsumed into the IDEAL GAS LAW.

IDEAL GAS LAW   The ideal gas law admits two modes of expression, *viz.*,

$$PV = nRT \qquad \text{OR} \qquad PV = N k_B T.$$

The LHS is the same in each case, and consists of the product of the pressure and the volume. It is not insignificant that the units associated with the ideal gas law are those of energy:

$$[P] \times [V] = \frac{\text{N}}{\text{m}^2} \times \text{m}^3 = \text{N} \cdot \text{m} = \text{J}.$$

On each RHS, T represents the absolute temperature [expressed in kelvin, K] of the system. The quantity n denotes the number of moles of material present in the gas comprising the system, while N represents the number of molecules in the same sample. The **Universal Gas Constant**, R, and the **Boltzmann Constant**, k_B, are determined to be

$$R = 8.31447 \frac{\text{J}}{\text{mole} \cdot \text{K}} \qquad \text{and} \qquad k_B = 1.3806 \times 10^{-23} \frac{\text{J}}{\text{molecule} \cdot \text{K}},$$

respectively. These quantities, $\{N, n\}$, and [empirical] constants, $\{R, k_B\}$, are related via AVOGADRO'S NUMBER,[6] $N_A \simeq 6.0225 \times 10^{23}$. *I.e.*,

$$n = \frac{N}{N_A} \qquad \text{and} \qquad R = N_A k_B.$$

A host of conditions must be met in order for the ideal gas law to be operative.

    o   The gaseous substance must be sufficiently rarefied or diffuse.

---

[3]Some historians ascribe this law to Guillaume Amontons (1663–1705). We refer to it as Gay-Lussac's.
[4]Avogadro's ideas did not gain widespread currency until after his death.
[5]Hey! What are molecules? Perhaps a better expression would be "... the same amount of chemical stuff in accord with the idea of invariant elemental proportions propounded by Dalton *et al.*," but such devotion to historicity is unnecessary since we have plenty of evidence for the existence of molecules.
[6]Accurate determinations of N_A were not made until the early years of the twentieth century.

    o   The system must be in thermal equilibrium for there to exist a unique characteristic temperature,[7] T.

    o   *etc.*

In the remainder of this chapter, the first of these quoted conditions shall be examined in greater detail. Discussions of thermal equilibrium and temperature scales shall await our introduction of Thermodynamics, starting in Chapter 37. Henceforth, our employment of the ideal gas law presupposes that all of the necessary conditions are met [to sufficient precision] in the physical system under consideration.

Two idealisations implicit in the ideal gas law which bear on the diffuseness of the gaseous state are that the molecular constituents of the gas are point particles and that they are non-interacting.

• **POINT PARTICLES** The molecules occupy no volume, and thus all of the space available to the gas can accommodate the molecules.

> ASIDE: This only seems arcane until a *Gedanken* experiment, involving several dozen small marbles in a shoebox, clarifies matters.
>
> Any particular marble may occupy any[8] position within the box.
>
> That the marbles do not overlap or interpenetrate diminishes the volume available to each member of the collection. As a matter of phenomenology, the marbles are dissuaded from overlapping through the agency of a [strongly repulsive and short-range] **contact force**.

• **NON-INTERACTING** To a first approximation, dilute gases are non-interacting. However, some degree of attraction among the gas molecules is absolutely necessary for there to be changes of phase [*cf.* Chapter 38] or chemical composition.

The situation is subtle, and edge effects play an important rôle.

    — Deep in the [homogeneous] fluid, the environs of any particular gas particle are isotropic on intermediate length scales.[9] Thus, any interaction between particles in this region will tend to produce ZERO net effect.   [Ensuring the veracity of this last statement may require averaging over "long-enough" time scales.]

    — Near the edges there is a relative paucity of interaction partners on the "wall side" *vis-à-vis* toward the interior of the box. This asymmetry allows a net intra-fluid force to accrue for particles near the edges.

> ASIDE: That gases spontaneously condense to liquid [under certain conditions], and liquids cohere, indicates that attractive forces must be dominant on the length-scales corresponding to interparticle spacings in the liquid phase. Nonetheless, the short-range repulsive forces are dominant at shorter scales, and intermolecular forces vanish for long length scales. Many

---

[7]This temperature must be expressed using an ABSOLUTE SCALE, *e.g.,* kelvin.

[8]CAVEAT: There are edge/boundary effects too. However, for boxes with all sides large on the scale set by the marbles, such *side-effects* are guaranteed to be small.

[9]Intermediate length scales are long compared to the size of the molecules and short with respect to the physical dimensions of the system.

models with these features are extant. The best known are the so-called "6–12 models" with potential energy functions given by

$$U_{6\text{--}12} = \frac{C_{12}}{r^{12}} - \frac{C_6}{r^6}\,,$$

where $\{C_{12}, C_6\}$ are phenomenological [measured] constants.

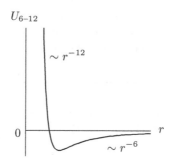

Relaxing the assumption of non-interacting point molecule constituents, in a particular way, yields an extension of the ideal gas law called the **Van der Waals equation of state**.

EQUATION of STATE   An equation of state [EoS] is any valid mathematical expression relating thermodynamic parameters. In modern parlance, Boyle's Law, Charles' Law, Gay-Lussac's Law, the ideal gas law, and the Van der Waals equation are all equations of state.

This definition, sufficient for now, will be embellished when we encounter the FIRST LAW OF THERMODYNAMICS in Chapter 42.

VAN DER WAALS EQUATION   The Van der Waals EoS reads:

$$\left(P + a\left(\frac{n}{V}\right)^2\right)\left(V - bn\right) = nRT\,,$$

where $\{a, b\}$ are two empirical [substance-specific] constant parameters. It is a refinement of the ideal gas law.

There are three things to note about the Van der Waals extension of the ideal gas law.

- The empirical parameters, $\{a, b\}$, arise from the finite size of, and the forces of interaction between, the molecular constituents of the gas:

    a   is a measure of the strength of the attraction between gas particles

    b   represents the volume excluded by a mole of gas particles

- The Van der Waals EoS is readily reformulated in terms of the MOLAR DENSITY[10] of the gas, $\eta = n/V$:

$$\left(P + a\eta^2\right)\left(1 - b\eta\right) = \eta RT\,.$$

---

[10]Molar density is equivalent to number density.

• The pressure appearing in the reformulated Van der Waals equation may be expanded perturbatively in terms of the molar density, about the value $\eta = 0$. The series expansion

$$P = P_0 + P_1\,\eta + P_2\,\eta^2 + \dots$$

has coefficients P_i, which are to be determined self-consistently. Substituting this series form of the pressure into the Van der Waals equation and solving order-by-order in η, one obtains the results displayed in the following table.

Order	Equation	Implication
0th	$P_0 = 0$	No gas ⇔ no pressure
1st	$\eta\,P_1 = \eta\,RT$	Ideal gas law
2nd	$\eta^2\,P_2 = \eta^2\,(b\,RT - a)$	Correction to ideal gas law
3rd	...	Higher-order corrections

The zeroth-order part confirms that in the absence of gas the pressure vanishes. The first-order piece reproduces the ideal gas law. Higher-order terms generate perturbative corrections to lower-order results.

A crucially important point of consideration for gaseous fluids is that:

**For a fixed amount of ideal gas at constant temperature,
the density varies in proportion to pressure.**

• Suppose that an object, *Bob*, floats[11] at a particular height, H_0, in a column of gas. According to Archimedes, the buoyant force provided by the surrounding substance, acting on Bob, exactly cancels his weight.

IF Bob were raised a small distance above H_0 and released, THEN he would suffer a net downward force on account of the diminution of the buoyant force[12] arising from the decrease in density of the fluid substance.

IF Bob were pushed a small distance below H_0 and let go, THEN he would experience a net upward force due to the enlarged buoyant force[13] owing to the increase in density of the fluid substance.

The significance of this analysis is the intimation that the weight and buoyant forces are combining to produce a **restoring** force acting on the object. Thus, Bob's motion might exhibit features of SIMPLE HARMONIC OSCILLATION, which were briefly discussed in VOLUME I and will be carefully examined in upcoming chapters.

• The local version of Pascal's formula reads: $dP = \rho\,g\,dh$. According to the ideal gas law EoS, at constant temperature, the density varies linearly with pressure. Hence,

$$dP \propto P\,dh\,,$$

characteristic of an exponential relation between pressure and height.

---

[11]We say, in this instance, that Bob possesses NEUTRAL BUOYANCY.
[12]One would say that Bob has NEGATIVE BUOYANCY.
[13]Bob's got POSITIVE BUOYANCY in this case.

As a final thought in this chapter, let's determine the amount of mechanical work done by an expanding ideal gas. The notion of a force acting through a distance generalises to a pressure acting throughout a change in volume. *Ergo*,

$$W_{if}[P] = \int_{V_i}^{V_f} P \, dV.$$

For a fixed amount of ideal gas held at constant temperature, the work done is

$$W_{if}[P] = \int_{V_i}^{V_f} P \, dV = \int_{V_i}^{V_f} \frac{nRT}{V} \, dV = nRT \ln [V] \Big|_{V_i}^{V_f} = nRT \ln \left[\frac{V_f}{V_i} \right].$$

IF the gas $\left\{ \begin{array}{c} \text{expands} \\ \text{Contracts} \end{array} \right\}$, THEN the mechanical work done by the gas is $\left\{ \begin{array}{c} \text{POSITIVE} \\ \text{NEGATIVE} \end{array} \right\}$.

Chapter 11

Through the Earth and Back

A tunnel has been dug from the Perimeter Institute for Theoretical Physics [located in Waterloo, Ontario, Canada] straight down to the centre of the Earth and through to the other side. The excavated material was used to build the island of *Oolretaw* [in the Indian Ocean, southwest of Perth, Australia]. The Diameter Institute for Experimental Physics is located at the Oolretaw terminus of the shaft.

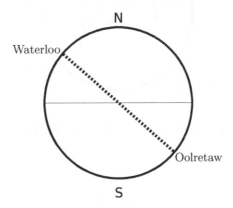

FIGURE 11.1   A Tunnel through the Earth

ASIDE: The Perimeter Institute has approximate latitude $43.4634°$ N and longitude $80.5208°$ W. Therefore, the Diameter Institute is found at $43.4634°$ S and $99.4792°$ E.

In this chapter, the Earth shall be modelled as a solid sphere with uniform mass density

$$\rho_\oplus = \frac{M_\oplus}{\frac{4\pi}{3}\,R_\oplus^3} \sim 5520 \text{ kg/m}^3 \,.$$

ASIDE: Owing to the extent and relative homogeneity of the Earth's mantle and off-setting corrections from the core and crust, it turns out that $\rho_\oplus$ lies in the range of silicate minerals.

Gauss's Law for Gravity [*cf.* Chapter 48 in VOLUME I] determines the net gravitational force exerted by a spherically symmetric body upon a point-like body of mass m, located at distance r from the centre of the spherical source, in direction $\hat{r}$, to be

$$\vec{F}_{\text{G}} = m\,\vec{g}(r) = m\,\frac{G\,M'}{r^2}\,[-\hat{r}] = -\frac{G\,M'\,m}{r^2}\,[\hat{r}]\,,$$

where M' is the mass of that portion of the spherical body nearer to the centre than the point-like object.

◯ · IF the field point is external to the distribution, **THEN** the entire body participates in the gravitational interaction and $M' = M_{\text{Tot}} = M_\oplus$.

☉ IF the field point is internal to the source, **THEN** the contributions to the effective mass come only from material located nearer to the centre than r. *I.e.*,

$$M' = \int_{\text{Volume within } r} \rho \, dV = \int_0^r \rho \, 4\pi r^2 \, dr \,,$$

assuming spherical symmetry.

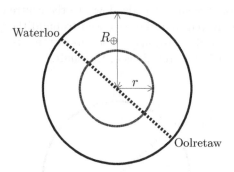

FIGURE 11.2 Gauss's Law for Gravity Applied to the Tunnel

When the density is constant [as assumed in this model of the Earth], the integral becomes

$$M' = \rho_\oplus \, \frac{4\pi}{3} r^3 \,.$$

In anticipation of the next step, the contributing mass may be rewritten as the appropriate
[volume] fraction of the total mass:

$$M' = M_\oplus \, \frac{r^3}{R_\oplus^3} \,.$$

Thus, the net force on the [point-like] object in the tunnel is

$$\vec{F}_{\text{G}} = -\frac{G M_\oplus \, m}{R_\oplus^3} \, r \, \hat{r} = -\omega^2 \, m r \, \hat{r} \,,$$

where, for notational simplicity,[1] some parameters have been subsumed into a [positive, real]
constant, ω^2. In this case,

$$\omega^2 = \frac{G M_\oplus}{R_\oplus^3} = \frac{6.67 \times 10^{-11} \times 6 \times 10^{24}}{(6.38 \times 10^6)^3} \simeq \frac{400}{(6.38)^3} \times 10^{-6} = 1.54 \times 10^{-6} \,,$$

with apparent units

$$[\omega^2] = \left[\frac{\frac{\text{N·m}^2}{\text{kg}^2} \, \text{kg}}{\text{m}^3} \right] = \frac{\frac{\text{kg·m}}{\text{s}^2}}{\text{kg} \cdot \text{m}} = \frac{1}{\text{s}^2} \,.$$

---

[1]There is much more to it than this, as we'll see shortly. The important bits here are that the force is
linear in both the mass of the particle and its radial coordinate.

In truth, ω is an angular frequency and its natural units are radians per second. Thus, the magnitude of the angular frequency is $\omega = \sqrt{\omega^2} = 1.24 \times 10^{-3}$ rad/s, and the corresponding oscillatory period is

$$\mathcal{T} = 2\pi/\omega \simeq 5058\,\text{s} = 84.3 \text{ minutes}.$$

EXAMPLE [*Express Male from Waterloo to Oolretaw*]

Suppose that PK were to step into the tunnel joining Waterloo with Oolretaw.

FIGURE 11.3   Freefalling in a Tunnel through the Earth

Q: What happens?

A: Let's employ Newton's Laws via the dynamics recipe in order to find PK's trajectory. Before starting, we must make some [simplifying] assumptions and approximations:

- ◊   PK acts as an effectively point-like particle with constant mass m.
- ◊   The Earth is spherically symmetric, with radius $R_\oplus$.
- ◊   The Earth has mass $M_\oplus \gg m$ and is approximately homogeneous.
- ◊   Friction and drag forces are inoperative or negligible.
- ◊   PK enters the mouth of the hole with ZERO initial velocity.

Here, we shall follow a slightly abbreviated form of the familiar dynamical recipe.

(1)   A sketch of the in-flight situation appears in Figure 11.3.

(2)   PK is the sole dynamical constituent and is acted upon by a single force, gravity.

$$\text{Gravity} \circ \vec{F_{\text{G}}} = -\omega^2\, m\, r\, \hat{r}, \text{ where } \omega^2 = G\, M_\oplus/R_\oplus^3.$$

(3)   PK's position is given by the magnitude of his radial distance from the centre, along with a discrete determination of whether he is in the northern or the southern hemisphere. The most natural coordinate to use is $x \in [-R_\oplus, +R_\oplus]$, where the northern/southern parts of the tunnel have positive/negative x. The position coordinate is denoted by "x" for several reasons:

- the motion occurs in 1-d,

- coordinates are arbitrary, as are the labels we give to them, and

- a true radial coordinate [the r in polar coordinates] is positive definite.

(4) Application of N2 to determine PK's acceleration while he is in the tunnel yields

$$m\,a = -\omega^2\,m\,x\,.$$

Thus, PK's trajectory satisfies the [second-order linear ordinary differential] equation of motion

$$\frac{d^2x}{dt^2} = -\omega^2\,x\,.$$

(5) Solutions to this equation of motion may be obtained by a variety of means.

> ASIDE: A qualitative feature of the dynamics is that the solution trajectories must be bending toward the origin, $x = 0$, at all times.

> ⌢ IF $x(t)$ is positive [*i.e.*, PK is in the northern hemisphere], THEN the acceleration is negative, AND the velocity becomes less positive [northward motion slowing] or increasingly negative [southward speeding up].

> ⌣ IF $x(t) < 0$ [*i.e.*, PK is in the southern half], THEN $a(t) > 0$, AND the velocity becomes less negative [southward slowing] or increasingly positive [northward speed increasing].

METHOD ONE: [*Rote Recognition*]

Two elementary mathematical facts,

$$\frac{d\cos(x)}{dx} = -\sin(x) \qquad \text{and} \qquad \frac{d\sin(x)}{dx} = \cos(x)\,,$$

iterated [in either order] reveal that both **sine** and **cosine** can be employed to construct putative dynamical trajectories:

$$x(t) = A\,\cos(\omega\,t) + B\,\sin(\omega\,t)\,.$$

The simplest form of the solution arises when $t = 0\,\mathrm{s}$ is deemed the instant at which PK steps into the hole, in which case

$$x(t) = A\,\cos(\omega\,t)\,,$$

for some constant A. The boundary/initial condition that PK starts from rest at the surface of the Earth establishes that $A = R_{\oplus}$.

METHOD TWO: [*Able Ansatz*]

Let us propose an *educated guess* for the sought-after solution,

$$x(t) = A \exp(r\,t)\,,$$

for some *a priori* unspecified parameters[2] A and r. This particular mathematical form for the solution stems from the property that the derivative of an exponential is itself an exponential. Since this attribute is preserved under iteration, the exponential function makes for a rather good guess.

The assumed form of $x(t)$ allows computation of its derivatives:

$$\frac{dx}{dt} = r\,A\,\exp(r\,t) \qquad \text{and} \qquad \frac{d^2x}{dt^2} = r^2\,A\,\exp(r\,t)\,.$$

Substitution of these into the dynamical equation yields algebraic constraints involving the parameters. In the present case,

$$\left.\begin{aligned} \text{LHS} &= \frac{d^2x}{dt^2} = r^2\,A\,\exp(r\,t) \\[4pt] \text{RHS} &= -\omega^2\,x = -\omega^2\,A\,\exp(r\,t) \end{aligned}\right\} \quad \Longrightarrow \quad r^2 = -\omega^2 \quad \Longrightarrow \quad r = \pm i\,\omega\,.$$

Insistence that the proposed solution satisfy the equation of motion has determined two [complex, imaginary] values for the parameter r. Given that there is no basis for preferring one over the other, we are obligated to accept an admixture of both, *viz.*,

$$x(t) = A_+\,\exp(r_+\,t) + A_-\,\exp(r_-\,t) = A_+\,\exp(+i\,\omega\,t) + A_-\,\exp(-i\,\omega\,t)\,.$$

> ASIDE: The existence of two r-solutions is fortuitous, as two constants of integration, $A_\pm$, are required in the general solution of a second-order differential equation. Mathematical theorems guaranteeing the EXISTENCE and UNIQUENESS of solutions of second-order linear ordinary differential equations with constant coefficients assure us that physically relevant solutions were not overlooked in this analysis.

That PK is [momentarily] at rest at $x = +R_\oplus$ when $t = 0$ must be incorporated into the trajectory. Setting $t = 0$ in $x(t)$ and in $\frac{dx}{dt}$ yields two equations,

$$R_\oplus = A_+ + A_- \qquad \text{and} \qquad 0 = i\,\omega\left(A_+ - A_-\right),$$

with unique and self-consistent solution

$$A_+ = A_- = \frac{R_\oplus}{2}\,.$$

Thus, in the present case with all of its assumptions, PK's trajectory,

$$x(t) = \frac{R_\oplus}{2}\left(\exp(+i\,\omega\,t) + \exp(-i\,\omega\,t)\right),$$

is [courtesy of Euler's formulae] precisely equivalent to that obtained in METHOD ONE,

$$x(t) = R_\oplus\,\cos(\omega\,t)\,.$$

---

[2]This particular 'r' has nothing to do with radial position. It's just a parameter.

METHOD THREE: [*Fabulous Factor*]

Assuming that a solution to the dynamical equation exists [and that its t derivative is well-defined], one can multiply both sides of the dynamical equation by twice the first derivative of the solution. This transforms the left and right hand sides of the equation into forms which are readily integrable.

$$\left. \begin{array}{l} \text{LHS} = 2\,\dfrac{dx}{dt}\,\dfrac{d^2x}{dt^2} = \dfrac{d}{dt}\left(\dfrac{dx}{dt}\right)^2 \\[3mm] \text{RHS} = -2\,\omega^2\,x\,\dfrac{dx}{dt} = \dfrac{d}{dt}\left(-\omega^2\,x^2\right) \end{array} \right\} \quad \Longrightarrow \quad \left(\dfrac{dx}{dt}\right)^2 = -\omega^2\,x^2$$

$$\Longrightarrow \quad \dfrac{dx}{dt} = \pm i\,\omega\,x\,.$$

A constant of integration was set to zero in obtaining the first integral of the equation of motion, just above. The remaining analysis proceeds apace in the manner of METHOD TWO.

(6)   PK's trajectory has the kinematic features sketched below.

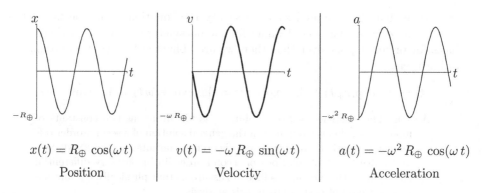

$x(t) = R_\oplus\,\cos(\omega\,t)$	$v(t) = -\omega\,R_\oplus\,\sin(\omega\,t)$	$a(t) = -\omega^2\,R_\oplus\,\cos(\omega\,t)$
Position	Velocity	Acceleration

FIGURE 11.4   Kinematic Aspects of the Express Male Trajectory

Chapter 12

Introduction to Simple Harmonic Oscillation

The topic of this chapter is **Simple Harmonic Oscillation**. Let's commence by parsing this three-word phrase.

OSCILLATION A particle undergoing oscillation [*a.k.a.* oscillatory motion] has a trajectory, $\vec{r}(t)$, that repeats exactly in time with period $\mathcal{T}$, *viz.*,

$$\vec{r}(t + n\,\mathcal{T}) \equiv \vec{r}(t)\,, \quad \text{for all times } t \text{ and integers } n.$$

A multifarious variety of physical phenomena possess [to varying degrees of approximation] this property of repetition. Of the superabundance of possibilities, four kinesthetic examples spring readily to mind.

* Pacing back and forth at a steady speed [making very sudden turns] is oscillatory.

  [This yields a repeating (isosceles) triangular spacetime trajectory, $x(t)$, and is symmetric under time reversal.]

* Repetitive exercise, such as bench press, often involves a two-count drive and a four-count return.

  [Such a spacetime trajectory, $x(t)$, is shaped like a sawtooth and is not time-reversal symmetric.]

* In a *quadrille* eight people perform five sets of similar movements and at the end each finds himself back in his original spot on the dance floor. In this instance the repeated pattern consists of sub-units which are repetitive [or quasi-repetitive].

  [Each dancer's trajectory is a path in a 3-d (t, x, y) space.]

* Running constant speed laps on a circular track is an exemplar of oscillation.

  [The trajectory in this instance is helical.]

An important, and oft overlooked, consideration is that **in order for oscillation to occur, the forces which act on the particle must vary in a regular manner which repeats with the same frequency as the motion itself.**

HARMONIC OSCILLATION The trajectories of particles undergoing harmonic oscillation are [infinitely] smooth, in that all time derivatives exist and are continuous.

[In mathematics jargon, functions possessing this property are dubbed "C^{∞}."]

The requirements of oscillation and smoothness can only be met by sinusoidally-based [`cosine` or `sine`] trajectories.

Of the four examples of oscillations enumerated above, only three can entertain any hope of being harmonic.

*   Pacing back and forth across a room at a steady speed [making very quick sudden turns] is inherently *jerky*, rather than harmonic.

*   Gracefully lifting weights or dancing involves smooth movement. Running circular laps at a constant speed is the epitome of *smooth*.

SIMPLE HARMONIC OSCILLATION   Simple Harmonic Oscillation, or SHO, is distinguished by having a single oscillatory mode, *i.e.*, being describable in terms of a single sinusoidal function rather than a sum or product of sinusoids.

Of the three examples of oscillations which might possibly be considered harmonic, listed above, only one can be SHO.

*   Lifting weights may entail smooth movements, but the forward/backward-in-time asymmetry ensures that the motion is not simple.

*   The manoeuvres involved in dancing exhibit periodicity on different time scales and therefore cannot possibly be simple in the sense of this chapter.

*   Constant-speed circular laps are simple, as shall be explicated in Chapter 15.

CYCLE   A cycle of the motion consists of one complete excursion through the [repetitive part of the] trajectory. The start time and place are not particularly germane except insofar as the motion must be continuous. In common practice, significant points [extrema or zero-crossings] are customarily employed as markers for determination of a cycle.

PERIOD   The cycle period is the duration of the time interval through which one oscillation occurs. The symbol employed throughout these volumes to denote period is $\mathcal{T}$. The dimension and SI unit associated with period are time and seconds, respectively.

FREQUENCY   The cycle frequency is the reciprocal of the period. It counts the number of oscillatory cycles which occur in a prescribed time interval. The SI unit for frequency is the hertz,[1] [Hz].

$$1 \text{ Hz} = \text{one cycle per second.}$$

---

[1]So named in honour of Heinrich Hertz (1857–1894).

KINEMATICS OF SHO

A certain particle, constrained to move in 1-d, has a sinusoidal trajectory. WLOG the x-axis may be aligned with the direction in which the particle is confined, and the origin of coordinates chosen to be at the midpoint of the range through which the particle may be found as it oscillates. Under these conditions, the particle's trajectory is

$$x(t) = A \cos(\omega t + \varphi).$$

The RHS of this expression is parsed as follows.

A The amplitude is equal to the maximum displacement [from the centre of symmetry] of the particle.

> ASIDE: The origin of coordinates was chosen to eliminate any constant term in $x(t)$. While this may be easily relaxed, an operational concern is:
>
> **Q:** How might the centre of the motion be reliably determined?
>
> **A1:** One approach is to *split the difference* between the locations of the forward and backward extrema.
>
> **A2:** An alternative method is to ascertain the *time-average position* of the particle over a timespan incorporating a large number of periods.
>
> **A3:** In Chapter 15, it shall be realised that the centre of symmetry, the origin, is the **equilibrium position** of the particle.

cos   **Cosine** is a sinusoidal function. Its argument is a linear function of time, and thus, by the ordinary laws of composition of functions, $x(t)$ is oscillatory [provided that the time domain incorporates at least two complete oscillation periods[2]].

ω The angular frequency of oscillation, ω, has SI units of radians per second. The angular frequency is proportional to the cycle frequency, ν, *i.e.*,

$$\omega = 2\pi\nu \qquad \Longleftrightarrow \qquad \nu = \frac{\omega}{2\pi},$$

as the argument of the sinusoidal function must pass through 2π radians in each and every complete oscillation. Since the period is the reciprocal of the cycle frequency,

$$\mathcal{T} = \frac{1}{\nu} = \frac{2\pi}{\omega}.$$

φ The phase angle provides the necessary shift to the sinusoidal function to bring the general oscillatory structure of $x(t)$ into conformity with the particulars of the physical system under consideration.

The three parameters, A, ω, and φ, are not all on equal footing. The angular frequency, ω, has its value fixed by physical aspects of the oscillating system. In order to change the value of ω, one must make essential modifications which amount to the creation of a new system. On the other hand, the amplitude and the phase are incidental parameters, whose values are chosen to yield conformity with the prescribed initial data or boundary conditions. These can be changed [almost] at will without affecting the identity of the physical system.

---

[2]This caveat is needed to placate the lawyers.

ASIDE: This inequivalence amongst the parameters is not so mysterious. The same effect was present in every encounter with the constant acceleration kinematical formulae,

$$v(t) = v_0 + a\,t \qquad \text{and} \qquad x(t) = x_0 + v_0\,t + \frac{1}{2}\,a\,t^2\,.$$

The acceleration is specified dynamically, via N2, while the initial position and velocity are contingent on the particular instant at which $t = 0$.

READING PERIOD, AMPLITUDE, AND PHASE FROM SHO TRAJECTORIES

Ascertaining empirical values of the SHO parameters is straightforward and algorithmic. A subset of data, extracted from long-term observations of an SHO system, appears in the plot displayed in Figure 12.1.

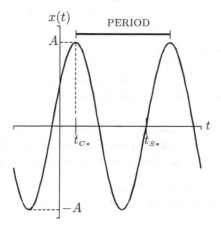

FIGURE 12.1   A Sketch of $x(t)$ *vs.* t for an SHO System

$\mathcal{T}$ The period is the time interval between successive recurrences in the trajectory. Knowledge of the period provides a determination of the angular frequency, via

$$\omega = \frac{2\,\pi}{\mathcal{T}}\,.$$

A The amplitude, A, can be read off the graph as indicated.

[It is assumed that A is constant or changing quite slowly.]

ϕ The phase angle is not uniquely determined. Foremost, it is only defined modulo $2\,\pi$, by the very nature of oscillatory behaviour. Second, the phase angle depends on whether one views the trajectory as a shifted **cosine** or a shifted **sine**. Third, the shift could be forward or backward. Fourth, an additional shift by $\pm\pi$ causes the sinusoids to flip, which can be annulled by formally resetting the amplitude to $-A$. Notwithstanding these ambiguities, in Figure 12.1,

$$\cos\left(\omega\, t_{C*} + \phi_C\right) = 1 \qquad \Longrightarrow \qquad \phi_C = -\omega\, t_{C*},$$
$$\sin\left(\omega\, t_{S*} + \phi_S\right) = 0 \qquad \Longrightarrow \qquad \phi_S = -\omega\, t_{S*}.$$

Suppose that two SHO systems have the same angular frequency and phase, with differing amplitudes, as illustrated in Figure 12.2.

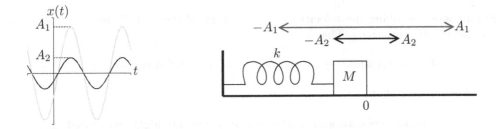

FIGURE 12.2   Two SHO Systems with the Same $\{\omega, \phi\}$ and Different A

The amplitude of the motion described by the grey trajectory is A_1, while the amplitude associated with the black trajectory is A_2. Perhaps these trajectories are particular to identical copies of the same system subject to [slightly] different initial conditions.

Suppose that two SHO systems differ in both angular frequency and amplitude, as illustrated in Figure 12.3.

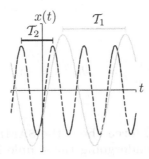

FIGURE 12.3   Two SHO Systems with Different $\{A, \omega\}$

The amplitudes of the motions described by the trajectories can be read from the graph. The period[3] of the oscillatory system whose motion is described by the grey trajectory is

---

[3] Recall that the period is the duration of the time interval in which the repeated motion occurs once.

longer than the period of the system associated with the black trajectory. The two SHO systems are necessarily different, since their angular frequencies are distinct.

$$T_1 > T_2 \qquad \Longleftrightarrow \qquad \omega_1 < \omega_2$$

FORMAL KINEMATICS OF THE SIMPLE HARMONIC OSCILLATOR

IF the precise form of the trajectory of a particle is known[4] to be

$$x(t) = A \cos(\omega t + \varphi),$$

THEN functions describing the velocity and acceleration of the particle are easily obtained by [successive] differentiations. That is,

$$v(t) = -\omega A \sin(\omega t + \varphi) \qquad \text{AND} \qquad a(t) = -\omega^2 A \cos(\omega t + \varphi).$$

Two final comments follow.

○   The position, velocity, and acceleration functions are highly correlated, as was illustrated in the figures concluding Chapter 11. In particular, $v(t)$ and $a(t)$ must oscillate with exactly the same frequency as $x(t)$.

○   At the start of this chapter, it was remarked that in an oscillating system the forces which give rise to the behaviour of the system must themselves oscillate with the same frequency. The following is a heuristic proof of this claim.

The net external force acting on the particle undergoing SHO obeys N2,

$$M a = F_{\text{NET}}, \quad \text{and therefore} \quad F_{\text{NET}} = -M \omega^2 x,$$

where the SHO equation of motion [defining the acceleration] has been invoked. The first and second time derivatives of the net external force are straightforwardly determined to be

$$\frac{dF_{\text{NET}}}{dt} = -M \omega^2 \frac{dx}{dt} = -M \omega^2 v$$

and

$$\frac{d^2 F_{\text{NET}}}{dt^2} = -M \omega^2 \frac{d^2 x}{dt^2} = -M \omega^2 a = -M \omega^2 \left(-\omega^2 x \right) = -\omega^2 F_{\text{NET}}.$$

Hence,

The net external force obeys the SAME dynamical equation
as the particle undergoing the simple harmonic oscillation.

Therefore, the net force must oscillate with the same frequency as the SHO system upon which the forces act.

> ASIDE: It is the net force which must oscillate. Individual forces need not. Consider instances in which a subset of the forces persistently sums to zero [*e.g.*, normal and weight forces], while the remaining forces oscillate.

---

[4]How we know this is irrelevant for the present purposes.

Chapter 13

SHO–Time

In this chapter, we shall investigate two instances of simple harmonic oscillation, as well as two other systems which are nearly so. Here, all of the components of the example systems shall be assumed to be ideal. Frictional and drag forces are assumed to vanish.

EXAMPLE [*Mass–Spring System*]

One end of an ideal spring with force constant k is attached to an anchoring wall, while the other end is affixed to a block of mass M, which is otherwise free to ride on a horizontal frictionless plane. The situation is depicted in Figure 13.1.

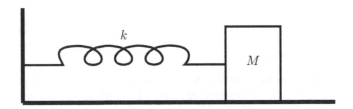

FIGURE 13.1   An Ideal Mass–Spring System

The block is displaced from its equilibrium position[1] and released.

Q: What happens to the block subsequent to its release?

A: Application of the dynamical recipe to this situation takes the following form.

1. The sketch of the system is in Figure 13.1. WLOG imagine that the block is to the right of its equilibrium position, and thus the spring is extended.

2. The block is the sole dynamical constituent. It is subject to three forces: its weight, the normal force of contact with the plane, and the spring force.

3. The interesting dynamics occur in the horizontal direction [x], so henceforth we'll specialise to 1-d. Setting $x = 0$ at the equilibrium position, along with the convention that $x > 0$ denotes extension, fixes the coordinates.

4. Application of N2 yields

$$M\,a = F_{\text{NET}} = -k\,x \quad \Longrightarrow \quad a = -\omega^2\,x\,, \quad \text{for } \omega = \sqrt{\frac{k}{M}}\,.$$

---

[1]That there is an equilibrium position and that it corresponds to the location of the block when the spring has no extension [or compression] from its natural length is worked out in excruciating detail in Chapters 15, 22, and 25 of VOLUME I.

(5 - 6)   This analysis shows that the mass–spring system instantiates SHO.

For the next example, let's add the ideal torsion fibre to our dynamics toolkit.

IDEAL TORSION FIBRE   An ideal torsion fibre is the rotational analogue of the ideal spring. As such, it is insensitive to size and shape effects, possesses no rotational inertia, and gives rise to a Hookian torque distinguished by the relation

$$\tau_F = -\kappa\,\Delta\theta\,.$$

The torque produced by the torsion fibre is

—   restorative, as evidenced by the overall minus sign

κ scaled by the **torsion fibre constant**, a phenomenological parameter

$\Delta\theta$ linear in the angle through which the fibre is twisted

The fibre constant depends on the composition and geometry [size and shape] of the fibre. The SI units of κ are Newton·metres per radian [$\mathrm{N\cdot m/rad}$].

EXAMPLE [*Torsional Pendulum*]

One end of an ideal torsion fibre with fibre constant κ is anchored, while the other end is affixed to a block of mass M and moment of inertia I, which is otherwise free to twist about the axis defined by the fibre itself. The situation is illustrated in Figure 13.2.

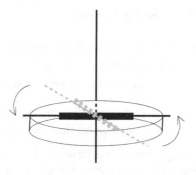

FIGURE 13.2   An Ideal Torsion Fibre System

The block is rotated from its equilibrium angular position[2] and then released.

Q: What happens to the block subsequent to its release?

A: Application of the dynamical recipe to this situation takes the following form.

1. The sketch of the system is in Figure 13.2. The bottom of the fibre has been twisted anticlockwise with respect to the top when looking down.

---

[2]That there is an equilibrium angle and that it corresponds to the orientation of the load when the fibre has no twist follows naturally by analogy with the mass–spring system.

2. The block is the dynamical constituent. The torque produced by the torsion fibre is the only non-zero torque acting to produce rotation about the axis defined by the fibre.

3. Looking down along the fibre toward the block, we take rotation in the anti-clockwise sense to be positively directed. The angular displacement of the block is measured from its equilibrium position.

4. Application of [the rotational version of] N2 yields

$$I\alpha = \tau_{\text{net}} = -\kappa\theta \quad\Longrightarrow\quad \alpha = -\omega^2\theta, \quad \text{for } \alpha = \frac{d^2\theta}{dt^2}, \quad \text{and } \omega = \sqrt{\frac{\kappa}{I}}.$$

(5 - 6) The torsion fibre system constitutes a less familiar realisation of SHO.

EXAMPLE [*Simple Pendulum*]

A **simple pendulum** consists of a point-like bob with mass M, attached to one end of a massless rigid rod[3] of length L. The other end of the rod is hinged in such a manner that it is able to swing freely in a plane, as depicted in Figure 13.3. In equilibrium, the pendulum is at rest, with the bob directly below the hinge point. The bob is set in motion by being first pulled to one side and then released.

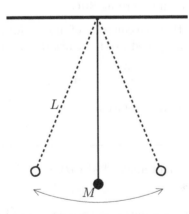

FIGURE 13.3   A Simple Pendulum System

Q: What happens to the bob?

A: The dynamical recipe applies to this situation also.

1. The sketch of the system is presented in Figure 13.3.

2. The bob is the dynamical constituent and experiences two forces: its weight, and "tension" [by which we mean a force internal to the rod which ensures that its length remains fixed].

----

[3]**Q:** Shouldn't this be a *string*? **A:** The rod ensures that the bob follows a segment of circular arc.

3. To perform the analysis, we shall employ instantaneous coordinates which are tangential and radial to the trajectory of the bob. The state of the bob is parameterised by the angle, θ, made by the rod with respect to the vertical [in the plane in which the rod swings].

4. Application of N2 yields

$$0 = M\,a_r = T - M\,g\,\cos(\theta) \qquad \text{and} \qquad M\,a_t = -M\,g\,\sin(\theta)\,,$$

where a_r and a_t are the instantaneous radial and tangential accelerations of the bob. Rigidity of the rod ensures that radial acceleration vanishes. The tangential acceleration is

$$a_t = -g\,\sin(\theta)\,,$$

which is the same as would be experienced by a block sliding down an inclined plane [*cf.* the rollercoaster loop-the-loop in VOLUME I].

The bob undergoes motion along a segment of circular arc. The tangential position, velocity, and acceleration of the bob have angular counterparts. These are related via

$$S = L\,\theta\,, \qquad v_t = L\,\frac{d\theta}{dt} = L\,\omega\,, \qquad \text{and} \quad a_t = L\,\frac{d^2\theta}{dt^2} = L\,\frac{d\omega}{dt} = L\,\alpha\,.$$

Substituting this last relation into N2 yields

$$L\,\alpha = -g\,\sin(\theta) \qquad \Longrightarrow \qquad \frac{d^2\theta}{dt^2} = -\omega^2\,\sin(\theta)\,, \quad \text{for } \omega = \sqrt{\frac{g}{L}}\,,$$

which is NOT the equation governing SHO!

(5 - 6) One's disappointment that the equation of motion for the simple pendulum is not [quite] of SHO form is deepened by the realisation that the differential equation

$$\frac{d^2 f}{dx^2} = -\sin(\,f(x)\,)$$

does not admit a closed-form solution.

All is not lost, however. For small angles,[4] $\sin(\theta) \sim \theta$. Let's not take my word for this but verify it by several means.

∠　　The claim is verified when the angle becomes vanishingly small [using L'Hôpital's Rule]:

$$\lim_{\theta \to 0} \frac{\sin(\theta)}{\theta} = \lim_{\theta \to 0} \frac{\frac{d\sin(\theta)}{d\theta}}{\frac{d\theta}{d\theta}} = \lim_{\theta \to 0} \frac{\cos(\theta)}{1} = 1\,.$$

∠　　The Maclaurin[5] series approximation to **sine** reads:

$$\sin(\theta) = \theta - \frac{\theta^3}{3!} + \frac{\theta^5}{5!} - \frac{\theta^7}{7!} + \ldots\,.$$

When θ is small, the linear term dominates and the higher-order terms may be safely and consistently neglected.

---

[4] In this analysis, angles are specified in radians, unless otherwise explicitly indicated.
[5] The Maclaurin series is a Taylor series in which the expansion is developed about the origin.

∠   Explicit calculation[6] supports the claim that $\sin(\theta) \sim \theta$ for small θ.

θ [rad]	$\sin(\theta)$		
.4	$0.3894\ldots$	$=$	$.4 - .0106$
.3	$0.29552\ldots$	$=$	$.3 - .00448$
.2	$0.19866\ldots$	$=$	$.2 - .00134$
.1	$0.099833\ldots$	$=$	$.1 - .000167$
.05	$0.0499791\ldots$	$=$	$.05 - .0000208$

An angle of 0.4 `radians` is roughly equivalent to $23°$. Even at this relatively large angular amplitude, the discrepancy between θ and $\sin(\theta)$ is only $2\frac{1}{2}\%$.

Thus, provided that the amplitude of the swing be not too large, the equation of motion for the pendulum assumes, to an excellent approximation, the SHO form:

$$\alpha \simeq -\frac{g}{L}\,\theta\,.$$

The [approximate] angular frequency of simple harmonic oscillation is $\omega = \sqrt{g/L}$.

. . . .

EXAMPLE [*Physical Pendulum*]

A **physical pendulum** consists of an extended rigid distribution of matter constrained to oscillate about a fixed axis of rotation, as shown in Figure 13.4. Two significant points are noted: the axis of rotation, ⊙, and the CofM of the body, +.

[Figure 13.4 is a projection into the plane perpendicular to the axis and containing the CofM.]

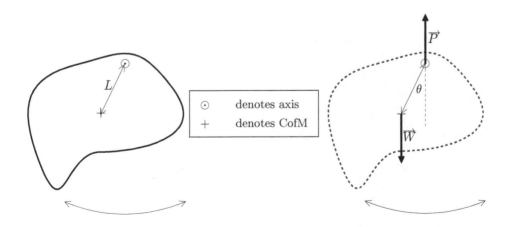

⊙	denotes axis
+	denotes CofM

FIGURE 13.4   A Physical Pendulum and Its FBD

---

[6]This ought to be one's last resort!

The weight force, deemed to act at the location of the CofM, provides a torque,

$$\tau_W = -M g L \sin(\theta),$$

about the axis of rotation. The forces which maintain the shape of the physical pendulum and hold it in place are not able to contribute to the net external torque. Thus,

$$I \alpha = -M g L \sin(\theta) \qquad \Longrightarrow \qquad \alpha = -\frac{M g L}{I} \sin(\theta).$$

Clearly, the physical pendulum is precisely analogous to the simple pendulum.

Chapter 14

Springs in Series and Parallel

For grins, let's investigate series and parallel arrangements of springs.

EXAMPLE [*Series Composition of Two Springs*]

A block of mass M riding on a frictionless horizontal plane is attached to a fixed wall by means of two ideal springs with respective spring constants k_1 and k_2, arranged in series as illustrated in Figure 14.1.

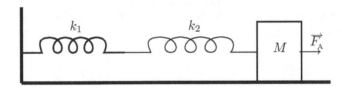

FIGURE 14.1   Two Ideal Springs Arranged in Series

An applied force F_A acts on the block, stretching the composite spring by an amount[1] $\Delta x_{(12)}$, and the system is in a state of static equilibrium. Since the system is not moving,[2] the two constituent springs experience the same amount of applied force, while each contributes to the net extension. That is:

$$\Delta x_{(12)} = \Delta x_1 + \Delta x_2 \qquad \text{and} \qquad F_1 = F_2 = F_A .$$

Three Hookian relations hold [the first for the composite system, the latter two for the constituent springs]:

$$F_A = k_{(12)} \, \Delta x_{(12)} , \qquad \begin{aligned} F_1 &= k_1 \, \Delta x_1 \\ F_2 &= k_2 \, \Delta x_2 \end{aligned} .$$

Combining all of these relations in light of $\Delta x_{(12)} = \Delta x_1 + \Delta x_2$ and $F_1 = F_2 = F_A$ yields

$$\frac{1}{k_{(12)}} = \frac{\Delta x_{(12)}}{F_A} = \frac{\Delta x_1 + \Delta x_2}{F_A} = \frac{\Delta x_1}{F_A} + \frac{\Delta x_2}{F_A} = \frac{\Delta x_1}{F_1} + \frac{\Delta x_2}{F_2} = \frac{1}{k_1} + \frac{1}{k_2} .$$

Thus we obtain the *snazzy* result that **the reciprocal of the effective spring constant for two springs connected in series is the sum of the reciprocals of the respective constants for the individual springs.**

Several comments are pertinent here.

---

[1] As in the discussion of effective elastic moduli, quantities proper to the series combination of springs i and j are labelled (ij).

[2] More precisely, it is not accelerating.

...   Going through the motions [cross-multiplying, *etc.*] produces an explicit formula for the effective spring constant for two springs in series:

$$k_{(12)} = \frac{k_1 \, k_2}{k_1 + k_2} \, .$$

$k_{(12)} < k_{1,2}$ The effective spring constant is less than each of the constituent spring constants.

$k_1 = k_2$ IF $k_1 = k_2 = k$, THEN

$$\frac{1}{k_{(12)}} = \frac{1}{k_1} + \frac{1}{k_2} = \frac{1}{k} + \frac{1}{k} = \frac{2}{k} \qquad \Longrightarrow \qquad k_{(12)} = \frac{k}{2} \, .$$

Since two identical springs will each stretch by a given amount under a fixed external force, subjecting the series combination to the same external force will double this extension. Therefore the effective spring constant of the series combination is half that of the constituent springs.

$k_1 \gg k_2$ If the spring constants for the constituent springs differ greatly, it is the smaller of the two which sets the scale for the effective spring constant of the series combination. Explicitly,

$$\lim_{k_1 \gg k_2} \frac{1}{k_{(12)}} = \lim_{k_1 \gg k_2} \frac{1}{k_1} + \frac{1}{k_2} = \lim_{k_1 \gg k_2} \left(\frac{1}{k_2} \left[1 + \frac{k_2}{k_1} \right] \right) \simeq \frac{1}{k_2} \qquad \Longrightarrow \qquad k_{(12)} \simeq k_2 \, .$$

This extreme form of scale separation can arise when EITHER $k_1 \to \infty$, OR $k_2 \to 0$. In the former case, one of the springs is stiff to the point of not stretching at all. In the latter case, one spring is very easily stretched.

EXAMPLE [$\mathcal{N}$ *Springs in Series*]

The same block of mass M is attached to the wall via $\mathcal{N}$ distinct springs combined purely in series, as shown in Figure 14.2.

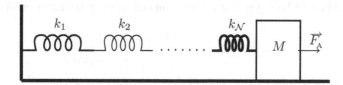

FIGURE 14.2 $\mathcal{N}$ Ideal Springs Arranged in Series

METHOD ONE: [*Extension*]

It is straightforward to derive the general formula for the composition of $\mathcal{N}$ springs by extending the argument made in the two-spring case. The resultant expression is

$$\frac{1}{k_{(1234\ldots\mathcal{N})}} = \frac{\Delta x_{(1234\ldots\mathcal{N})}}{F_{\mathrm{A}}} = \frac{\sum_{i=1}^{\mathcal{N}} \Delta x_i}{F_{\mathrm{A}}} = \sum_{i=1}^{\mathcal{N}} \frac{\Delta x_i}{F_i} = \sum_{i=1}^{\mathcal{N}} \frac{1}{k_i} \, .$$

METHOD TWO: [*Recursion*]

Recursive application of the two-spring formula also leads to the general result:

$$\frac{1}{k_{(1234...\mathcal{N})}} = \frac{1}{k_1} + \frac{1}{k_{(2345...\mathcal{N})}}$$

$$= \frac{1}{k_1} + \left(\frac{1}{k_2} + \frac{1}{k_{(345...\mathcal{N})}} \right)$$

$$= \frac{1}{k_1} + \frac{1}{k_2} + \left(\frac{1}{k_3} + \frac{1}{k_{(45...\mathcal{N})}} \right)$$

$$\vdots$$

$$= \frac{1}{k_1} + \frac{1}{k_2} + \frac{1}{k_3} + \ldots + \frac{1}{k_{\mathcal{N}-2}} + \frac{1}{k_{\mathcal{N}-1}} + \frac{1}{k_{\mathcal{N}}}$$

Three of the comments made in the example above merit amplification here.

$k_{(...)} < k_i$ The effective spring constant is less than every one of the constituent spring constants.

$k_i = k$ IF $k_1 = k_2 = \ldots = k_{\mathcal{N}} = k$, THEN

$$\frac{1}{k_{(1...\mathcal{N})}} = \frac{1}{k_1} + \frac{1}{k_2} + \ldots + \frac{1}{k_{\mathcal{N}}} = \frac{\mathcal{N}}{k} \qquad \Longrightarrow \qquad k_{(1...\mathcal{N})} = \frac{k}{\mathcal{N}}.$$

SCALE   The scale of the effective coupling is set by the weakest of the springs.

EXAMPLE [*Parallel Composition of Two Springs*]

A block of mass M riding on a frictionless horizontal plane is attached to a fixed wall by means of two ideal springs with respective spring constants k_1 and k_2, arranged in parallel as illustrated in Figure 14.3.

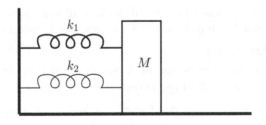

FIGURE 14.3   Two Ideal Springs Arranged in Parallel

An applied force, F_A, acts on the block, extending the composite spring by the amount Δx, and the system is in static equilibrium. The two constituent springs undergo the same extension, while each contributes to the overall cancelling of the net force acting on the block. That is:

$$\Delta x_1 = \Delta x_2 = \Delta x \qquad \text{and} \qquad F_1 + F_2 = F_A.$$

The following three Hookian relations hold [for the assemblage and each constituent]:

$$F_A = k_{[12]} \, \Delta x_{[12]} \,, \qquad \begin{aligned} F_1 &= k_1 \, \Delta x_1 \\ F_2 &= k_2 \, \Delta x_2 \end{aligned}.$$

Combining all of these relations in light of $F_A = F_1 + F_2$ and $\Delta x_{[12]} = \Delta x_1 = \Delta x_2 = \Delta x$ yields

$$k_{[12]} = \frac{F_A}{\Delta x_{[12]}} = \frac{F_1 + F_2}{\Delta x} = \frac{F_1}{\Delta x} + \frac{F_2}{\Delta x} = \frac{F_1}{\Delta x_1} + \frac{F_2}{\Delta x_2} = k_1 + k_2 \,.$$

Thus we obtain the *eminently cool* result that **the effective spring constant for two springs arranged in parallel is the sum of the individual spring constants**.

Several comments are pertinent here.

$k_{[12]} > k_{1,2}$ The effective spring constant is greater than either of the constituent spring constants.

$k_1 = k_2$ IF $k_1 = k_2 = k$, THEN

$$k_{[12]} = k_1 + k_2 = 2\,k \,.$$

Pithily: It takes twice the applied force to extend two identical springs, arranged in parallel, by a fixed amount, as it does to extend one alone. Therefore the effective spring constant of the parallel combination is twice k.

$k_1 \gg k_2$ When the spring constants for the constituent springs differ greatly, the scale for the effective spring constant of the parallel combination is set by the larger of the two. Explicitly,

$$\lim_{k_1 \gg k_2} k_{[12]} = \lim_{k_1 \gg k_2} k_1 + k_2 \simeq k_1 \,.$$

The two most extreme forms of scale separation arise when EITHER $k_1 \to \infty$, OR $k_2 \to 0$. In the former case, one of the springs is so stiff that by barely stretching it cancels the applied force. In the latter case, one spring is so weak that it cancels very little of the applied force.

EXAMPLE   [$\mathcal{N}$ *Springs in Parallel*]

Suppose that the block is attached to the wall with $\mathcal{N}$ distinct springs [all having the same equilibrium length] combined purely in parallel, as shown in Figure 14.4.

METHOD ONE: [*Extension*]

The general formula for the composition of $\mathcal{N}$ springs is derived by generalising and extending the argument made above in the two-spring case.

$$\Delta x_{[1...\mathcal{N}]} = \Delta x_1 = \Delta x_2 = \ldots = \Delta x_{\mathcal{N}} = \Delta x \qquad \text{and} \qquad F_1 + F_2 + \ldots + F_{\mathcal{N}} = F_A \,.$$

From the Hookian relations for the parallel system and each constituent,

$$F_A = k_{[1...\mathcal{N}]} \, \Delta x_{[1...\mathcal{N}]} \qquad \text{and} \qquad F_i = k_i \, \Delta x_i \,, \ \forall \, i \in 1 \ldots \mathcal{N} \,,$$

one readily obtains that

$$k_{[1...\mathcal{N}]} = k_1 + k_2 + \ldots + k_{\mathcal{N}} = \sum_{i=1}^{\mathcal{N}} k_i \,.$$

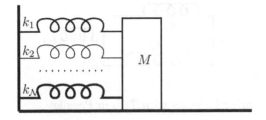

FIGURE 14.4 $\mathcal{N}$ Ideal Springs Arranged in Parallel

METHOD TWO: [*Recursion*]

Recursive application of the two-spring relation yields the $\mathcal{N}$-spring result.

$$
\begin{aligned}
k_{[1234...\mathcal{N}]} &= k_1 + k_{[234...\mathcal{N}]} \\
&= k_1 + \left(k_2 + k_{[34...\mathcal{N}]}\right) \\
&= k_1 + k_2 + \left(k_3 + k_{[4...\mathcal{N}]}\right) \\
&\ \ \vdots \\
&= \sum_{i=1}^{\mathcal{N}} k_i \,.
\end{aligned}
$$

Here are a few more germane comments.

$k_{[...]} > k_i$ The effective spring constant is greater than each and every one of the constituent spring constants.

$k_i = k$ IF $k_1 = k_2 = \ldots = k_{\mathcal{N}} = k$, THEN

$$
k_{[1...\mathcal{N}]} = k_1 + k_2 + \ldots = \mathcal{N}\,k\,.
$$

SCALE   The scale of the effective coupling is set by the strongest of the springs.

EXAMPLE [*Mixed Combinations of Three Springs*]

Three springs can be arranged in four distinct ways. One of these is purely series and another is purely parallel. The remaining two combinations are mixed, as illustrated in Figures 14.5 and 14.7. To determine an overall spring constant for these networks, one must *iteratively* replace local sets of springs occurring in pure series or parallel subnetworks with single effective springs. We'll do this for the distinct A and B configurations below.

A: ONE IN SERIES WITH TWO IN PARALLEL

Springs 1 and 2 form a parallel subnetwork with an effective spring constant of

$$
k_{[12]} = k_1 + k_2 \,.
$$

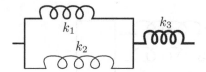

FIGURE 14.5 One Spring in Series with Two in Parallel

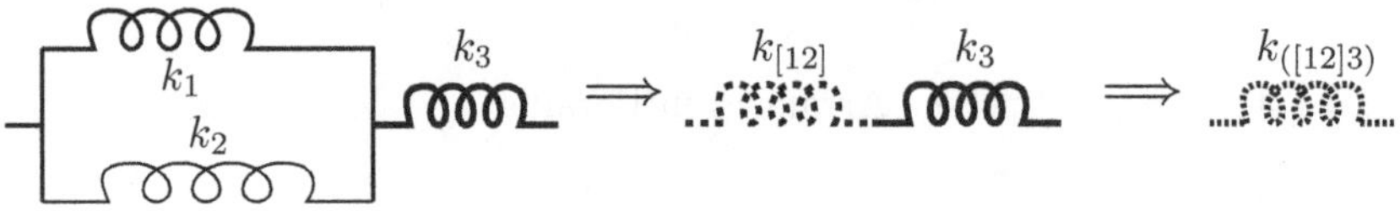

FIGURE 14.6 Iterative Solution to Spring Network A

The [12] combination is in series with spring 3, and hence

$$\frac{1}{k_{([12]3)}} = \frac{1}{k_{[12]}} + \frac{1}{k_3} = \frac{1}{k_1 + k_2} + \frac{1}{k_3} = \frac{k_1 + k_2 + k_3}{(k_1 + k_2)k_3}.$$

Thus, for this combination of the three springs, the effective spring constant is

$$k_{([12]3)} = \frac{(k_1 + k_2)k_3}{k_1 + k_2 + k_3}.$$

B: One in parallel with two in series

Springs 1 and 3, appearing in Figure 14.7, form a series subnetwork with an effective spring constant of

$$\frac{1}{k_{(13)}} = \frac{1}{k_1} + \frac{1}{k_3} \qquad \Longrightarrow \qquad k_{(13)} = \frac{k_1 k_3}{k_1 + k_3}.$$

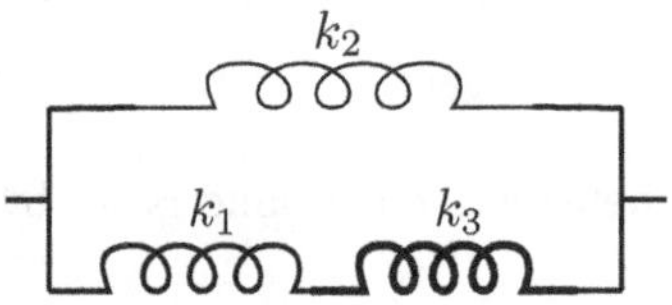

FIGURE 14.7 One Spring in Parallel with Two in Series

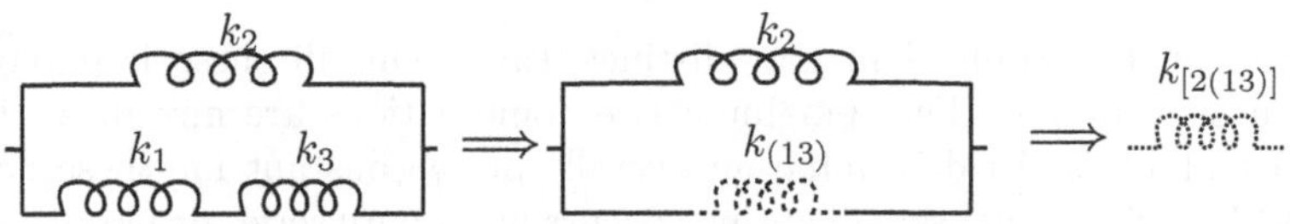

FIGURE 14.8 Iterative Solution to Spring Network B

The (13) combination is in parallel with spring 2, and thus the effective spring constant is

$$k_{[2(13)]} = k_2 + k_{(13)} = k_2 + \frac{k_1 k_3}{k_1 + k_3} = \frac{k_1 k_2 + k_2 k_3 + k_3 k_1}{k_1 + k_3}.$$

Chapter 15

SHO: Kinematics, Dynamics, and Energetics

The equation of motion characteristic of SHO is

$$a(t) = -\omega^2 \, x(t) \, .$$

The general solutions are sinusoidal:

$$x(t) = A \, \cos(\omega \, t + \varphi) \, .$$

In consideration of the bloke from Chapter 12, [who is still] running at a constant speed around a circular track, we shall make explicit the connection between SHO and uniform circular motion. First recall [VOLUME I, Chapter 8] a few properties of uniform circular motion:

○ The trajectory **covers**[1] a circle of radius R. The distance run is equal to the arc length, $S = R \, \theta$ [up to an additive constant which can be absorbed into the angular position, θ].

○ The constant tangential speed, v_t, is equal to the time rate of change of the arc length:

$$v_t = \frac{dS}{dt} = R \, \frac{d\theta}{dt} = R \, \omega \, .$$

Constancy of the tangential speed is concomitant with constant angular speed.

○ The instantaneous tangential acceleration,

$$a_t = \frac{dv_t}{dt} = R \, \frac{d\omega}{dt} = R \, \alpha \, ,$$

vanishes, and thus so too must the angular acceleration, α.

○ For uniform circular motion, the magnitude of the centripetal acceleration,

$$a_c = \frac{v_t^2}{R} \, ,$$

is constant, while its direction is constantly changing.[2]

The upshot of all of this is that the angular position of the runner is given by

$$\theta(t) = \omega \, t + \theta_0 \, ,$$

---

[1] The trajectory wraps around the circle repeatedly; it constitutes a multiple cover. [The map from time, t, onto the locus of points comprising the entirety of the trajectory is SURJECTIVE; it is not INJECTIVE.]

[2] Recall from VOLUME I's discussion of circular motion, and of rotation in general, that centripetal means "centre-seeking" and the "radially inward direction" depends on one's position.

where the constant angular speed, ω, and the initial angle, θ_0, are chosen so as to conform to the initial data.

Polar coordinates, with origin at the centre of the circular path, provide the most natural framework for description and analysis of the trajectory. Naturalness, however, does not preclude transformation to a Cartesian basis [with common origin], via

$$x(t) = R \cos\big(\theta(t)\big) \qquad\qquad y(t) = R \sin\big(\theta(t)\big)$$
$$\qquad\qquad\text{and}$$
$$= R \cos\big(\omega\,t + \theta_0\big) \qquad\qquad = R \sin\big(\omega\,t + \theta_0\big)\;.$$

The angular argument of the x- and y-component sinusoids is changing [linearly] in time, while the xy-phase difference remains constant at $\pi/2$.

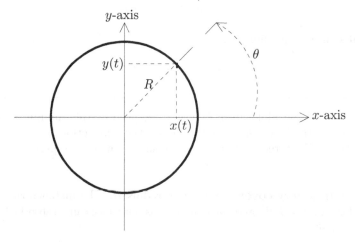

FIGURE 15.1   Uniform Circular Motion in Polar and Cartesian Coordinates

In a nutshell, **uniform circular motion has the appearance of simple harmonic oscillation when projected onto an axis**, *i.e.*, viewed *edge on*.

This discursion provides an insight into uniform circular motion beyond that available in VOLUME I and better equips us to ask:

Q: What generalisations of uniform circular motion exist?

Au: Dropping the condition of uniformity was explored somewhat in VOLUME I.

Ac: There are several avenues by which the circularity restriction might be relaxed. One of these is allowing the two components to possess unequal amplitudes while retaining the $\pi/2$ phase difference [*cf.* Kepler's elliptical orbits, discussed in VOLUME I, Chapters 49 and 50]. In the left panel of Figure 15.2, the two curves,

$$\vec{R}_{\text{A1:2}} = \big(\cos(t),\, 2\sin(t)\big) \qquad\text{and}\qquad \vec{R}_{\text{A2:1}} = \big(2\cos(t),\, \sin(t)\big),$$

illustrate this. A second approach involves equal-amplitude components with relative phase other than $\pi/2$. The two curves found in the right panel of Figure 15.2,

$$\vec{R}_{\pi/4} = \big(\cos(t + \pi/4),\, \sin(t)\big) \qquad\text{and}\qquad \vec{R}_{-3\pi/7} = \big(\cos(t - 3\,\pi/7),\, \sin(t)\big),$$

illuminate this second means. A third avenue, not shown, is the combination of unequal amplitudes and non-$\pi/2$ relative phase. Yet another possibility, also not explicitly considered here, is allowing the components to oscillate with different frequencies. These loci form **Lissajous Patterns**.

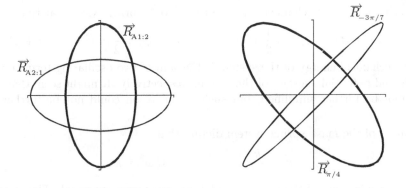

FIGURE 15.2 SHO-Inspired Generalisations of Uniform Circular Motion

---

ENERGETICS OF SIMPLE HARMONIC OSCILLATION

A mass–spring system is pictured in Figure 15.3. The usual assumptions about the spring being Hookian, the horizontal plane frictionless, the air drag-free, *etc.*, are made.

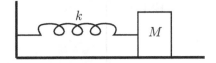

FIGURE 15.3 A Mass–Spring SHO System

From VOLUME I, we recollect that the spring force is counted among the conservative forces. The spring potential energy function is[3]

$$U_{\mathrm{s}}(x) = \frac{1}{2}\, k\, x^2\,.$$

This system experiences SHO, and thus the trajectory of the block can be expressed as

$$x(t) = A\,\cos(\omega\, t + \varphi) \qquad \text{and} \qquad v(t) = -\omega\, A\,\sin(\omega\, t + \varphi)\,.$$

Hence, the potential energy stored in the block–spring system at time t is

$$U_{\mathrm{s}}(t) = U_{\mathrm{s}}\big(x(t)\big) = \frac{1}{2}\, k\, x^2(t) = \frac{1}{2}\, k\, A^2\,\cos^2(\omega\, t + \varphi)\,.$$

Meanwhile, the kinetic energy of the block is

$$K(t) = K\big(v(t)\big) = \frac{1}{2}\, M\, v^2 = \frac{1}{2}\, M\, \omega^2\, A^2\,\sin^2(\omega\, t + \varphi)\,.$$

---

[3]This form of U_{s} presupposes that the origin of coordinates is located at the equilibrium point, and that the arbitrary constant additive piece is set to zero.

Consequently, the total mechanical energy of the block–spring system, at time t, is

$$E(t) = K(t) + U_\mathrm{s}(t) = \frac{1}{2} M \omega^2 A^2 \sin^2(\omega t + \varphi) + \frac{1}{2} k A^2 \cos^2(\omega t + \varphi).$$

The total mechanical energy of the system is the sum of two oscillating terms representing the kinetic and potential energies of the system, respectively. It might seem that this latter result cannot be further simplified, but such a conclusion could not be further from the truth.

The dynamics of the mass–spring system dictate that

$$\omega^2 \equiv \frac{k}{M} \quad \Longrightarrow \quad M \omega^2 \equiv k,$$

and thus the coefficients of the two sinusoidal-squared terms are equal. The expression for the total mechanical energy now reads

$$E = \frac{1}{2} k A^2 \left[\cos^2(\omega t + \varphi) + \sin^2(\omega t + \varphi) \right].$$

Despite the explicit time dependence of the angular arguments, the quantity within the square brackets is equal to 1, because at each instant the angles are exactly the same. As a consequence, **the total mechanical energy**,

$$E = \frac{1}{2} k A^2,$$

is a constant of the motion. This result is best conveyed by the *iconic* Figure 15.4.

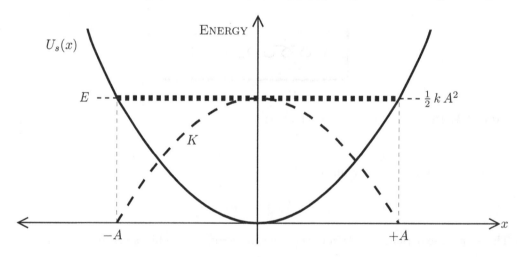

FIGURE 15.4 Energetics of an SHO

In Figure 15.4, the potential energy function is $U_\mathrm{s} = (1/2) k x^2$. The total mechanical energy, $E = (1/2) k A^2$, intersects the potential energy function at the extrema of the motion, *i.e.*, at $x = \pm A$. The extrema are the points at which the block is instantaneously at rest.

There is more to it than this. Constancy of the total mechanical energy dictates that the kinetic energy function acquire a dependence on position,[4] *viz.*,

$$K(x) = E - U_\mathrm{s}(x).$$

---

[4]This notion was alluded to in Chapter 27 of VOLUME I, where it was shown that, under certain circumstances, trajectories could be obtained by direct integration of the velocity, expressed in terms of E and $U(x)$.

Negative values of kinetic energy, $K = \frac{1}{2} M v^2$, are precluded.[5] Manifest positivity of K imparts energetic force to an argument that the block must not venture forth beyond its prescribed range. The block stops at $\pm A$ and is unable to proceed past these points, so as to ensure that the kinetic energy remains non-negative.

Consideration of Figure 15.4 also reveals a pattern of energy oscillation concomitant with the oscillation of the block, but at twice the frequency.

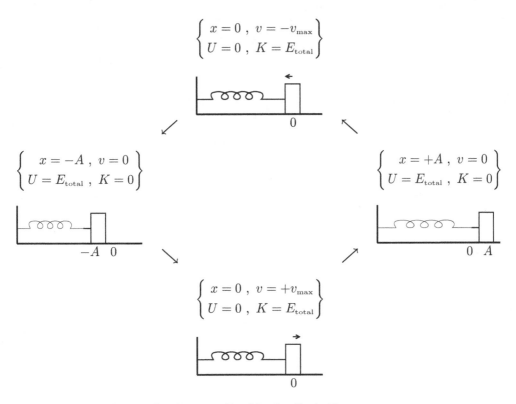

FIGURE 15.5 Energy Oscillates at Double the Cycle Frequency

---

[5]Negative K would require that the mass be negative OR the speed be imaginary-valued. Negative mass is problematic [putting it mildly] in that such a particle's acceleration response occurs anti-parallel to the net force acting upon it. Notions of imaginary velocity require the consistent imposition of a complex structure on the spatial coordinates. No empirical evidence exists to justify such a move in classical physics. [Speculative models which harmoniously incorporate SUPERSYMMETRY in Grand Unified (Quantum Field) Theories propose such a complexification of space(time). Theorists who work on such models must ensure that no vestiges of this enlargement of space appear in the classical regime.]

Chapter 16

Damped Oscillation: Qualitative

Our enthusiasm for the study of ideal SHOs is tempered by the discomfiting realisation that such beasties are not to be found in nature. Actual oscillating systems [almost[1]] never follow precisely repeating trajectories. Instead, they are observed to endure a relentless diminution of their range of motion until, finally, they come to a stop.

The aim of this chapter is to augment the SHO model by incorporating linear drag, thus beginning the investigation of **damped harmonic oscillation** [DHO].

> ASIDE: The monikers "D'OH" and "DUH" (Damped Harmonic Undulation) were passed over in favour of DHO.

In the next chapter we shall solve, completely and generally, the equations of motion for a damped harmonic oscillator. Our goal here is to infer the physical properties of a class of solutions via a heuristic approach.

Consider the usual mass–spring system:

- □   a block of [constant] mass M, riding upon a horizontal frictionless plane, and

- □   a spring with force constant k, one end anchored, the other affixed to the block.

To this we add

- □   a [resistive] drag force, linear in the velocity of the particle,

$$\vec{F_{\text{D}}} = -b_1\,\vec{v}\,.$$

The details pertaining to the fluid medium [density, pressure, viscosity, *etc.*] and the block [size, shape, surface properties] are subsumed into the drag coefficient, b_1.

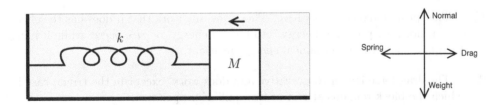

FIGURE 16.1   A Damped Mass–Spring System

Suppose that the mass–spring system is only slightly damped, in that it is nearly oscillatory, with a slowly decaying amplitude. Such a system possesses dynamical behaviour on two markedly different time scales. The short time scale is set by the period of the

---

[1] CAVEAT: In the presence of an external driving force all bets are off.

quasi-oscillation, $\mathcal{T}$. The long time scale is characterised by an appreciable decrease in the amplitude of the motion. For example, the "half-time," $\tau_{0.5}$, is the time interval through which the amplitude is reduced to $1/\sqrt{2}$ of its original value. That the two time scales are widely disparate ensures that $\tau_{0.5} \gg \mathcal{T}$.

That the total mechanical energy of a pure SHO [undamped] system is a constant of the motion was shown in Chapter 15. This constancy cannot be maintained when drag is incorporated into the model. The reasons for this bear some examination.

DHO   Foremost is that the drag force is non-conservative. Thus, no potential energy
               function can be associated with this force.

> **Hence, the mechanical work performed by the drag force will**
> [unless cancelled exactly by work from other non-conservative forces]
> **produce a change in the total mechanical energy.**

DHO   In addition,[2] the resistive nature of drag forces ensures that negative work is
               done, thereby reducing the total mechanical energy of the system.

The TOTAL ENERGY–WORK THEOREM[3] can be paraphrased as: The change in the total mechanical energy of the system throughout the time interval Δt, extending from some initial time t_i to final time $t_f = t_i + \Delta t$, is equal to the net work done, during the time interval, by non-conservative forces acting on the system. Thus,

$$\Delta E = W_{if}\left[\vec{F}_{\text{NET, non-C}}\right].$$

Enumeration of the forces acting on the block: weight, normal, spring, and drag, reveals that

$\vec{W}$ The weight force is conservative, but this is moot, since the gravitational force
               does no work on a horizontally sliding block, anyway.

$\vec{N}$ The normal force is decidedly non-conservative. However, it also does no work,
               since the block rides on the planar surface.

$\vec{F}_{\text{s}}$ The spring force is conservative. Therefore, any work that it does has the effect
               of transforming potential energy into kinetic energy, or *vice versa*, while leaving
               the quantity of total mechanical energy invariant.

$\vec{F}_{\text{D}}$ The drag force is non-conservative and does work, except in the trivial case in
               which the block remains at rest at its equilibrium position.

The drag force performs the net non-conservative work on the system.

---

[2]The snarky among us might think it more appropriate to write, "In subtraction, ..."
[3]This theorem was introduced and proven in VOLUME I, Chapter 28.

Rather than consider the amount by which the energy changes through a time interval of finite duration,[4] one might instead investigate the instantaneous power produced by the drag force. It is

$$P_D = \vec{F_D} \cdot \vec{v} = -b_1 \, \vec{v} \cdot \vec{v} = -b_1 \, v^2 \,,$$

where $\vec{v}$ is the [instantaneous] velocity of the particle and v is its speed. That the power is negative is consistent with the resistive/dissipative nature of the drag force. The implications of this result are reasonable: more power is lost to drag when the object is moving quickly, and less when it is moving slowly, and none at all when it is stopped.

There are two different paths along which one might continue this analysis. Naturally, we'll take both!

METHOD ONE: [*Energetics Approach*]

The rate of power loss in this instance is proportional to the kinetic energy of the particle:

$$P_D = -b_1 \, v^2 = -b_1 \left(\frac{2}{M} \frac{1}{2} M \right) v^2 = -\frac{2 \, b_1}{M} \, K \,.$$

Were this an SHO, the kinetic energy would be changing [throughout the oscillatory cycle] in the manner described in the iconic Figure 15.4 in Chapter 15, which is reproduced here as Figure 16.2.

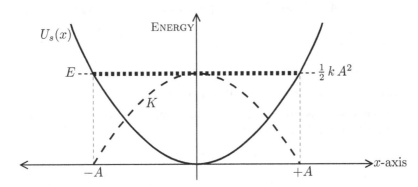

FIGURE 16.2   Energetics of an SHO

The total energy is the sum of the kinetic and potential energies, both of which are changing rapidly [on the short time scale] in opposite ways so as to conserve the value of the total mechanical energy. Inspection of the diagram suggests that the average values of the kinetic and the potential energy over one complete cycle [or equivalently, over a long time[5]] are equal.[6]

  ASIDE: There exists a general result, the VIRIAL THEOREM, which asserts: IF the potential energy function has the form

$$V(x) = x^n \,, \quad \text{for some } n \,,$$

  THEN the average kinetic energy is related to the average potential energy by

$$\langle K \rangle = \frac{n}{2} \, \langle V \rangle \,.$$

  For the harmonic oscillator, $n = 2$, and the Virial Theorem assures us that the average kinetic energy and average potential energies are equal.

---

[4]This is usually quite hard to determine.
[5]Remember that a dissipationless SHO is under consideration here.
[6]The equality is only weakly suggested by the [spatial] energy curves because the average is temporal.

Ergo, the average kinetic energy of an SHO must also be equal to one half of its [constant] total mechanical energy. We are labouring under the assumption that the total mechanical energy is changing slowly [on the long time scale]. Thus, the average power dissipated by the drag force throughout one entire oscillation is

$$\langle P_{\mathrm{D}} \rangle = \left\langle -\frac{2\,b_1}{M}\,K \right\rangle = -\frac{2\,b_1}{M}\,\langle K \rangle = -\frac{2\,b_1}{M}\,\frac{1}{2}\,\langle E \rangle = -\frac{b_1}{M}\,\langle E \rangle \simeq -\frac{b_1}{M}\,E\,.$$

The final [approximate] equality holds provided that the oscillation time scale, over which the average is taken, is much shorter than the longer time scale on which the energy changes by an appreciable amount. This result, that the average time rate of change of the total mechanical energy of the particle is proportional to the value of the energy itself, is characteristic of exponential functions. Integrating yields

$$E = E_0 \exp\left(-\frac{b_1}{M}\,t\right),$$

where E_0 is the total mechanical energy of the system at $t = 0$.

The final connection to *reality*™ comes from recognition that the total mechanical energy is related to the amplitude, via $E = (1/2)\,k\,A^2$. Hence, the amplitude varies with time as

$$A = A_0 \exp\left(-\frac{b_1}{2\,M}\,t\right) = A_0 \exp\left(-\gamma\,t\right) = A_0\,e^{-\gamma\,t}\,,$$

where the **decay constant** $\gamma = b_1/(2\,M)$.

METHOD TWO: [*Shoulder to the Wheel Approach*]

As the damped system is presumed to be only slightly perturbed from an ideal SHO, the particle's trajectory can be modelled as

$$x(t) = A(t)\,\cos(\omega\,t + \varphi)\,,$$

where the amplitude is now taken to vary according to a long time scale [*i.e.,* slowly], while the sinusoid varies on a short time scale, $T = 2\,\pi/\omega$ [*i.e.,* quickly]. The angular frequency, ω, is assumed to not differ appreciably from its unperturbed value, $\omega = \sqrt{k/M}$. The velocity is, of course, the time rate of change of the position, and thus

$$v(t) = \frac{dx(t)}{dt} = \frac{dA(t)}{dt}\,\cos(\omega\,t + \varphi) - \omega\,A(t)\,\sin(\omega\,t + \varphi) \simeq -\omega\,A(t)\,\sin(\omega\,t + \varphi)\,.$$

In the final [approximate] equality, the slow, long-time-scale, changing-amplitude contribution to the velocity was deemed to be much smaller than the quick, short-time-scale, oscillation part and could thus be safely neglected. Formally, this is akin to the approximation $\frac{dA(t)}{dt} \sim 0$, *i.e.,* that the amplitude remains effectively constant when viewed on sufficiently short time scales. Therefore, the instantaneous power dissipated by the drag force is

$$P_{\mathrm{D}} = -b_1\,v^2 \simeq -b_1\,\omega^2\,A^2(t)\,\sin^2(\omega\,t + \varphi) \simeq -b_1\,\frac{k}{M}\,A^2(t)\,\sin^2(\omega\,t + \varphi)\,.$$

This relation states the [approximate] time dependence of the flow of energy out of the system with more detail than is necessary to understand the behaviour of the system on longer time scales. To distill this superabundance of information into a single meaningful quantity, we shall compute the time average of the power throughout a complete oscillation. This will lead to the determination of the average power through one complete cycle.

ASIDE: This process is often referred to as "coarse graining." The notion is that one can better expose the long-term behaviour by "integrating out" or "averaging over" short-term choppiness. Careful attention to the paintings of the so-called *Group of Seven*[7] makes it abundantly clear that, paradoxically, blurring the trunks and branches of individual trees shows the forest more clearly.

Prior experience with averaging various quantities dictates the first few steps:

$$\left\langle P_{\mathrm{D}} \right\rangle_{T} \simeq \left\langle -b_1 \frac{k}{M} A^2(t) \sin^2(\omega t + \varphi) \right\rangle_{T} \simeq -b_1 \frac{k}{M} A_0^2 \left\langle \sin^2(\omega t + \varphi) \right\rangle_{T},$$

where the slowly changing value of the amplitude has been replaced by a [constant] representative value A_0. The remaining task is to accurately determine the average value of the squared sinusoid over one complete cycle of oscillation. There are a variety of ways[8] in which one may perform this computation. In this instance, we shall employ the most direct means.

$$
\begin{aligned}
\left\langle \sin^2(\omega t + \varphi) \right\rangle_{T} &= \frac{1}{T} \int_0^T \sin^2(\omega t + \varphi)\, dt \\
&= \frac{1}{2\pi} \int_0^{2\pi} \sin^2(\theta)\, d\theta = \frac{1}{2\pi} \int_0^{2\pi} \left[\frac{1 - \cos(2\theta)}{2} \right] d\theta \\
&= \frac{1}{2\pi} \left[\frac{2\pi - 0}{2} - \frac{0 - 0}{4} \right] = \frac{1}{2}.
\end{aligned}
$$

A glance at the graph of $\sin^2(\theta)$ over a complete cycle corroborates this result. Insertion of this relation into the expression for the average power yields

$$\left\langle P_{\mathrm{D}} \right\rangle_{T} \simeq -b_1 \frac{k}{M} A_0^2 \left\langle \frac{1}{2} \right\rangle = -\frac{b_1}{M} \frac{1}{2} k A_0^2.$$

Given our assumptions, the energy of the system throughout the cycle is effectively constant at a value not far removed from that associated with an SHO whose amplitude is equal to A_0, and therefore the average time rate of change of energy over the short time scale of a single cycle is

$$\frac{dE}{dt} \simeq \left\langle P_{\mathrm{D}} \right\rangle_{T} \simeq -\frac{b_1}{M} E.$$

Here we have recaptured the manifestly exponential behaviour of the total energy[9] as obtained just previously.

All the results of the previous method are obtained here, too.

---

[7]This name is given to a group of early twentieth century Canadian painters: F. Carmichael, L. Harris, A.Y. Jackson, F. Johnston, A. Lismer, J.E.H. MacDonald, and F. Varley, who adopted and adapted late nineteenth century impressionistic styles to create powerful images of rugged boreal landscapes. Five other painters often associated with this minor art movement are A.J. Casson, E. Holgate, L. Fitzgerald, Tom Thomson, and [a bit of a stretch] Emily Carr.

[8]Recall the three means employed in computing a similar integral in the determination of the moment of inertia of a uniform hoop rotating about its diameter in VOLUME I, Chapter 37.

[9]We even obtained the same decay constant and everything!

EXAMPLE [*Half-Time Sho(w)*]

At the start of this chapter, the half-time was defined to be the time elapsed during the amplitude's reduction to $1/\sqrt{2}$ of its original value. Now that an expression for the time dependence of the amplitude has been found, we may write

$$A(t)\Big|_{t=0} = A_0\, e^{-\gamma \times 0} = A_0 \qquad \text{and} \qquad A(t)\Big|_{t=t_{0.5}} = A_0\, e^{-\gamma \times t_{0.5}} = \frac{A_0}{\sqrt{2}}\,.$$

Together, these relations imply that

$$\tau_{0.5} = \frac{\ln\left(\sqrt{2}\right)}{\gamma} = \frac{2\,M}{b_1}\,\ln\left(\sqrt{2}\right) = \frac{M\,\ln(2)}{b_1}\,.$$

This provides a predictive formula for the [long] time scale on which the amplitude decreases by a factor of $\sqrt{2}$. The significance of $\tau_{0.5}$ is that the total mechanical energy of the DHO is reduced to one-half of its original value after this[10] elapse of time.

───────────────────────────────

───────────────────────────────

[10]Owing to the self-similarity property of exponentials, it takes $2 \times \tau_{0.5}$ for the initial energy to drop to one-quarter of its initial value, $3 \times \tau_{0.5}$ to drop to one-eighth, *etc.*

Chapter 17

Damped Oscillation: Explicitly

A mass–spring system in which the mass rides on a frictionless plane and is immersed in a draggy medium appears in Figure 17.1 below.

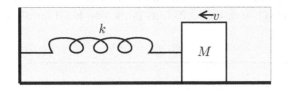

FIGURE 17.1   Mass–Spring System in a Draggy Medium

The assumptions underlying the model of this DHO system are listed below.

- ☐   The particle has constant mass M, and it moves in 1-d only.

- ☐   The plane on which the particle rides is horizontal and frictionless.

- ☐   The spring force is $F_s = -k\,x$, with force constant k. The equilibrium position of the block is at $x = 0$.

- ☐   The block experiences linear drag, $F_D = -b_1\,v$, owing to its motion through the fluid medium within which it is immersed. The linear drag coefficient, b_1, is constant.

The FBD for the block, when it lies to the right of its equilibrium position, $x > 0$, and is moving leftward, $v < 0$, appears in Figure 17.1. The normal force of contact with the plane cancels the weight of the block, which makes trivial the vertical equation of motion. In contrast, the horizontal component of the equation of motion reads:

$$M\,a_x = -k\,x - b_1\,v\,.$$

[The signs are consistent with the conventions adopted for the description of the system.]

The DHO equation of motion is a **second-order homogeneous linear ordinary differential equation with constant coefficients**:

$$v = \frac{dx}{dt} \quad \text{and} \quad a = \frac{d^2x}{dt^2} \quad \implies \quad 0 = M\,\frac{d^2x}{dt^2} + b_1\,\frac{dx}{dt} + k\,x\,.$$

SECOND ORDER   The maximum number of derivatives appearing in any single [additive] term in the expression is two.

HOMO-GENEOUS   Terms independent of x or its derivatives are NOT present.
Homogeneity $\iff$ LHS = 0.

LINEAR   IF $x_1(t)$ and $x_2(t)$ are distinct solutions of the equation of motion, THEN any linear combination

$$x_{1,2} = C_1\, x_1(t) + C_2\, x_2(t)\,,$$

with constant coefficients, $\{C_1, C_2\}$, is also a solution.

ORDINARY   All of the derivatives appearing in the equation of motion are with respect to a single parameter, the time.

DIFF EQN   The equation of motion constitutes a relation among the position, velocity, and acceleration of the block.

CONST COEFFS   There is no other time dependence, aside from that which is inherent in $x(t)$ and its derivatives.   [As an added bonus, the coefficients in the DHO EofM are all positive.]

The immediate task is to solve the equation of motion.

$$0 = M\,\frac{d^2x}{dt^2} + b_1\,\frac{dx}{dt} + k\,x \qquad \Longleftrightarrow \qquad 0 = \left[M\,\frac{d^2}{dt^2} + b_1\,\frac{d}{dt} + k\right]x(t)\,.$$

Q: How does one even begin this task?

A: With an *Ansatz*[1] for the trajectory of the particle.

Suppose that the trajectory, $x(t)$, has the form of an exponential function,

$$x(t) = e^{r\,t}\,,$$

where r is an as yet undetermined parameter.

> ASIDE: It is reasonable to inquire, "Where did this *Ansatz* come from?" Two answers spring to mind. The first is that a feature of the exponential function, *i.e.*, that all of its derivatives are proportional to the original function, makes it an ideal candidate for the solution of linear differential equations. The second is that, since *Ansatze* are "educated guesses," it is unlikely that we would report on an unsuccessful attempt.

The application of the differential operator to the *Ansatz* solution yields

$$0 = \left[M\,\frac{d^2}{dt^2} + b_1\,\frac{d}{dt} + k\right]e^{r\,t} = \left[M\,r^2 + b_1\,r + k\right]e^{r\,t}\,.$$

Thus,

EITHER $0 = M\,r^2 + b_1\,r + k\,,$ OR $r \to -\infty\,.$

The latter option leads to the extraneous solution $x(t) = 0$. Henceforth, our attention shall be focussed on the former with the intention of determining the particular value(s) of r for which the *Ansatz* satisfies this EofM.

Q: Wait! Isn't this a quadratic equation [*i.e.*, algebraic rather than differential]?

A: Yep! This is precisely the virtue of the *Ansatz* method.

---

[1] Recall that an *Ansatz* is a trial solution containing one or more free parameters. Particular values of these parameters are determined by substituting the *Ansatz* directly into the equation it is intended to satisfy and solving the constraints which then arise.

Application of the quadratic formula generates the solutions for r [*a.k.a.* the roots]:

$$r_\pm = \frac{-b_1 \pm \sqrt{b_1^2 - 4\,M\,k}}{2\,M} = -\frac{b_1}{2\,M} \pm \sqrt{\left(\frac{b_1}{2\,M}\right)^2 - \frac{k}{M}} = -\gamma \pm \sqrt{\gamma^2 - \omega_0^2}\,.$$

For concision, the roots are written in terms of two particular combinations[2] of $\{M, b_1, k\}$,

$$\gamma = \frac{b_1}{2\,M} \quad \text{and} \quad \omega_0^2 = \frac{k}{M}\,.$$

The γ term is precisely the amplitude decay constant [revealed in the qualitative analysis in Chapter 16], while ω_0 is the angular frequency that the mass–spring SHO system would possess were the drag force not present.

Next come two sticky wickets, along with their resolutions.

TWO   There are two roots of the quadratic equation, $r = r_\pm$, suggesting that there are two *Ansatz* solutions.

Resolution: Form an admixture of both solutions, *viz.*,

$$x(t) = C_+\, e^{r_+ t} + C_-\, e^{r_- t}\,,$$

where $C_\pm$ are constants whose values are determined by the initial or boundary conditions of the system.

> ASIDE: In retrospect, one realises that coefficients might well have been included in the original *Ansatz*. However, the linearity and homogeneity of the dynamical equation made it possible to elide them in the initial analysis.

An unforeseen bonus is that [owing to certain existence and uniqueness theorems applicable to second-order ordinary differential equations] **the unconstrained superposition of two distinct solutions is a fully general solution of the equation of motion.**

COMPLEX   Although the physical parameters $\{M, k, b_1\}$ and the combinations $\{\omega_0, \gamma\}$ are all manifestly positive constants characteristic of the physical mass–spring–damping system, the roots can be real or complex, depending on the sign of the term appearing under the square root.

Resolution: Consider all of the mathematically consistent possibilities. These are distinguished by the sign of the DISCRIMINANT: $\mathcal{D} = \gamma^2 - \omega_0^2$. There are three cases. Two of these, **overdamped** $[\mathcal{D} > 0]$ and **underdamped** $[\mathcal{D} < 0]$, exist as ranges, while the third, **critically damped** $[\mathcal{D} = 0]$, is the boundary between the first two.

OVERDAMPED   IF the value of the discriminant is positive,

$$\gamma^2 - \omega_0^2 > 0\,,$$

THEN the DHO system is overdamped. As a consequence, the square root term is real-valued, and thus $r_\pm$ are both real-valued and distinct. As γ and ω_0 are positive, it must be the case that

$$r_- < r_+ < 0\,.$$

---

[2]These also appear when the quadratic equation is scaled so as to set the coefficient of the r^2 term equal to unity.

Thus, the general solution of the overdamped equation of motion is

$$x(t) = C_+ \, e^{(-\gamma+\delta)\,t} + C_- \, e^{-(\gamma+\delta)\,t} \,,$$

for constant coefficients $C_\pm$, and $\delta = +\sqrt{\gamma^2 - \omega_0^2} < \gamma$, a system-dependent constant. Both of the temporal coefficients appearing in the exponents are negative.

There are several important comments that must be made.

O-D   Both terms contributing additively to $x(t)$ are decaying exponentials, with differing rates of decay. *Ergo*, the trajectory tends [asymptotically[3]] toward $x = 0$, smoothly but non-trivially.

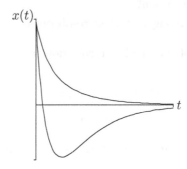

FIGURE 17.2   Trajectories of Two Overdamped Oscillators

O-D   After a somewhat long time has elapsed, the term with the more negative exponential factor, $-\gamma - \delta$, will become insignificant in comparison with the less quickly attenuating term. When this occurs, the trajectory of the block will exhibit [approximate] exponential decay [toward its equilibrium position] at rate $-\gamma + \delta$.

O-D   The relative sizes and signs of the coefficients $C_\pm$ determine whether the block goes monotonically or overshoots [once], *en route* to its final position.

O-D   Consider the factorisation:

$$-\gamma + \delta = -\gamma \left[1 - \sqrt{1 - \frac{\omega_0^2}{\gamma^2}} \right] \,.$$

In the limit $\frac{\omega_0^2}{\gamma^2} \to 0$, in which the effects of damping are certain to dominate over the system's attempt to oscillate, the argument of the more slowly decaying exponential becomes

$$-\gamma + \delta \simeq -\gamma \left[\frac{1}{2} \frac{\omega_0^2}{\gamma^2} \right] = -\frac{\omega_0^2}{2\,\gamma} = -\frac{k}{b_1} \,.$$

For this type of DHO system, the asymptotic rate at which the block approaches the equilibrium position is independent of its mass.

---

[3] Asymptotically, used in this context, means "in the limit that the time parameter grows large."

CRITICALLY DAMPED   IF the value of the discriminant is zero,

$$\gamma^2 - \omega_0^2 = 0\,,$$

THEN the DHO system is critically damped. Here, $r_\pm = -\gamma$ is a real double root. One might be concerned that the double root might vitiate the claims made above about generality. However, the general solution[4] of the critically damped equation of motion is

$$x(t) = (A + B\,t)\,e^{-\gamma\,t}\,,$$

for constant coefficients A and B.

The critically damped case has the peculiarities explored below.

C-D   The exponential suppresses both the constant and the linear terms, leading to a smooth and monotonic[5] [asymptotic] descent toward $x = 0$.

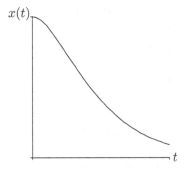

FIGURE 17.3   The Trajectory of a Critically Damped Oscillator

C-D   At criticality, the discriminant vanishes, and yet it remains factorisable:

$$0 = \gamma^2 - \omega_0^2 = \left(\gamma + \omega_0\right)\left(\gamma - \omega_0\right).$$

Since both γ and ω_0 are positive, it must be that $\gamma = \omega_0$.

---

[4]Despite our not proving that this is the general solution, it is easily verified to be a valid solution.
[5]It can be proven that, when the block starts from rest, it does not overshoot its equilibrium position.

UNDERDAMPED   IF the value of the discriminant is negative,

$$\gamma^2 - \omega_0^2 < 0 \,,$$

THEN the DHO system is underdamped. The square root term is imaginary-valued, and the $r_\pm$ are complex conjugates. Thus, the general solution of the underdamped equation of motion is

$$x(t) = A_+\, e^{(-\gamma + i\widetilde{\omega})\, t} + A_-\, e^{(-\gamma - i\widetilde{\omega})\, t} = e^{-\gamma t} \left[A_+\, e^{+i\widetilde{\omega}\, t} + A_-\, e^{-i\widetilde{\omega}\, t} \right] ,$$

for constant coefficients $A_\pm$, and $\widetilde{\omega} = \sqrt{|\gamma^2 - \omega_0^2|} = +\sqrt{\omega_0^2 - \gamma^2} < \omega_0$, a system-dependent constant.

There are several important aspects of the underdamped case deserving of mention.

U-D   An overall exponential envelope, with decay constant γ, suppresses the two terms in the bracket. The exponentials with imaginary arguments are equivalent [via Euler's Theorem] to a linear combination of sinusoids. Therefore, the trajectory oscillates within the confines of the [asymptotically] shrinking envelope, as illustrated in Figure 17.4.

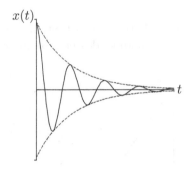

FIGURE 17.4   The Trajectory of an Underdamped Oscillator

The underdamped case corresponds to that exposed in the qualitative analysis of Chapter 16. The exponential falloff in amplitude predicted by the energetics arguments is precisely realised in the underdamped solutions.

U-D   Consider the not-so-obvious expansion of the oscillatory frequency:

$$\widetilde{\omega} = \omega_0 \sqrt{1 - \frac{\gamma^2}{\omega_0^2}} \,.$$

In the limit $\frac{\gamma^2}{\omega_0^2} \to 0$, in which the effects of damping are small in comparison to the underlying SHO dynamics, this becomes

$$\widetilde{\omega} \simeq \omega_0 \left[1 - \frac{1}{2} \frac{\gamma^2}{\omega_0^2} \right] = \omega_0 \left[1 - \frac{b_1^2}{8\, M\, k} \right] .$$

The effective frequency of the oscillation within the envelope is slightly diminished from what it would be were damping not present. Conversely, the period is increased. One interpretation of this effect is that it takes longer for the system to complete an oscillation owing to the resistive action of the drag force.

Chapter 18

Forced Oscillations

In the presentation of simple and damped harmonic oscillators thus far, the dynamical evolution of the system proceeded without external interference. That is, any direct external action on the system was limited to the imposition of initial conditions. *E.g.*, the block–spring system was displaced from its equilibrium position by some specified amount and then let go, or was given an impulsive kick to set it in motion.

There exist infinitely many ways in which an external agent might interact with the system. So as to make this chapter somewhat less than infinitely long,[1] we'll consider only the application of two classes of force: constant and sinusoidally varying.

In every case, the dynamical equation is obtained by including the external force in the expression of N2 for the DHO system. By appeal to results obtained in previous chapters one may immediately write:

$$F_{\mathrm{A}}(t) = M\,a + b_1\,v + k\,x\,.$$

This second-order linear ordinary differential equation with constant coefficients is NOT **homogeneous**, owing to the presence of the external force. Inhomogeneity is not an insurmountable obstacle, because **linearity** ensures that general solutions to the inhomogeneous equation can be obtained by superposition of general solutions of the homogeneous equation, $x_H(t)$, and any particular solution $x_P(t)$ incorporating the effects of the driving force.

This is subtle, and although we shan't actually prove this, we shall briefly argue for its consistency and mention its most profound consequence.

H   The solution of the homogeneous [unforced] DHO equations possesses two integration constants, and thus the superposition described above has precisely the degrees of freedom needed to accommodate any initial conditions. In all instances this solution is exponentially damped. This is termed the **transient** part, as it mediates the evolution of the system from its initial conditions to its asymptotic behaviour.

P   **The asymptotic behaviour of the trajectory is determined entirely by the particular solution, which in turn is dictated by the forcing function.**

   [Colloquially, the driven DHO system is said to "forget" its initial conditions.]

To see how all of this works, consider the case of a constant force suddenly applied to a DHO originally at rest at its equilibrium position.

---

[1] It may only seem so to the beleaguered reader.

EXAMPLE [*A Constantly Driven DHO*]

The simplest driving force is the sudden-onset[2] constant[3] force:

$$F_{\mathrm{A}}(t) = \begin{cases} 0, & t < 0 \\ F_0, & t > 0 \end{cases}.$$

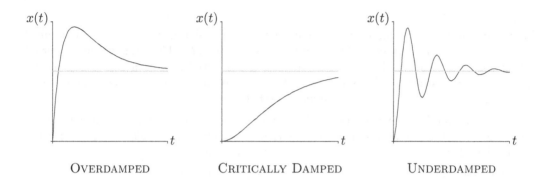

OVERDAMPED   CRITICALLY DAMPED   UNDERDAMPED

FIGURE 18.1   Trajectories of DHOs Suddenly Subjected to a Constant Applied Force

The homogeneous solutions are exactly those explicitly worked out in Chapter 17. These exist in three classes, OVER-/CRITICALLY/UNDERDAMPED, determined by the value of the discriminant, $\gamma^2 - \omega_0^2$. When the inhomogeneous term, F_{A}, is constant, a particular solution[4] of the equation of motion is

$$x_P(t) = x_\infty = \frac{F_{\mathrm{A}}}{k},$$

as may be verified by substitution. Figure 18.1 reveals that the range of expected behaviours is captured by the superposition of the transients and the particular solution. That is, in all three cases the trajectory approaches x_∞ in the limit $t \to \infty$, irrespective of the precise form of the initial conditions and the transient dynamics.

The most fruitful case, and thus the one considered in greatest detail, arises when the force itself oscillates harmonically. The additional demand of simplicity constrains the forcing function to be sinusoidal with fixed values of frequency, amplitude, and phase. WLOG, the forcing function phase may be chosen to vanish by felicitous assignment of the instant $t = 0$, in which case

$$F_{\mathrm{A}} = F_0 \cos(\omega\, t).$$

In this expression, the amplitude, F_0, and the driving frequency, ω, are constant parameters. Recall that the underlying mass–spring system[5] has mass M, linear damping coefficient b_1, and spring constant k. Therefore, the driven DHO requires a total of five parameters for its physical description:

$$\{M, b_1, k\} \quad \text{and} \quad \{F_0, \omega\}.$$

---

[2] A natural refinement of this model is to have the applied force grow continuously from zero to its ultimate [constant] value within a narrow temporal window.

[3] A constant force may be thought of as oscillating with zero frequency, *i.e.*, having infinite period.

[4] The snarky among us might quip that this is a particularly fine solution.

[5] Although our study of forced oscillations shall explicitly presuppose the existence of a mass–spring system, the analysis generalises to other realisations of driven DHOs.

o   The reduced variables, $\{\omega_0, \gamma\}$, are sufficient to describe the DHO dynamics. The most important upcoming results shall be reformulated in these terms.

o   No *a priori* relation between the angular frequency of the applied force, ω, and the natural frequency of the SHO/DHO, $\omega_0 = \sqrt{k/M}$, is posited.

[In Chapter 20, we'll explicate the consequences of choosing $\omega \sim \omega_0$.]

That [in the constant-force context] the long-time behaviour of the system is uncorrelated with its initial conditions, and that the driving frequency is independent of the natural frequency of the [underlying] SHO, taken together, suggest that the particular solution of the dynamical equations, $x(t)$, is oscillatory with angular frequency equal to that of the driving force, yet phase-shifted.

ASIDE: To appreciate the soundness of this proposal, imagine assisting a very small child on a playground swing set. When the child is continually held, his back-and-forth motion is completely controlled by you, irrespective of the natural frequency of the swing.

The phase shift issue is subtle. The sensations that you experience while guiding the child lead to the realisation that the child's motion is [generally] not quite in sync with the natural motion of the swing. Ordinarily, there is an [unconscious] inclination on your part to adjust the pushing frequency so as to eliminate these odd sensations and enter into the more pleasurable state of **resonance**. [We shall indulge this urge in Chapter 20.]

An *Ansatz* for the position response [the trajectory] incorporating the proposal reads:

$$x(t) = A_0 \sin(\omega t - \varphi),$$

where ω is the driving frequency.

ASIDE: An equivalent *Ansatz* is $x(t) = A_0 \cos(\omega t - \tilde{\varphi})$, where $\tilde{\varphi}$ differs from φ by $\pi/2$.

The amplitude, A_0, and phase angle, φ, are parameters to be determined by demanding that the dynamical equations be satisfied. It must be remarked that an implicit phase offset of $\pi/2$ is built into the trajectory on account of its being expressed in terms of sine [rather than cosine like the forcing function].

ASIDE: **Q:** WHY write the *Ansatz* in this form?

A1: The additional phase offset inherent in the use of sine brushes aside discontinuities which might otherwise infect a mild physical regime. Such discontinuous behaviour, while not pathological, requires extra care in the analysis.

A2: The **momentum response** of a system happens to be a better characteristic than its position response.

A certain harmonically oscillating forcing function and the phase-shifted position response that it elicits from a DHO are illustrated together[6] in Figure 18.2.

Given that the *Ansatz* for the position response is $x(t) = A_0 \sin(\omega t - \varphi)$,

$$v(t) = \frac{dx(t)}{dt} = \omega A_0 \cos(\omega t - \varphi) \quad \text{and} \quad a(t) = \frac{dv(t)}{dt} = -\omega^2 A_0 \sin(\omega t - \varphi).$$

Upon substitution of these results into the dynamical equation, it reads

$$\begin{aligned} F_0 \cos(\omega t) &= M a + b_1 v + k x \\ &= -M \omega^2 A_0 \sin(\omega t - \varphi) + b_1 \omega A_0 \cos(\omega t - \varphi) + k A_0 \sin(\omega t - \varphi). \end{aligned}$$

---

[6] Each curve has its proper scale and units. They are overlaid to show their mutual correlation.

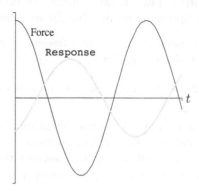

FIGURE 18.2   Sinusoidal Driving Force and Phase-Shifted Dynamical Response

Solving for the *Ansatz* parameters requires invocation of the identities governing trigonometric functions of differences of angles:

$$\sin(\omega\,t - \varphi) = \sin(\omega\,t)\,\cos(\varphi) - \cos(\omega\,t)\,\sin(\varphi)$$
$$\cos(\omega\,t - \varphi) = \cos(\omega\,t)\,\cos(\varphi) + \sin(\omega\,t)\,\sin(\varphi)$$

Substitution of these formulae into the dynamical equation, expansion, and collection of terms with common time-dependent parts leads to

$$0 = \Big[F_0 - \left(M\,\omega^2 - k\right)A_0\,\sin(\varphi) - \omega\,b_1\,A_0\,\cos(\varphi)\Big]\sin(\omega\,t)$$
$$+ \Big[\left(M\,\omega^2 - k\right)\cos(\varphi) - \omega\,b_1\,\sin(\varphi)\Big]\cos(\omega\,t)$$

In order for this equality to persist throughout a time interval, [we make the claim that[7]] the bracketed terms must each separately vanish. *Ergo*, a system of two equations ensues:

$$\begin{cases} \#1 & 0 = F_0 - \left(M\,\omega^2 - k\right)A_0\,\sin(\varphi) - \omega\,b_1\,A_0\,\cos(\varphi) \\ \#2 & 0 = \left(M\,\omega^2 - k\right)\cos(\varphi) - \omega\,b_1\,\sin(\varphi) \end{cases}$$

The latter equation, #2, is readily solved for the tangent of the phase angle,

$$\tan(\varphi) = \frac{\sin(\varphi)}{\cos(\varphi)} = \frac{M\,\omega^2 - k}{\omega\,b_1} = \frac{M\left(\omega^2 - \omega_0^2\right)}{\omega\,b_1}.$$

Solving the former equation for the amplitude requires re-expression of $\cos(\varphi)$ and $\sin(\varphi)$ in terms of the SHO and driver parameters. The first step is the realisation that

$$\cos(u) = \frac{1}{\sec(u)} = \frac{1}{\sqrt{1 + \tan^2(u)}} \quad \text{and} \quad \sin(u) = \frac{\tan(u)}{\sec(u)} = \frac{\tan(u)}{\sqrt{1 + \tan^2(u)}}.$$

The second step involves substituting the expression for $\tan(\varphi)$ obtained just above and

---

[7]See the parallel discussion of this claim in VOLUME III, Chapter 44, and its justification found just ahead in Chapter 22.

simplifying to yield

$$\cos(\varphi) = \frac{1}{\sqrt{1 + \left(\frac{M\,\omega^2 - k}{\omega\,b_1}\right)^2}} = \frac{\omega\,b_1}{\sqrt{M^2\left(\omega^2 - \omega_0^2\right)^2 + \omega^2\,b_1^2}}$$

and

$$\sin(\varphi) = \frac{\frac{M\,\omega^2 - k}{\omega\,b_1}}{\sqrt{1 + \left(\frac{M\,\omega^2 - k}{\omega\,b_1}\right)^2}} = \frac{M\left(\omega^2 - \omega_0^2\right)}{\sqrt{M^2\left(\omega^2 - \omega_0^2\right)^2 + \omega^2\,b_1^2}}.$$

Third, inserting these expressions into the former equation, #1, determines the amplitude of the response function:

$$F_0 = A_0\left((M\,\omega^2 - k) \times \frac{M\left(\omega^2 - \omega_0^2\right)}{\sqrt{M^2\left(\omega^2 - \omega_0^2\right)^2 + \omega^2\,b_1^2}} + \omega\,b_1 \times \frac{\omega\,b_1}{\sqrt{M^2\left(\omega^2 - \omega_0^2\right)^2 + \omega^2\,b_1^2}}\right)$$

$$= A_0\left(\frac{M^2\left(\omega^2 - \omega_0^2\right)^2 + \omega^2\,b_1^2}{\sqrt{M^2\left(\omega^2 - \omega_0^2\right)^2 + \omega^2\,b_1^2}}\right) = A_0\,\sqrt{M^2\left(\omega^2 - \omega_0^2\right)^2 + \omega^2\,b_1^2}$$

$$\implies \quad A_0 = \frac{F_0}{\sqrt{M^2\left(\omega^2 - \omega_0^2\right)^2 + \omega^2\,b_1^2}}.$$

EXAMPLE [*A Sinusoidally Driven DHO*]

A particular block–spring system has parameters $\{m = 4, k = 16\}$ [the units shall remain implicit throughout this example]. The system is immersed in a fluid medium and the drag coefficient for the block under the prevailing conditions is $b_1 = 5$. An external driving force, $F_A = 13\,\cos(4\,t)$, is applied to the block and acts for a long time.

(a) Compute the natural frequency of the block–spring SHO.

(b) Compute the damping constant of the block–spring–drag system.

(c) Is the unforced system under-, critically, or overdamped?

(d) Under the influence of the external (driving) force the block will oscillate.   (i) Ascertain the angular frequency of the motion of the block.   Determine   (ii) the amplitude and   (iii) relative phase shift   of the block's response to the driving force.

This system is a driven damped oscillator.

(a) $\omega_0 = \sqrt{k/m} = 2\,\mathbf{rad/s}$.

(b) $\gamma = b_1/(2\,M) = \frac{5}{8}\,\mathbf{Hz}$.

(c) $\gamma^2 - \omega_0^2 = \left(\frac{5}{8}\right)^2 - 2^2 = -\frac{231}{64} < 0$, and therefore the DHO is underdamped.

(d) (i) $4\,\mathbf{rad/s}$, (ii) $A_0 = \dfrac{F_0}{\sqrt{M^2(\omega^2 - \omega_0^2)^2 + \omega^2\,b_1^2}} = \dfrac{13}{\sqrt{16\,(16 - 4)^2 + (16)(25)}} = \dfrac{1}{4}$,

(iii) $\tan(\varphi) = \dfrac{M\left(\omega^2 - \omega_0^2\right)}{\omega\,b_1} = \dfrac{4\,(16 - 4)}{(4)(5)} = \dfrac{12}{5}$. Therefore, $\varphi \simeq 1.176\,\mathbf{rad} \simeq 67.38°$.

Having successfully obtained the particular solution of the equation of motion governing the sinusoidally driven DHO, our attention must be turned to issues of economy and uniqueness. That is, although the physical DHO is specified by $\{M, b_1, k\}$, its dynamical behaviour is completely characterised by

$$\omega_0 = \sqrt{\frac{k}{M}} \qquad \text{and} \qquad \gamma = \frac{b_1}{2M}.$$

Rewriting the expressions for $\tan(\varphi)$ and A_0 in terms of these parameters yields

$$\tan(\varphi) = \frac{\omega^2 - \omega_0^2}{2\,\gamma\,\omega} \qquad \text{and} \qquad A_0 = \frac{F_0/M}{\sqrt{\left(\omega^2 - \omega_0^2\right)^2 + 4\,\gamma^2\,\omega^2}}.$$

We finish this chapter with a couple of quick observations and the promise to review and systematise these results in the next chapter.

φ The phase angle can be POSITIVE, ZERO, or NEGATIVE depending on whether the driving frequency, ω, is GREATER THAN, EQUAL TO, or LESS THAN the SHO frequency, ω_0. These possible behaviours and their primary physical implications are summarised in the table which follows.

	$\omega < \omega_0$	$\omega = \omega_0$	$\omega > \omega_0$
φ	$\varphi < 0$	$\varphi = 0$	$\varphi > 0$
PHASE RESPONSE VS DRIVING	LEADS	IN PHASE	LAGS
POSITION RESPONSE			
MOMENTUM RESPONSE			

In the sketches, the forcing function is denoted by the darker line, while typical [POSITION and MOMENTUM] responses are shown in grey. The standard nomenclature, LEADS/IN PHASE/LAGS, is seen to match the momentum responses.

A_0 Underlying the DHO system is an SHO whose defining condition is that the acceleration and position are directly proportional at all times. It is not surprising then that the maximum displacement is proportional to F_0/M, an acceleration.

Chapter 19

Impedance and Power

Let's recapitulate the most important aspects of the previous chapters.

SHO   The SHO modelled only the effects of the spring force and the inertia of the block. All dissipative effects, *e.g.*, friction and drag, were assumed to vanish, as were any other forces which might have done a net amount of work on the block. Under these conditions, the total mechanical energy is conserved, and thus it is [almost] inevitable that the motion be oscillatory. The dynamical behaviour of the system is governed by its natural frequency, ω_0.

DHO   More realistic, and yet tractable, models incorporate the effects of [linear] drag. That these models are a substantial extension of the SHO model is evident. First, the solutions are NOT oscillatory! Second, an additional parameter, γ, is needed to characterise the solutions.

DRIVEN DHO   IF a DHO is subjected to a sinusoidal driving force,

$$F_{\text{A}}(t) = F_0 \cos(\omega t) \,,$$

THEN the steady–state [post-transient] response of the system is harmonic oscillation at the driving frequency,

$$x(t) = A_0 \sin(\omega t - \varphi) \,.$$

The parameters $\{A_0, \varphi\}$ in the response function admit expression in terms of the physical parameters of the DHO and the driving force:

$$\varphi = \arctan\left(\frac{M\left(\omega^2 - \omega_0^2\right)}{\omega\, b_1}\right) \qquad \text{and} \qquad A_0 = \frac{F_0}{\sqrt{M^2\left(\omega^2 - \omega_0^2\right)^2 + \omega^2\, b_1^2}} \,.$$

Alternatively, $\{A_0, \varphi\}$ can be written [almost] exclusively in terms of the reduced set of parameters $\{\omega_0, \gamma, \omega, F_0\}$:

$$\varphi = \arctan\left(\frac{\omega^2 - \omega_0^2}{2\,\gamma\,\omega}\right) \qquad \text{and} \qquad A_0 = \frac{F_0/M}{\sqrt{\left(\omega^2 - \omega_0^2\right)^2 + 4\,\gamma^2\,\omega^2}} \,.$$

Now that the existence of these solutions of the harmonically driven DHO equations of motion is beyond doubt, a discussion of the impedance triangle framework is warranted.

IMPEDANCE TRIANGLE   The impedance triangle is a mnemonic device expressing the amplitude and phase angle appearing in the position response *Ansatz*.

Consider a right triangle with sides $M\left(\omega^2 - \omega_0^2\right)$, $\omega\, b_1$, and Z, as in Figure 19.1.

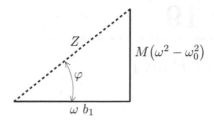

FIGURE 19.1   The Impedance Triangle for the Damped DHO

φ It is certainly the case that

$$\tan(\varphi) = \frac{\text{rise}}{\text{run}} = \frac{M\left(\omega^2 - \omega_0^2\right)}{\omega\, b_1},$$

reproducing the previously obtained characterisation of the phase angle.

Z The sum of the squares of the sides is equal to the square of the denominator appearing in the expression for the maximum amplitude.

PHASE ANGLE   The phase angle quantifies the temporal relation between the driver STIMULUS and the [momentum] RESPONSE of the DHO. This phase shift is physical and cannot be cavalierly done away with by split timing, since $t = 0$ was set in framing the parameterisation of the driving force.

LAG   IF $\varphi > 0$, THEN the response lags the driver. That is, the oscillatory features[1] exhibited by the driving force are later echoed in the momentum response.

IN PHASE   IF $\varphi = 0$, THEN the response exactly mimics the driver. Oscillatory features of the driving force and momentum response occur simultaneously.

LEAD   IF $\varphi < 0$, THEN the response leads the driver. It is as though the response anticipates the future changes in the driving force.

IMPEDANCE   The impedance, Z, of a driven damped mechanical oscillator is the factor relating the amplitude of the position response of the system to that of the harmonic driving force, *i.e.*,

$$A_0 = \frac{F_0}{Z} \quad \Longrightarrow \quad Z = \sqrt{M^2\left(\omega^2 - \omega_0^2\right)^2 + \left(\omega\, b_1\right)^2}.$$

The impedance depends on properties of both the DHO and the driving force.

---

[1]Such features might include: peaks, valleys, central crossings in a particular direction, *etc.*

Next, we consider the impedance triangle in a special case, and for limiting values of the DHO and forcing function parameters.

$\omega = \omega_0$ IF the driving force oscillates at the natural frequency of the underlying SHO, THEN the phase angle vanishes, $\varphi = 0$, AND the impedance is $Z = \omega_0\, b_1 = \sqrt{k/M}\; b_1$. This situation is called **resonance**.

$b_1 \to 0$ In the limit that the drag coefficient approaches zero, the underlying SHO re-emerges and the phase angle, φ, tends toward $\pm\pi/2$, with its sign contingent on that of $\omega - \omega_0$. The identities

$$\sin\left(\alpha + \frac{\pi}{2}\right) = \cos(\alpha) \qquad \text{and} \qquad \sin\left(\alpha - \frac{\pi}{2}\right) = -\cos(\alpha)$$

reveal that at low driving frequencies the position response is in phase with the driver, while at high frequencies it is completely out of phase. Surprisingly, the impedance does not vanish when the damping coefficient does:

$$\lim_{b_1 \to 0} Z = M\left|\omega - \omega_0\right|,$$

provided that there is a mismatch between the driving and SHO frequencies.

$\omega \to 0$ As the driving frequency vanishes, *i.e.*, $\omega \to 0$, the time scale over which the forcing function oscillates becomes increasingly long. The force then is effectively constant on the shorter time scales characteristic of the underlying SHO/DHO. The phase angle tends to $-\pi/2$, and as a consequence the trajectory is in phase with the [almost constant] forcing function. The impedance does not vanish when $\omega \to 0$. Instead,

$$\lim_{\omega \to 0} Z = M\,\omega_0^2 = k\,,$$

in consonance with the results obtained in the constant force example in Chapter 18.

$\omega \to \infty$ When the driver oscillates much faster than the SHO natural frequency, the phase angle approaches $+\pi/2$, indicating that the response is completely out of phase with the driver. In this instance, the impedance becomes unboundedly large, and thus the amplitude of the position response becomes very small.

Let's examine the flows of energy[2] into and out of the DHO system.

$k\,x$ The spring force is conservative. **Work done by the spring does not add to, nor subtract from, the total mechanical energy of the system.**

[The spring can store energy and release it back to the system.]

$b_1\,v$ The drag force is dissipative. **Work done by drag diminishes the total mechanical energy of the system.**

[Drag almost[3] always converts mechanical energy to heat.]

---

[2]It is a truism that, to understand how and why events unfold as they do in the political and economic spheres, one should *follow the money*. In the physical realm, one should *follow the energy*.

[3]This qualifier is needed because when the block is at rest, as it is at the extremal points of its trajectory, the drag power vanishes.

F_A The applied force is non-conservative. **Work done by an applied force adds to[4] the total mechanical energy of the system.**

[At different times, the applied force may transfer energy into or out of the system.]

The instantaneous flows of energy, *i.e.*, the applied and drag powers, need not always sum to zero, because the spring is able to store and release energy. Eschewing the [inessential] complexity of the instantaneous power, we consider instead the average power throughout each cycle.

> ASIDE: Equivalently, one may consider the net amount of energy passing into, or out of, the driven DHO in each complete cycle.

[Net energy transferred per cycle = the average input/output power × the period of oscillation.]

There is more to this than a desire on our parts to avoid having to deal with complicated time-dependence. In order for the mass–spring system to oscillate, its total mechanical energy must either oscillate in a synchronised manner or remain constant. In either case, zero net energy accrues to the system throughout any number of complete cycles. As a consequence, **the net work input by the applied driving force acting through some number of complete cycles must be exactly equal to the amount of energy dissipated by the drag force during the same time interval.**

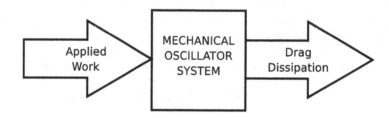

FIGURE 19.2   The DHO Converts Input Applied Work to Output Dissipated Energy

This idea is schematically illustrated in Figure 19.2. The DHO is portrayed as the mechanism by which energy supplied by the applied force is dissipated. The average power flowing through the DHO may be computed by calculating the applied power [input] OR the drag power [output]. As is our custom, we'll do both.

METHOD ONE: [*Average Power Input Through Each Cycle*]

The instantaneous applied power is

$$P_A = \vec{F_A} \cdot \vec{v} = \left[F_0 \cos(\omega\, t) \right] \cdot \left(\omega\, A_0 \cos(\omega\, t - \varphi) \right) = \frac{\omega\, F_0^2}{Z} \cos(\omega\, t) \cos(\omega\, t - \varphi)\,.$$

In the final equality, the response amplitude, A_0, was expressed in terms of the amplitude of the driving force and the impedance of the system.

The time average of the input power is obtained by integrating P_A through the duration of one period, *i.e.*, from $t = 0$ to $t = T = 2\,\pi/\omega$, and dividing the result by T. That is,

$$\langle P_A \rangle = \frac{1}{T} \int_0^T \frac{\omega\, F_0^2}{Z} \cos(\omega\, t) \cos(\omega\, t - \varphi)\, dt\,.$$

---

[4]Of course, for negative external work this means "subtracts from."

To effect this integration, we change to a new variable, $u = \omega t$, in which case

$$t = \frac{u}{\omega}, \qquad dt = \frac{du}{\omega}, \qquad \text{and} \qquad 0 \le t \le T \implies 0 \le u \le \omega T = 2\pi.$$

The average input power, expressed in terms of u, reads

$$\langle P_A \rangle = \frac{\omega F_0^2}{Z} \frac{1}{2\pi} \int_0^{2\pi} \cos(u) \cos(u - \varphi) \, du.$$

Evaluation of the integral relies on the identity[5] $\cos(u - \varphi) = \cos(u)\cos(\varphi) + \sin(u)\sin(\varphi)$, along with the following elementary integrations:

$$\int_0^{2\pi} \cos^2(u) \, du = \int_0^{2\pi} \frac{1 + \cos(u)}{2} \, du = \frac{1}{2}\left[u + \sin(u)\right]_0^{2\pi} = \frac{2\pi + 0}{2} - 0 = \pi, \quad \text{and}$$

$$\int_0^{2\pi} \cos(u) \sin(u) \, du = 0 \qquad \text{[by symmetric integration]}.$$

Therefore, the average power input to the mass–spring system is

$$\langle P_A \rangle = \frac{\omega F_0^2}{Z} \frac{1}{2\pi} \left[\pi \times \cos(\varphi) + 0 \times \sin(\varphi)\right] = \frac{\omega F_0^2}{2Z} \cos(\varphi).$$

In this formulation of the average power, the $\cos(\varphi)$ factor is dubbed the **power factor**. The evident simplicity of the above expression notwithstanding, further distillation is both possible and preferred. According to the impedance triangle, $\cos(\varphi) = \omega b_1 / Z$, while $Z^2 = M^2 \left(\omega^2 - \omega_0^2\right)^2 + \omega^2 b_1^2$. By incorporating these facts into the expression for the power, the following result is obtained.

$$\boxed{\langle P_A \rangle = \frac{F_0^2 \, \omega^2 \, b_1 / 2}{M^2 \left(\omega^2 - \omega_0^2\right)^2 + \omega^2 b_1^2}}$$

There are two things to note about this result before proceeding to METHOD TWO.

≥ 0 The average input power is positive, as expected, despite the possibility that there are moments when the instantaneous power is negative.

SHAPE   Some representative sketches of $\langle P_A \rangle$ vs. ω appear in Figure 19.3.

METHOD TWO: [*Average Power Dissipated Through Each Cycle*]

The instantaneous drag power is [*cf.* Chapter 16 for the undriven[6] case],

$$P_D = \vec{F_D} \cdot \vec{v} = -b_1 v^2 = -b_1 \left(\omega A_0 \cos(\omega t - \varphi)\right)^2 = -\frac{b_1 \omega^2 F_0^2}{Z^2} \cos^2(\omega t - \varphi).$$

[Recall that the instantaneous drag power is never positive.]

---

[5] The time-dependence of the angles does not preclude the imposition of this identity.
[6] A snarky person might dub this the "*slacker* DHO."

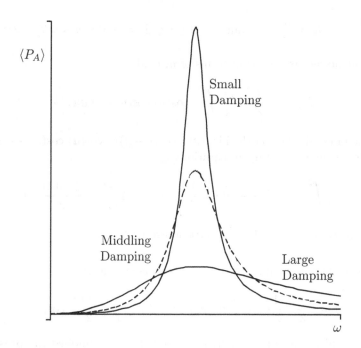

FIGURE 19.3 Input Power *vs.* Driving Frequency for Fixed $\{M, k, F_0\}$

The time-average of the power associated with the drag force is constructed in the same manner as the average input power. This analysis is slightly more straightforward, relying on a simple rescaling and shift of the integration variable, *i.e.*, $u = \omega t - \varphi$, to convert the integral into a standard form amenable to evaluation:

$$\frac{1}{T} \int_0^T \cos^2(\omega t - \varphi)\,dt = \frac{1}{2\pi} \int_{-\varphi}^{2\pi - \varphi} \cos^2(u)\,du = \frac{1}{2\pi} \int_0^{2\pi} \cos^2(u)\,du = \frac{1}{2}.$$

Therefore the average drag power during a cycle of oscillation is:

$$\langle P_D \rangle = -\frac{b_1\,\omega^2\,F_0^2}{2\,Z^2} = \frac{-b_1\,\omega^2\,F_0^2/2}{M^2\left(\omega^2 - \omega_0^2\right)^2 + \omega^2\,b_1^2}.$$

In steady state, the average power output from the DHO system is exactly equal to the average power flowing into it!

Chapter 20

Resonance

Careful examination of the flows of energy into and out of harmonically driven DHO systems led to expressions for the input [applied] and output [drag] average powers:

$$\langle P_A \rangle = \frac{F_0^2 \, \omega^2 \, b_1/2}{M^2 \left(\omega^2 - \omega_0^2\right)^2 + \omega^2 \, b_1^2} = -\langle P_D \rangle \, .$$

Rather than perpetuate the *faux* distinction between the input and drag powers, we'll just call this the average power transferred *through* the DHO. In Figure 19.3 of Chapter 19, reproduced here as Figure 20.1, the mass, the spring constant, and the amplitude of the driving force are held fixed while the average power flowing through the system is plotted *vs.* the frequency of the driving force for a representative sample of damping constants b_1.

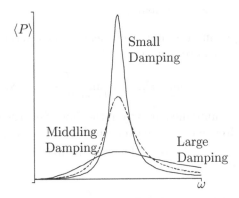

FIGURE 20.1   Average Power Transferred *vs.* Driving Frequency for Fixed $\{M, k, F_0\}$

The power spectra displayed in Figure 20.1 reveal that the flow of energy through the driven DHO system depends strongly on the driving frequency.

PEAK   The partial derivative of the average input/output power with respect to the frequency of the driving force, ω, is

$$\frac{\partial \langle P \rangle}{\partial \omega} = F_0^2 \, b_1 \left[\frac{\omega}{M^2 \left(\omega^2 - \omega_0^2\right)^2 + \omega^2 \, b_1^2} - \frac{\omega^2 \left(2 \, M^2 \left(\omega^2 - \omega_0^2\right) \omega - b_1^2 \, \omega\right)}{\left[M^2 \left(\omega^2 - \omega_0^2\right)^2 + \omega^2 \, b_1^2\right]^2} \right]$$

$$= -F_0^2 \, b_1 \, M^2 \, \frac{\omega \left(\omega^4 - \omega_0^4\right)}{\left[M^2 \left(\omega^2 - \omega_0^2\right)^2 + \omega^2 \, b_1^2\right]^2} \, .$$

This expression vanishes, *i.e.*,

$$\frac{\partial \langle P \rangle}{\partial \omega} \bigg|_{\omega = \omega_{max}} \equiv 0 \,,$$

when the power through the DHO is maximised. Simplification reveals that

$$0 = \omega_{max}^4 - \omega_0^4 = \left(\omega_{max}^2 + \omega_0^2 \right) \left(\omega_{max} + \omega_0 \right) \left(\omega_{max} - \omega_0 \right) \,,$$

which has the sole physical solution

$$\omega_{max} = \omega_0 \equiv \sqrt{\frac{k}{M}} \,.$$

The maximal average power transfer through the driven DHO occurs when the system is driven at the natural frequency of its underlying SHO. This is another aspect of the phenomenon called **resonance**.

RESONANCE   The MAXIMUM rate at which power flows through the DHO occurs when $\omega = \omega_0$. Under this condition,

$$\langle P \rangle_{max} = \frac{F_0^2}{2 \, b_1} \,.$$

LOW DRAG   The maximum average power occurring in the drag-free limit, *i.e.*, $b_1 \to 0$, depends crucially on whether the system is being driven at its natural frequency. That is, whether or not the system is in resonance.

IN   Curiously, making the drag coefficient smaller actually increases the rate of energy transfer in resonance, *i.e.*,

$$\lim_{b_1 \to 0} \langle P \rangle_{max} = \lim_{b_1 \to 0} \frac{F_0^2}{2 \, b_1} \longrightarrow \infty \,.$$

An offset to this unbridled growth in the flow of energy is a concomitant narrowing of the frequency range over which the rate of transfer is appreciable.

OUT   Out of resonance, the rate of power dissipation vanishes along with the drag coefficient. *I.e.*,

$$\lim_{b_1 \to 0} \langle P_A \rangle = \lim_{b_1 \to 0} \frac{F_0^2 \, \omega^2 \, b_1 / 2}{M^2 \left(\omega^2 - \omega_0^2 \right)^2 + \omega^2 \, b_1^2} = 0 \,,$$

provided that $\omega \neq \omega_0$.

The counterpoint to all of this is that increased damping tends to decrease the peak rate at which energy is dissipated while making the system less sensitive to the driver frequency.

LOW-FREQUENCY DRIVING   IF $\omega \to 0$, THEN $\langle P \rangle \to 0$.

In the limit that the force is constant [or very nearly so, owing to a very slow rate of change], we have seen that the system asymptotically comes to rest and thereby ceases to both absorb work from the applied force and dissipate energy via the linear drag force.

HIGH-FREQUENCY DRIVING   IF $\omega \to \infty$, THEN $\langle P \rangle \to 0$.

When the applied force oscillates much faster than the system can respond, the amplitude of the response approaches zero, on account of the unbounded increase of the impedance. Thus, the motion of the particle effectively ceases, and the system's ability to absorb and re-emit work is curtailed.

RESONANCE   Resonance occurs when the driving force oscillates with the frequency characteristic of the underlying SHO system, $\omega = \omega_0$. In resonance, the [momentum] response of the system is phase-aligned with the driving force, and the rate at which energy flows through the system is maximised.

> ASIDE: It must be remarked that the impedance is not minimised at resonance, owing to the presence of the power factor. In fact, for fixed $\{M, b_1, k\}$, the minimum impedance and maximum amplitude occur when $\omega = \omega_0 \sqrt{1 - 2\gamma^2/\omega_0^2}$.

The beneficient properties of resonance occur for driving frequencies in a range near ω_0, as is nicely illustrated by the power spectra curves exhibited in Figure 20.1. The question which must then arise is:

Q: How might this range of frequencies be characterised?

A: The WIDTH of the peak in the power spectrum is its second-most important aspect. Precisely how this width is measured is a matter of convention.

FULL-WIDTH AT HALF-MAXIMUM   The full-width at half-maximum[1] of the resonant peak in the power spectrum is a convenient proxy for the breadth of the resonance. The details are outlined below and illustrated in Figure 20.2.

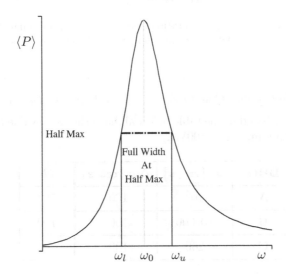

FIGURE 20.2   Full-Width at Half-Maximum for a Driven DHO Resonant Power Curve

---

[1] One must exercise care because another oft-used measure is the **half-width at half-maximum**.

The lower and upper frequencies, $\{\omega_l, \omega_u\}$, are those at which the power is reduced to one-half that of the peak power.

$$\langle P \rangle\Big|_{\omega=\omega_{l/u}} = \frac{1}{2}\,\langle P \rangle_{\max} = \frac{F_0^2}{4\,b_1}\,.$$

The full-width at half-maximum is the [positive] difference between the two half-power frequencies:

$$\Delta\omega = \omega_u - \omega_l\,.$$

While the full-width at half-maximum is a [dimensionful] measure of the shape of the power spectrum of the oscillator, it does not quite describe the relative sharpness of the resonance peak. To overcome this particular limitation, we recognise that the natural frequency establishes the scale to which $\Delta\omega$ should be referred.

QUALITY FACTOR The quality factor of a DHO is the ratio of its natural frequency to its full-width at half-maximum:

$$Q = \frac{\omega_0}{\Delta\omega}\,.$$

WHEN the quality factor is $\begin{Bmatrix} \text{large} \\ \text{small} \end{Bmatrix}$, THEN the resonance is $\begin{Bmatrix} \text{narrow and sharp} \\ \text{broad and flat} \end{Bmatrix}$.

High-Q means:

narrow taper'd peak
small damping parameter
slow energy loss.

ASIDE: The quality factor admits interpretation in terms of the approximate fraction of energy loss per cycle of an undriven DHO, *cf.* the energetic analysis of damping in Chapter 16.

EXAMPLE [*Peaks, Widths, and Quality Factors for a Triplet of Driven DHOs*]

The three different DHOs listed in the table below all have the same value for their respective full-widths at half-maximum, $\Delta\omega = 100\,\textbf{rad/s}$.

DHO	$\omega_0\,[\textbf{rad/s}]$	$\Delta\omega\,[\textbf{rad/s}]$	Q
A	10^6	100	10^4
B	10,000	100	100
C	200	100	2

Comparing these DHO systems to the spectra in Figure 20.1 suggests a very rough correspondence between **A**, **B**, and **C** and the cases labelled Small, Middling, and Large Damping. The upshot of this is that, although all three DHOs share the same full-width at half-maximum, they possess radically different spectral shapes. The quality factor distinguishes them.

Chapter 21

The First Wave

Q: What else is there besides oscillation?

A: Waves!

MECHANICAL WAVE   A mechanical wave is a propagating disturbance.

- PROPAGATING implies movement, transmission, and even causality.

- DISTURBANCE is sufficiently generic to accommodate many situations.

The existence of a medium in which the disturbance resides and through which it propagates is implicitly assumed. **The medium through which the wave moves need not experience [bulk] motion.**

WAVE SPEED   The speed at which the wave propagates, c, is determined primarily by the medium in which it moves. The wave speed exhibits dependence on certain aspects of the wave itself through the phenomenon of **dispersion**.

DISPERSIONLESS   Two related consequences of the assumption of dispersionless propagation [in 1-d] are listed below.

SHAPE   The shape of the disturbance remains unchanged.

SPEED   All waves in a particular medium travel with precisely the same speed.

All of the mechanical waves considered hereafter shall be taken to be dispersionless. On account of the constancy of the shapes of these waves, there can be no **attenuation**, *i.e.*, loss of mechanical energy. Generally speaking, attenuation leads to diminishment of some aspect of the disturbance.

A mechanical wave may be solitary in nature, like a tsunami or a pulse on a string, as illustrated in Figure 21.1, or it may consist of many repeated disturbances, like ocean waves or a succession of string pulses, as shown in Figure 21.2.

ASIDE: In Figures 21.1, 21.2, 21.3, and the like, the wave form is plotted *vs.* position in space. Such a plot of the spatial variation of the wave represents a "snapshot," or a frozen instant in time. In contrast, Figure 21.4 displays temporal variation of the wave disturbance viewed at a particular, fixed, point in space. These two points of view are **dual** [complementary].

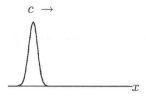 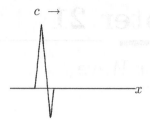

FIGURE 21.1   Solitary Waves

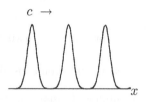 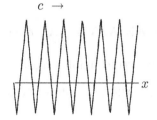

FIGURE 21.2   Repeated Waves

When a wave form consists of many repeated disturbances, the notions of **wavelength**, **period**, **frequency**, **wave speed**, **wave number**, and **angular frequency** become increasingly[1] well-defined.

WAVELENGTH   The wavelength, λ, associated with a repeating wave is equal to the distance between successive identical wave features[2] taken at the same instant in time. The dimension of λ is length, and its SI unit is the metre.

[It is essential that the position measurements occur at the same time!]

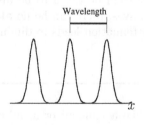 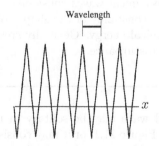

FIGURE 21.3   Wavelengths of Various Repeated Waves

---

[1]Strictly speaking, these notions are meaningful only in the limit of infinite repetition.
[2]These precisely repeating features might be crests, troughs, oriented zero-crossings, or any such distinguishable aspect of the wave disturbance.

WAVE NUMBER   The wave number, k, of a repeating wave is

$$k = \frac{2\pi}{\lambda}.$$

The units of wave number are $\mathtt{rad/m}$, or more simply $\mathtt{m^{-1}}$ *a.k.a.* "inverse metres."

PERIOD   The period, T, associated with a repeating wave is the time interval between the passage of successive identical wave features beyond a particular point in space. The dimension of T is time, and its SI unit is the second.

[It is essential that the time measurements occur at the same position!]

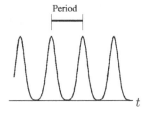 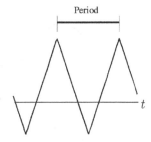

FIGURE 21.4   Periods of Various Repeated Waves

FREQUENCY   The frequency, ν, of a repeating wave is the rate at which elementary copies of the wave disturbance pass by a particular position. The frequency is the reciprocal of the period, *i.e.*,

$$\nu = \frac{1}{T}.$$

The SI unit associated with wave frequency is the inverse-second, *a.k.a.* hertz, $\mathtt{Hz}$ [introduced previously in Chapter 12 of this volume, and Chapter 34 of VOLUME I].

ANGULAR FREQUENCY   The angular frequency, ω, of a wave with frequency ν, and period T, is

$$\omega = 2\pi\nu = \frac{2\pi}{T}.$$

The unit of angular frequency is the radian per second, $\mathtt{rad/s}$.

TRAVELLING WAVE   A travelling wave is a repeating [mechanical[3]] wave propagating in a homogeneous medium and possessing well-defined frequency and wavelength.

HARMONIC WAVE   A harmonic wave is a travelling wave with an [infinitely] smooth[4] spatio-temporal shape.

WAVE SPEED   A travelling wave propagates forward a distance of one wavelength in the span of each one-period time interval, as illustrated in Figure 21.5. In the figure, a portion of the wave as it appeared at the instant t has been tagged. One period later, at $t + T$, the wave is indistinguishable from how it appeared at t except that the tagged region has shifted forward by an amount equal to one wavelength.

> ASIDE: IF one waits through some number of complete periods, THEN the tagged region will have advanced through an equal number of wavelengths.

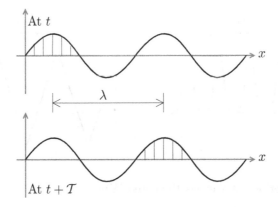

FIGURE 21.5   A Travelling Wave Moves Forward at the Rate of λ per T

The speed of propagation of the wave is the distance through which the phase has advanced, divided by the time it took:

$$c = \frac{\lambda}{T} = \lambda \nu .$$

This is the speed with which the **phase** of the wave propagates. At least three important points follow below.

- The units and dimensions are completely consistent, *i.e.*, $[c] = \mathtt{m/s}$.

- Although expressed in terms of wave parameters, $\{\nu, \lambda\}$, the speed is determined by properties of the medium in which the wave resides. Generally speaking, the frequency is fixed by the source of the wave, and the wavelength adjusts to accommodate changes in wave speed.

---

[3]Other types, *e.g.*, electromagnetic waves, exist but are not explicitly considered here.
[4]The smoothness criterion strongly restricts the allowed mathematical form of these waves.

- The wave speed, in terms of the angular frequency, ω, and wave number, k, is

$$c = \lambda \nu = \frac{2\pi}{k} \frac{\omega}{2\pi} = \frac{\omega}{k}.$$

- $c = \lambda \nu$ inspires the classic physics joke recounted in the Epilogue.

EXAMPLE [*A Harmonic Wave*]

Figure 21.6 displays the harmonic wave described by the function $\psi(t,x) = \sin(\pi x - 2\pi t)$. To interpret the figure, it is helpful to know that the domain is $\{t \in [-1, 2.5],\ x \in [-1, 2.0]\}$ and that the [LIGHT/DARK] shading is proportional to the height/depth of ψ above/below the (t, x) plane. The dark lines denote the portions of the x and t axes which are "visible" from above the ψ–surface. The two plots below the spacetime diagram illustrate the restrictions of the wave to the x- and t-axes. That is, the first shows

$$\psi(0, x) = \sin(\pi x - 2\pi t)\Big|_{t=0,\ -1 \le x \le 2} = \sin(\pi x), \quad -1 \le x \le 2,$$

while the second displays

$$\psi(t, 0) = \sin(\pi x - 2\pi t)\Big|_{-1 \le t \le 5/2,\ x=0} = \sin(-2\pi t) = -\sin(2\pi t), \quad -1 \le t \le 5/2.$$

Squinting at the curves and the shaded 3-d image is rewarded by recognition of their mutual consistency.

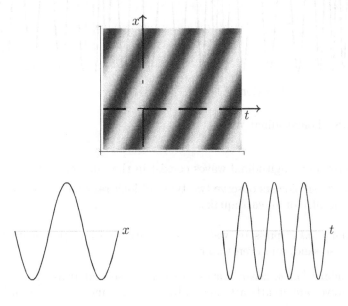

FIGURE 21.6 Combined Spatial and Temporal Aspects of a Repeated Wave

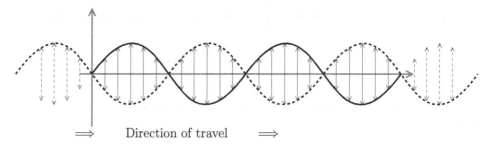

FIGURE 21.7   Transverse Waves

Two archetypes of **travelling waves** exist.

TRANSVERSE   A transverse wave is a propagating disturbance for which the motions of particles in the medium [through which the wave moves] are perpendicular to the motion of the wave. Such a wave is illustrated in Figure 21.7.

LONGITUDINAL   A longitudinal wave is a propagating disturbance for which the motions of particles in the medium are parallel to the motion of the wave, as indicated in Figure 21.8.

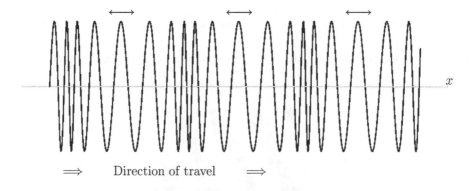

FIGURE 21.8   Longitudinal Waves

Q: Can tranverse and longitudinal waves coexist in the same medium?

A: You betcha! Seismologists observe two types of long-range disturbances [waves] emanating from the epicentre of an earthquake.

> **P**   Primary [*a.k.a.* pressure] waves are longitudinal and propagate through solid rock, loose sand, and even water.

> **S**   Secondary [*a.k.a.* shear] waves are transverse and travel through solid rock. These waves are greatly attenuated by sandy ground and are completely stymied by bodies of water.

It is also observed that P-waves typically travel faster than do S-waves. How the speeds of transverse and longitudinal waves are determined by mechanical aspects of the media in which they propagate will be studied in Chapters 26 and 27.

Chapter 22

Wave Dynamics and Phenomenology

The propagation of a wave in 1-d, $\psi(t,x)$, is governed by local dynamics in which the instantaneous acceleration depends on the local curvature of the wave form, *i.e.,*

$$a \propto \frac{\partial^2 \psi}{\partial x^2}.$$

Identifying the acceleration as $a = \frac{\partial^2 \psi}{\partial t^2}$, and realising that the wave speed [squared] is the sole dimensional parameter extant which can fulfill the rôle of proportionality constant, one obtains the canonical form of the **wave equation**:

$$0 = \frac{\partial^2 \psi}{\partial x^2} - \frac{1}{c^2} \frac{\partial^2 \psi}{\partial t^2}.$$

Mathematically, the wave equation is a **second-order homogeneous linear partial differential equation with constant coefficients**. A rewriting of the wave equation:

$$\frac{1}{c^2} \frac{\partial^2 \psi}{\partial t^2} = \frac{\partial^2 \psi}{\partial x^2},$$

puts it into a form resembling $m\,a = F_{\text{NET}}$. The factor of c^{-2} ensures consistency of the units [*cf.* the introduction of inertia in Chapter 10 of VOLUME I]. Evidently, the curvature [the second-derivative] of the wave function is the dynamical *cause* of its time dependence. Physically, this reinterpretation of the wave equation is consistent with one's intuitive sense of the essential dynamics of wave propagation. For instance, the leading and trailing edges of waves are usually less strongly curved than regions near peaks and troughs. The instantaneous accelerations [associated with the disturbance in the medium] are typically greatest near the peaks and troughs.

[Think about how ships are tossed at sea.]

It is surprising, but true, that any function of the form

$$\psi(t,x) = \psi_-(x - c\,t) + \psi_+(x + c\,t)$$

is a solution of the wave equation. Prior to proving this, four things are noted.

LINEAR SUPERPOSITION   IF ψ_- and ψ_+ are solutions of the wave equation, THEN their sum, weighted by arbitrary constant coefficients, is also a solution.

SINGLE VARIABLE   The wave $\psi(t,x)$ varies both in time and space and so is, *a priori*, a function of two variables. Each of $\psi_\pm$ is a function of a single variable [itself expressed as a linear combination of t and x].

COMBO   Writing $x \pm ct$ is advantageous from the "snapshot" point of view, but has the drawback of being dimensional [with units of length]. Adoption of $kx \pm \omega t$, via

$$kx \pm \omega t = k\big(x \pm (\omega/k)\,t\big) = k\big(x \pm ct\big),$$

remedies this minor deficiency.

RELATIVE   The wave is characterised by relative differences in ψ [*i.e.*, its shape]. Adding an arbitrary constant to ψ shifts its value, but has no effect on the dynamics.

PROOF that $\psi_{\pm}(x \pm ct)$ Satisfies the Wave Equation

Consider the first partial derivatives of $\psi_{\pm}(x \pm ct)$:

$$\frac{\partial \psi_{\pm}}{\partial t} = \frac{d\psi_{\pm}(z)}{dz}\,\frac{d(x \pm ct)}{dt} = (\pm c)\,\psi'_{\pm} \qquad \text{and} \qquad \frac{\partial \psi_{\pm}}{\partial x} = \frac{d\psi_{\pm}(z)}{dz}\,\frac{d(x \pm ct)}{dx} = \psi'_{\pm},$$

where $\psi'_{\pm}$ denotes the derivative of $\psi_{\pm}(\alpha)$ with respect to its sole argument. The needed second partials are obtained in like fashion:

$$\left\{ \begin{array}{l} \dfrac{\partial^2 \psi_{\pm}}{\partial t^2} = (\pm c)\,\dfrac{d\psi'_{\pm}(z)}{dz}\,\dfrac{d(x \pm ct)}{dt} = (\pm c)^2\,\psi''_{\pm} = c^2\,\psi''_{\pm} \\[1.5em] \dfrac{\partial^2 \psi_{\pm}}{\partial^2 x} = \dfrac{d\psi'_{\pm}(z)}{dz}\,\dfrac{d(x \pm ct)}{dx} = \psi''_{\pm} \end{array} \right\}.$$

These second-order partial derivatives exist whenever the single-variable function $\psi_{\pm}$ is twice differentiable. When this condition holds, $\psi_{\pm}(x \pm ct)$ is guaranteed to satisfy the wave equation, as can be readily seen by direct substitution.

WAVES PROPAGATING IN A ONE DIMENSIONAL MEDIUM

Physical situations which admit representation by 1-d wave models include transverse waves on a stretched string and longitudinal waves produced in an organ pipe.

The solutions of the 1-d wave equation, discussed above, come in two forms, $\psi_{\pm}$, distinguished by the relative signs of the spatial and temporal terms appearing in the arguments $x - ct$ or $kx \pm \omega t$. By convention one says:

$$\left\{ \begin{array}{l} \psi_{-}(x - ct) \\ \psi_{+}(x + ct) \end{array} \right\} \quad \text{is} \quad \left\{ \begin{array}{l} \text{RIGHTMOVING} \\ \text{LEFTMOVING} \end{array} \right\}.$$

RIGHT-
MOVING   That ψ_{-} is rightmoving shall be demonstrated by three distinct means.

METHOD ONE: [*Finite time intervals and their corresponding phase shifts*]

A snapshot of a "−" type wave, taken at some particular time t_1, i.e., $\psi_-(x - ct_1)$, is a function of spatial position, x, only. Let's take note of the location, x_1, of a particular wave feature [a distinctive crest or trough, perhaps] at this instant. The assumption of dispersionless propagation ensures that at any later time $t_2 > t_1$, the distinctive feature is perfectly preserved, except that it will have migrated to another location, x_2. The only way to ensure accurate tracking of all such wave features is to insist that the argument of the function be invariant, *i.e.*,

$$\psi_-(x_1 - ct_1) \equiv \psi_-(x_2 - ct_2) \qquad \text{IFF} \qquad x_1 - ct_1 = x_2 - ct_2\,.$$

Rearranging the latter condition yields

$$x_2 - x_1 = c\,(t_2 - t_1) > 0\,,$$

and thus the future loci of the marked point are **to the right of** all past loci.

METHOD TWO: [*Constant phase approach*]

The argued-for invariance of $x - ct$ [associated with the specific wave feature whose motion is being tracked] has instantaneous and local implications. Let τ represent a parameter employed as a proxy for, or alternative to, the time. Constancy of $z = x - ct$ entails

$$0 = \frac{dz}{d\tau} = \frac{d}{d\tau}\,(x - ct) = \frac{dx}{d\tau} - c\frac{dt}{d\tau} \qquad \Longrightarrow \qquad \frac{dx}{d\tau} = c\frac{dt}{d\tau}\,.$$

Provided that the parameterisation is monotonic and free from singularities,

$$\frac{dx}{dt} = +c\,,$$

by implicit differentiation. Thus, the rightmoving wave disturbance propagates in the direction of increasing x with speed c.

METHOD THREE: [*Comoving perspective*]

One can choose to occupy an inertial reference frame moving with speed $+c$ alongside the medium in which the wave resides, while retaining the original time parameter. Expressing the original coordinates in terms of the new set reveals that

$$t = \tilde{t} \qquad \text{and} \qquad x = \tilde{x} + c\tilde{t} = \tilde{x} + ct \qquad \Longrightarrow \qquad \tilde{x} = x - ct\,.$$

The wave appears to be unchanging as both the wave and the observer are moving to the right at the same [constant] speed.

Therefore, $\psi_-(x - ct)$ waves **are** rightmoving. Henceforth, the "−" subscript shall be replaced by "R."

LEFT-MOVING   The same three lines of reasoning applied above yield corresponding results in the leftmover case. We'll limit ourselves to reproducing the METHOD ONE analysis.

A snapshot of this particular type of wave taken at time t_1 is $\psi_+(x + ct_1)$, which depends solely on the spatial position, x. A distinguishable wave feature is found at x_1 at the instant t_1. At any later time $t_2 > t_1$, the wave feature will have propagated to x_2. Feature/phase preservation,

$$\psi_+(x_1 + ct_1) \equiv \psi_+(x_2 + ct_2) \qquad \text{IFF} \qquad x_1 + ct_1 = x_2 + ct_2\,,$$

holds for all future t_2 [and past ones too!]. Therefore,

$$x_2 - x_1 = c\,(t_1 - t_2) < 0\,,$$

and thus the future loci of the marked point are **to the left of** all past loci. The leftmovers will be denoted by a subscripted L from this point on.

WAVES PROPAGATING IN A TWO-DIMENSIONAL MEDIUM

The situation in 2-d is more complicated than that in 1-d. There exist two archetypical forms: **circular waves** and **plane waves**. These are illustrated in Figure 22.1.

[Waves in 2-d may be of any reasonable shape, *e.g.*, elliptical, parabolic, *etc.*]

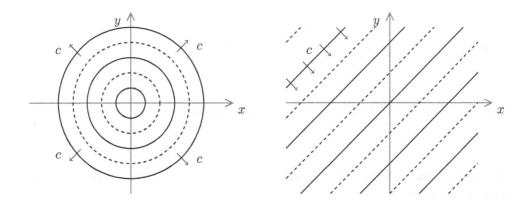

FIGURE 22.1   Circular and Plane Two-Dimensional Waves

⊙   Circular waves propagate radially with respect to a particular point.

   [This point is often chosen to be the origin of Cartesian or polar coordinates.]

   There are two distinct "species" of circular waves: **incoming** and **outgoing**.[1] These correspond to the left-/rightmovers in 1-d. An instance of outgoing 2-d circular waves occurs when a stone is dropped into a still pond and ripples propagate outward from the splashdown point. More complicated circular waveforms arise on a circular drum head.

‖   The local isotropy of a homogeneous medium guarantees that plane waves propagate equally well in any direction. In any particular direction, both left- and rightmoving waves may occur. Instances of approximate plane waves are found in the midst of oceans or lakes, far from shore.

   ASIDE: Water waves are a formidable system to model.

---

[1] *"Hi! I'm an outgoing wave. How're ya doin'? Nice day, eh?"*

WAVES PROPAGATING IN A THREE-DIMENSIONAL MEDIUM

Such waves surpass our limited ability to render them plainly on two-dimensional pages. Notwithstanding this, one can recognise that among the vast plethora of possible wave forms, there are again two limiting species: spherical and plane waves. These shall not be investigated further at this time but will be revisited briefly when the propagation of polarised light is considered in Chapter 33.

The waves that have been considered thus far have resided in homogeneous media.

> ASIDE: Wave propagation in a continuous but inhomogeneous medium is too daunting a subject for us at this juncture. [It will be touched upon in later chapters.]

The next question is inspired by the realisation that all good things must come to an end.

Q: What happens when a [1-d] wave runs out of medium?

A: There are two different behaviours, FIXED and FREE, distinguished by the constraints under which the wave impinges on the boundary. Fixed boundary conditions occur in the context of stringed musical instruments when a transverse wave moving in a string encounters the bridge of the instrument. Free boundary conditions are found when a longitudinal sound wave reaches the end of an open organ pipe.

FIXED   The material comprising the medium is held immobile at its boundary.

Consider a rightmoving pulse, ψ_R, propagating in a 1-d medium toward a fixed endpoint, as shown in the sequence portrayed in Figure 22.2.

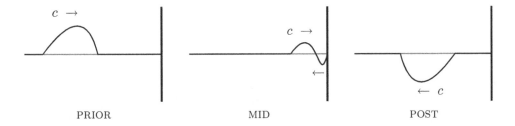

PRIOR   MID   POST

FIGURE 22.2   Fixed Endpoint Reflection

Initially, the pulse is moving to the right with speed c. Upon encountering the boundary, the pulse is **reflected** and **inverted**. Afterward, the inverted pulse moves to the left with speed c.

- IF the wave is sinusoidal, THEN the inversion amounts to a phase shift of π radians.

- IF dispersion, attenuation, and dissipation do not occur, THEN the reflected pulse has the same amplitude and shape [aside from inversion] as the incident pulse.

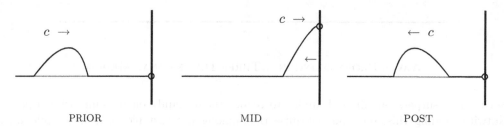

PRIOR   MID   POST

FIGURE 22.3   Free Endpoint Reflection

FREE   The material everywhere in the medium, including its peripheral points, is able to respond to the presence of the wave.

In Figure 22.3, a horizontal string is linked to a fixed vertical rod by means of a ring which is able to slide without friction. This arrangement enables tension to be maintained in the string while its end point moves freely in the transverse [vertical] direction.

An initially rightmoving pulse is **reflected** and **remains erect** upon reaching the sliding ring boundary. Both the incident and reflected waves move with speed c.

- IF the wave is sinusoidal, **THEN** the phase is continuous at the boundary. There is no additional shift.

- IF dispersion, attenuation, and dissipation do not occur, **THEN** the reflected pulse has the same shape and amplitude as the incident pulse.

Chapter 23

Linear Superposition of Waves

IF $\psi_1(t, x)$ and $\psi_2(t, x)$ are travelling waves residing in the same 1-d medium, THEN, according to our model of dispersionless propagation, each moves independently [unperturbed by the other's presence] AND the net disturbance produced in the medium is obtained by linear superposition:

$$\psi_{\text{net}}(t, x) = \psi_1(t, x) + \psi_2(t, x).$$

There are several things to note about linear superposition of waves.

- The superposition is local in both time, t, and space, x.

- Any number $\mathcal{N}$ of waves may be superposed:

$$\psi_{\text{net}}(t, x) = \sum_{i=1}^{\mathcal{N}} \psi_i(t, x).$$

- The constituent waves and their composite share a common wave speed.

- Very little restriction is imposed on the forms of the constituent waves.

  - These may themselves be superpositions of more primitive waves. Linear superposition is COMMUTATIVE and ASSOCIATIVE.

  - The wave may have any shape. Common idealisations are "triangle," "square," and "sawtooth" waves, and each of these admit many variations.

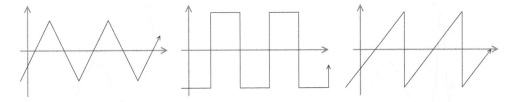

FIGURE 23.1   Three Archetypes of Repeated Waves

  - Notwithstanding the triangle, square, and sawtooth archetypes illustrated in Figure 23.1, cusps [sharp peaks or corners] are never[1] found in physical waves. Equivalently, the wave functions have continuous second partial derivatives, $\psi \in C^2$, everywhere within the medium [excluding the boundary]. Harmonic wave disturbances are C^∞ throughout the bulk of the medium.

---

[1]This "never" is to be interpreted [as in Gilbert and Sullivan's *H.M.S. Pinafore*] as "hardly ever," since kinks in the waveform may arise at the boundaries of the medium.

— Boundedness of the total energy [a technical consideration with physical implications] militates for square-integrability of all types of waves, and for compact support for solitary waves.

SQUARE-
INTEGRABLE   A real-valued function of one variable, $f(z)$, is square integrable provided that $\int f^2(z)\,dz$, taken over all of z-space, exists and is finite.

COMPACT
SUPPORT   Compact support means that the entirety of the wave disturbance at any instant is confined within a finite coordinate interval.

> ASIDE: This does not preclude the possibility of a pulse travelling to asymptotically far distances as long as, at any instant, the pulse has finite spatial extent.

Square-integrability is the weaker of these two conditions. The assumption that the wave is at least C^2 ensures that any pulse with compact support must also be square integrable. There exist many wave functions whose domain is infinite in extent and yet are square integrable.

> ASIDE: Harmonic waves are not square integrable, nor do they possess compact support.

SUPERPOSITION OF TWO RIGHTMOVING SOLITARY WAVES

Two distinct rightmoving pulses, ψ_{1R} and ψ_{2R}, superpose to form a compound rightmoving disturbance as shown in Figure 23.2.

$$\begin{aligned}\psi_{1R} &= \psi_{1R}(x - ct) \\ \psi_{2R} &= \psi_{2R}(x - ct)\end{aligned} \quad \Longrightarrow \quad \psi_{\text{net}} = \psi_{1R}(x - ct) + \psi_{2R}(x - ct) = \psi_{\text{net},R}(x - ct).$$

> ASIDE: A solitary pulse cannot possibly possess well-defined and unique wave number or angular frequency. Hence, the arguments of these wave functions are written as $x - ct$ rather than $k\,x - \omega\,t$.

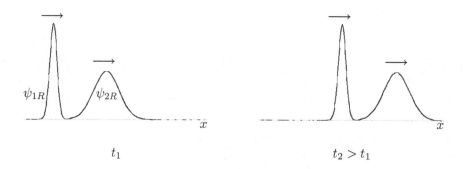

t_1 $t_2 > t_1$

FIGURE 23.2   Superposition of Two Rightmoving Pulses

In Figure 23.2, ψ_{1R} is narrow and tall and lies to the left of ψ_{2R}, which is broader and shorter. Both move to the right at the same speed, and thus at later times the disturbance will have moved to the right with its shape unchanged.

SUPERPOSITION OF RIGHTMOVING AND LEFTMOVING SOLITARY WAVES

The superposition of a rightmoving pulse and a leftmoving pulse exhibits non-trivial time dependence, as illustrated in Figure 23.3.

$$\begin{aligned} \psi_{1R} &= \psi_1(x - ct) \\ \psi_{2L} &= \psi_2(x + ct) \end{aligned} \quad \implies \quad \psi_{\text{net}}(t, x) = \psi_{1R}(x - ct) + \psi_{2L}(x + ct).$$

The superposition is a function of x and ct separately and cannot be reduced to dependence on a single variable due to the mixture of rightmoving/leftmoving constituents.

[In some circumstances, it proves advantageous to write: $\psi_{\text{net}}(z_+, z_-)$, where $z_\pm = x \pm ct$.]

In the first panel of Figure 23.3, *i.e.*, at time t_1, tall and narrow ψ_{1R} and the shorter, wider ψ_{2L} are well-separated and moving toward one another. At t_2, the two pulses have begun to merge. At t_3, they are completely merged, and it takes some effort to recognise the single compound pulse for what it truly is: the superposition of the incident pulses. The independent propagation of the waves ensures that they pass through one another and they are seen to have almost emerged at time t_4. At still later times, t_5, ψ_{1R} is to the right of ψ_{2L}, and both continue propagating in their respective directions.

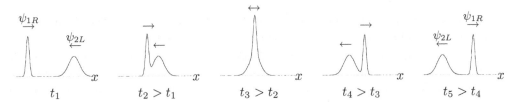

$$t_1 \qquad t_2 > t_1 \qquad t_3 > t_2 \qquad t_4 > t_3 \qquad t_5 > t_4$$

FIGURE 23.3 Superposition of Rightmoving and Leftmoving Pulses

This was so much fun, let's do it again, with the leftmover flipped!

With reference to Figure 23.4, ψ_{1R} and $-\psi_{2L}$ are initially well-separated and moving toward one another at time t_1. At t_2, the two pulses have begun to merge and partially cancel. For a brief interval including t_3, the pulses completely overlap and their combined disturbance exhibits three pulses. Propagating independently, the two waves pass through one another and begin to separate at time t_4. At still later times, *e.g.*, t_5, ψ_{1R} is to the right of $-\psi_{2L}$, and both continue propagating in their respective directions.

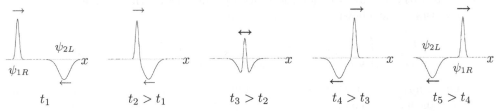

$$t_1 \qquad t_2 > t_1 \qquad t_3 > t_2 \qquad t_4 > t_3 \qquad t_5 > t_4$$

FIGURE 23.4 Superposition of Rightmoving and [Flipped] Leftmoving Pulses

Superposition of Repeated Waves

Myriad possible cases arise, as the repeated waves can be rightmoving or leftmoving, and can have all manners of shape, amplitude, frequency and wavelength, and relative phase. Two special cases involving the superposition of two rightmoving harmonic waves are examined in detail in Chapter 24.

φ The first case is the most restrictive: the wavelength, frequency,[2] and amplitude are all the same, while the relative phase of the two waves is permitted to vary.

ν , λ In the second case the amplitudes are chosen to be equal and the waves are in phase, while the frequencies and wavelengths differ.

Waves with differing amplitudes, wavelengths and frequencies, phases, and directions of travel can be superposed. However, the technical complexity of the analysis makes these somewhat arcane. The superposition of leftmoving and rightmoving waves possessing equal frequencies [wavelengths] and amplitudes is taken up in Chapter 25.

The Joy of Fourier Analysis

A rationale for the degree to which we have studied harmonic waves, to the exclusion of all other repeating and solitary waves, is provided by Fourier analysis.

Fourier (1768–1830) showed that certain collections of **sine** and **cosine** functions provide an orthonormal basis[3] for the infinite-dimensional space of [repeating] real-valued functions on an interval. The following is a brief précis of the Fourier series representation of functions.

$\sim$ Suppose that a function, $f(z)$, has support on the interval of width $2\,L$ centred on the origin *i.e.,* $z \in [-L, L]$.

• The function may exist only in this region, or perhaps this region may contain one complete wavelength [or multiple wavelengths] of a repeating function.

• Functions whose domains are not centred on the origin can easily be shifted so as to have this form. Alternatively, the analysis can readily be adapted.

$\sim$ The natural INNER PRODUCT on this function space [generalising the DOT PRODUCT on (ordinary) finite-dimensional vector spaces] is integration over the domain extending from $z = -L$ to $z = L$.

$\sim$ An orthonormal basis for the space of functions on the interval $z \in [-L, L]$ consists of the following:

$$\frac{1}{\sqrt{2L}} \cup \left\{ \frac{1}{\sqrt{L}} \sin\left(\frac{n\pi z}{L}\right), \frac{1}{\sqrt{L}} \cos\left(\frac{n\pi z}{L}\right) \right\}, \quad \text{where } n = 1, 2, \dots .$$

Orthonormality, in this context, is defined using the inner product [integration over the finite interval].

---

[2] On account of the speed relation, $c = \lambda\,\nu$, with c determined by properties of the medium, the wavelength and the frequency cannot both be independently chosen.

[3] Chapter 5 in VOLUME I contains a rudimentary discussion of orthonormal bases in finite-dimensional vector spaces.

Elementary integrations reveal that, for any pair of positive distinct integers $\{m, n \mid m \neq n\}$,

$$\frac{1}{L} \int_{-L}^{L} \sin\left(\frac{n\pi z}{L}\right) \sin\left(\frac{m\pi z}{L}\right) dz = 0$$

and

$$\frac{1}{L} \int_{-L}^{L} \cos\left(\frac{n\pi z}{L}\right) \cos\left(\frac{m\pi z}{L}\right) dz = 0 \,.$$

By symmetric integration, it is evidently the case that

$$\frac{1}{L} \int_{-L}^{L} \sin\left(\frac{n\pi z}{L}\right) \cos\left(\frac{m\pi z}{L}\right) dz = 0 \,,$$

for all positive integers m and n. Furthermore,

$$\frac{1}{\sqrt{2}\,L} \int_{-L}^{L} \sin\left(\frac{n\pi z}{L}\right) dz = 0 = \frac{1}{\sqrt{2}\,L} \int_{-L}^{L} \cos\left(\frac{n\pi z}{L}\right) dz \,,$$

for all $n \neq 0$.

NORMAL These basis vectors are normalised in the sense that the dot product of a basis vector with itself yields a squared magnitude which is exactly equal to 1. Explicitly performing the integrals, one cannot fail to obtain:

$$\frac{1}{L} \int_{-L}^{L} \sin^2\left(\frac{n\pi z}{L}\right) dz = 1$$

and

$$\frac{1}{L} \int_{-L}^{L} \cos^2\left(\frac{n\pi z}{L}\right) dz = 1 \,,$$

for all n among the positive integers. In addition,

$$\frac{1}{2L} \int_{-L}^{L} 1 \, dz = 1$$

shows that the "constant function" basis vector is normalised as well.

~ The projections of $f(z)$ onto the nth **sine**, nth **cosine**, and constant basis vectors are

$$f_{s,n} = s_n = \frac{1}{\sqrt{L}} \int_{-L}^{L} f(z) \sin\left(\frac{n\pi z}{L}\right) dz \,,$$

$$f_{c,n} = c_n = \frac{1}{\sqrt{L}} \int_{-L}^{L} f(z) \cos\left(\frac{n\pi z}{L}\right) dz \,,$$

$$f_0 = c_0 = \frac{1}{\sqrt{2L}} \int_{-L}^{L} f(z) \, dz \,,$$

respectively. These are the **components** of $f(z)$ in the respective "directions" of the basis functions.

$$f(z) = \begin{cases} 0, & z < 0, \\[2mm] \dfrac{4z}{L}, & 0 < z < \dfrac{L}{4}, \\[2mm] 2\left(1 - \dfrac{2z}{L}\right), & \dfrac{L}{4} < z < \dfrac{3L}{4}, \\[2mm] 4\left(\dfrac{z}{L} - 1\right), & \dfrac{3L}{4} < z < L. \end{cases}$$

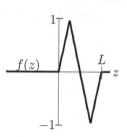

FIGURE 23.5 A Piecewise Function on a Bounded Interval

EXAMPLE [*Fourier Series Approximation to a Function on a Bounded Interval*]

Consider the piecewise continuous function shown in Figure 23.5, which is zero to the left of the origin and has a triangular wave cycle to the right.

The first order of business is to compute the projections onto [a subset of] the basis functions. That the average value of the function $f(Z)$ is zero suggests [correctly] that the projection onto the $1/\sqrt{2L}$ basis vector vanishes, *i.e.*, $c_0 = 0$. The only other pertinent observation that can be made is that, owing to the asymmetry of $f(z)$ under $z \leftrightarrow -z$, a mixture of `sine` and `cosine` terms is required. The non-zero projections, for basis vectors labelled 1 through 6, are:

$$\left\{ c_1 = \frac{8\sqrt{L}\left(\sqrt{2}-1\right)}{\pi^2} \; , \quad c_3 = -\frac{8\sqrt{L}\left(\sqrt{2}+1\right)}{9\pi^2} \; , \quad c_5 = -\frac{8\sqrt{L}\left(\sqrt{2}+1\right)}{25\pi^2} \; , \quad \cdots \right\}$$

$$\left\{ s_2 = \frac{4\sqrt{L}}{\pi^2} \; , \quad s_6 = -\frac{4\sqrt{L}}{9\pi^2} \; , \quad \cdots \right\} .$$

$c_1 \dfrac{1}{\sqrt{L}} \cos\left(\dfrac{\pi z}{L}\right) = \dfrac{8\left(\sqrt{2}-1\right)}{\pi^2} \cos\left(\dfrac{\pi z}{L}\right)$	
$s_2 \dfrac{1}{\sqrt{L}} \sin\left(\dfrac{2\pi z}{L}\right) = \dfrac{4}{\pi^2} \sin\left(\dfrac{2\pi z}{L}\right)$	
$c_3 \dfrac{1}{\sqrt{L}} \cos\left(\dfrac{3\pi z}{L}\right) = -\dfrac{8\left(\sqrt{2}+1\right)}{9\pi^2} \cos\left(\dfrac{3\pi z}{L}\right)$	

To reconstruct the original function, we merely multiply each basis function by its projection coefficient. The results of this operation for the three lowest-order non-zero terms are found in the nearby table in both algebraic and graphical form. With just these three terms an approximate version of the function, $f(z)$, is

$$\tilde{f}_3(z) = \frac{8\left(\sqrt{2}-1\right)}{\pi^2} \cos\left(\frac{\pi z}{L}\right) + \frac{4}{\pi^2} \sin\left(\frac{2\pi z}{L}\right) - \frac{8\left(\sqrt{2}+1\right)}{9\pi^2} \cos\left(\frac{3\pi z}{L}\right).$$

This approximation and the original function are plotted together on the left in Figure 23.6.

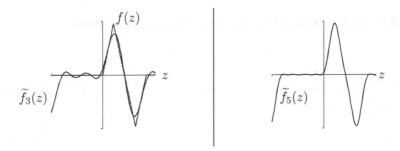

FIGURE 23.6 Three- and Five-Component Approximations to the Original Function

The five-term series approximation, illustrated on the right in Figure 23.6, is obtained by addition of the two terms in the table below to the three-term approximation.

$c_5 \dfrac{1}{\sqrt{L}} \cos\left(\dfrac{5\pi z}{L}\right) = -\dfrac{8\left(\sqrt{2}+1\right)}{25\pi^2} \cos\left(\dfrac{5\pi z}{L}\right)$	
$s_6 \dfrac{1}{\sqrt{L}} \sin\left(\dfrac{6\pi z}{L}\right) = -\dfrac{4}{9\pi^2} \sin\left(\dfrac{6\pi z}{L}\right)$	

EXAMPLE [*Fourier Series Approximation to a Repeating Wave*]

A repeating wave consisting of a succession of rectangular pulses is shown in Figure 23.7.

A single repetition unit (pulse) is described by the piecewise function expressed and illustrated in Figure 23.8.

The projections of the single pulse onto the basis functions are straightforwardly obtained.

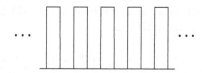

FIGURE 23.7 A Wave Comprised of a Train of Rectangular Pulses

$$f(z) = \begin{cases} 0, & -L < z < -\dfrac{L}{2}, \\ 1, & -\dfrac{L}{2} \leq z \leq \dfrac{L}{2}, \\ 0, & \dfrac{L}{2} < z < L. \end{cases}$$

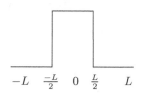

FIGURE 23.8 One Pulse from the (Infinite) Train

CONSTANT

$$c_0 = \frac{1}{\sqrt{2L}} \int_{-L}^{L} f(z)\,dz = 0 + \frac{1}{\sqrt{2L}} \int_{-L/2}^{L/2} dz + 0 = \frac{1}{\sqrt{2L}} L = \sqrt{\frac{L}{2}}.$$

Therefore the constant function contribution to the Fourier series representation of the pulse is

$$c_0 \frac{1}{\sqrt{2L}} = \frac{1}{2}$$

which is the average value of the impulse over the domain consisting of one complete repetition unit of the wave train.

COSINE

$$c_n = \frac{1}{\sqrt{L}} \int_{-L}^{L} f(z) \frac{1}{\sqrt{L}} \cos\left(\frac{n\pi z}{L}\right) dz = \frac{1}{\sqrt{L}} \int_{-L/2}^{L/2} \cos\left(\frac{n\pi z}{L}\right) dz$$

$$= \frac{\sqrt{L}}{n\pi} \left[\sin\left(\frac{n\pi}{2}\right) \times 2\right].$$

Consideration of this expression reveals that the cosine components are:

$$c_n = \begin{cases} 0 & , \; \forall n = \{0,2\} \mod (4) \\ \dfrac{2\sqrt{L}}{n\pi}(-1)^{\frac{n-1}{2}}, & \forall n = \{1,3\} \mod (4) \end{cases}.$$

SINE

$$s_n = \frac{1}{\sqrt{L}} \int_{-L}^{L} f(z) \frac{1}{\sqrt{L}} \sin\left(\frac{n\pi z}{L}\right) dz = \frac{1}{\sqrt{L}} \int_{-L/2}^{L/2} \sin\left(\frac{n\pi z}{L}\right) dz = 0.$$

All of the **sine** components vanish, since the pulses are symmetric [EVEN] within the domain.

The original function is approximately reconstructed by multiplying each basis function by its projection coefficient. The results of this operation for the lowest-order non-zero terms are found in the table below in both algebraic and graphical form.

$c_0 \dfrac{1}{\sqrt{2L}} + c_1 \dfrac{1}{\sqrt{L}} \cos\left(\dfrac{\pi z}{L}\right) = \dfrac{1}{2} + \dfrac{2}{\pi} \cos\left(\dfrac{\pi z}{L}\right)$	
$c_3 \dfrac{1}{\sqrt{L}} \cos\left(\dfrac{3\pi z}{L}\right) = -\dfrac{2}{3\pi} \cos\left(\dfrac{3\pi z}{L}\right)$	
$c_5 \dfrac{1}{\sqrt{L}} \cos\left(\dfrac{5\pi z}{L}\right) = \dfrac{2}{5\pi} \cos\left(\dfrac{5\pi z}{L}\right)$	
$c_7 \dfrac{1}{\sqrt{L}} \cos\left(\dfrac{7\pi z}{L}\right) = -\dfrac{2}{7\pi} \cos\left(\dfrac{7\pi z}{L}\right)$	

The five-component Fourier series approximation to the pulse, $\tilde{f}_5(z)$, is

$$\tilde{f}_5(z) = \frac{1}{2} + \frac{2}{\pi} \cos\left(\frac{\pi z}{L}\right) - \frac{2}{3\pi} \cos\left(\frac{3\pi z}{L}\right) + \frac{2}{5\pi} \cos\left(\frac{5\pi z}{L}\right) - \frac{2}{7\pi} \cos\left(\frac{7\pi z}{L}\right) + \cdots .$$

The approximation to the single-pulse is shown in Figure 23.9.

FIGURE 23.9 A Five-Component Approximation to the Single Pulse

Properties of Materials

Without doing a scintilla of additional work, we have in our possession the Fourier approximation to the train consisting of regular repetitions of the single pulse. Figure 23.10 takes the same sum of terms as was employed above for the single pulse and extends the domain of the function beyond $|z| \leq L$.

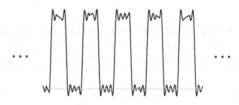

FIGURE 23.10   A Five-Component Approximation to the Entire Train of Pulses

This is ever so cool, eh?

Chapter 24

Linear Superposition of Rightmoving Harmonic Waves

Our task in this chapter is to study the superposition of rightmoving harmonic waves:

$$\psi_{1R} = A_1 \sin(k_1\,x - \omega_1\,t + \varphi_1) \qquad \text{and} \qquad \psi_{2R} = A_2 \sin(k_2\,x - \omega_2\,t + \varphi_2)\,.$$

Although this seems like an overly restricted case, the brief exposition on Fourier Analysis in Chapter 23 suggests that all manner of waves and pulses may be built using [perhaps infinitely many] harmonic waves.

The number of possible cases is too great for us to be exhaustive, so we shall restrict ourselves to two which most clearly exhibit the sorts of phenomena which can arise. These are:

φ EQUAL wave number/frequency and amplitude, DIFFERING phase,
$$k_1 = k_2 = k, \quad \omega_1 = \omega_2 = \omega, \quad A_1 = A_2 = A, \quad \text{while} \quad \varphi_2 = \varphi_1 + \varphi, \text{ and}$$

$\nu\,,\lambda$ EQUAL amplitude and phase, DIFFERING wave number/frequency,
$$A_1 = A_2 = A, \quad \varphi_2 = \varphi_1 = 0, \quad \text{while} \quad k_1 \neq k_2 \text{ and } \omega_1 \neq \omega_2.$$

TWO RIGHTMOVING HARMONIC WAVES: EQUAL AMPLITUDES AND FREQUENCIES

Two nearly identical rightmoving harmonic waves, differing only in phase, superpose to yield a rightmoving wave whose form depends crucially on their relative phase.

$$\psi_{1R} = A \sin(k\,x - \omega\,t)$$
$$\psi_{2R} = A \sin(k\,x - \omega\,t + \varphi)$$
$$\implies \quad \psi_{\text{net},R}(k\,x - \omega\,t) = A\Big[\sin(k\,x - \omega\,t) + \sin(k\,x - \omega\,t + \varphi)\Big]\,.$$

The trigonometric identity

$$\sin(a) + \sin(b) = 2\,\cos\left[\frac{a-b}{2}\right] \sin\left(\frac{a+b}{2}\right)$$

enables progress to be made in constructing the superposition. The identifications

$$a = k\,x - \omega\,t \qquad \text{and} \qquad b = k\,x - \omega\,t + \varphi$$

determine the arguments of the **sine** and **cosine** factors to be

$$\frac{a-b}{2} = -\frac{\varphi}{2} \qquad \text{and} \qquad \frac{a+b}{2} = k\,x - \omega\,t + \frac{\varphi}{2},$$

respectively. Realisation that **cosine** is an even function finally yields

$$\psi_{\text{net},R}(t,x) = 2\,A\,\cos\left[\frac{\varphi}{2}\right]\,\sin\left(k\,x - \omega\,t + \frac{\varphi}{2}\right)\,.$$

This is a harmonic wave with exactly the same wavelength and frequency as the two input waves, ψ_{1R} and ψ_{2R}, phase-shifted so as to lie precisely midway between them. The amplitude of the resultant wave scales with the amplitude of the input waves and depends crucially on the phase difference, *viz.*,

$$A_{\text{net}} = 2\,A\,\cos\left[\frac{\varphi}{2}\right].$$

Two final comments finish off this discussion.

- The relative phase determines the nature of the superposition.

- It is only necessary to consider relative phases $\varphi \in [-\pi, \pi]$, because of the $2\,\pi$ periodicity of harmonic waves.

EXAMPLE [*Equal Amplitude and Frequency Superposition with Five Relative Phases*]

Let's determine the superposition of $\psi_{1R} = A\,\sin(k\,x - \omega\,t)$ and $\psi_{2R} = A\,\sin(k\,x - \omega\,t + \varphi)$, for the five distinct relative phase shifts: $\varphi \in \{0, \pi/4, \pi/2, 3\,\pi/4, \pi\}$. As an aid in visually adding the waveforms, ψ_{1R} appears twice in each line [explicitly on the left, in black, and then again on the right where the superposition is shown, in grey].

0   IF $\varphi = 0$, THEN $A_{\text{net}} = 2\,A\,\cos(0) = 2\,A$, and thus

$$\psi_{\text{net}}(t, x) = 2\,A\,\sin(k\,x - \omega\,t).$$

The two input waves are in phase and their superposition yields completely **constructive interference**.

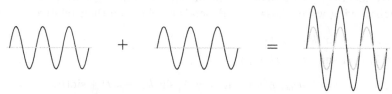

$$A\,\sin(k\,x - \omega\,t) + A\,\sin(k\,x - \omega\,t) = 2\,A\,\sin(k\,x - \omega\,t)$$

$\pi/4$ IF $\varphi = \pi/4$, THEN $A_{\text{net}} = 2\,A\,\cos(\pi/8) \simeq 1.85\,A$, and thus

$$\psi_{\text{net}}(t, x) = 2\,\cos(\pi/8)\,A\,\sin(k\,x - \omega\,t + \pi/8).$$

These input waves experience **partial constructive interference**.

$$A\,\sin(k\,x - \omega\,t) + A\,\sin(k\,x - \omega\,t + \pi/4) = 2\,\cos(\pi/8)\,A\,\sin(k\,x - \omega\,t + \pi/8)$$

$\pi/2$ IF $\varphi = \pi/2$, *i.e.,* these waves are orthogonal in the Fourier sense, THEN $A_\text{net} = 2\,A\,\cos(\pi/4) = \sqrt{2}\,A$, and thus

$$\psi_\text{net}(t, x) = \sqrt{2}\,A\,\sin(k\,x - \omega\,t + \pi/4)\,.$$

This is also an instance of **partial constructive interference**.

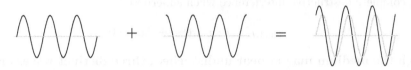

$$A\,\sin(k\,x - \omega\,t) + A\,\sin(k\,x - \omega\,t + \pi/2) = \sqrt{2}\,A\,\sin(k\,x - \omega\,t + \pi/4)$$

$3\pi/4$ IF $\varphi = 3\,\pi/4$, THEN $A_\text{net} = 2\,A\,\cos\left(\frac{3\pi}{8}\right) \simeq 0.765\,A$, and thus

$$\psi_\text{net}(t, x) = 2\,\cos(3\,\pi/8)\,A\,\sin(k\,x - \omega\,t + 3\,\pi/8)\,.$$

These waves suffer **partial destructive interference**.

$$A\,\sin(k\,x - \omega\,t) + A\,\sin(k\,x - \omega\,t + 3\,\pi/4) = 2\,\cos(3\,\pi/8)\,A\,\sin(k\,x - \omega\,t + 3\,\pi/8)$$

π   IF $\varphi = \pi$ [*i.e.*, the two waves are completely out of phase], THEN $A_\text{net} = 2\,A\,\cos(\pi/2) = 0$, and thus

$$\psi_\text{net}(t, x) = 0\,.$$

Complete cancellation of the waves is termed **destructive interference**.

$$A\,\sin(k\,x - \omega\,t) + A\,\sin(k\,x - \omega\,t + \pi) = 0$$

For another approach to constructing combinations of harmonic waves with equal frequency and wavelength, see Chapter 32.

EXAMPLE [*The Three-Phase Miracle*]

Three otherwise identical harmonic waves whose phases differ by $2\pi/3$ [*a.k.a.* 120°],

$$\psi_{1R} = A\sin(k\,x - \omega\,t)$$
$$\psi_{2R} = A\sin(k\,x - \omega\,t + 2\pi/3)$$
$$\psi_{3R} = A\sin(k\,x - \omega\,t + 4\pi/3)$$

endure complete destructive interference when superposed:

$$\psi_{\text{net},R} = \psi_{1R} + \psi_{2R} + \psi_{3R} \equiv 0.$$

Though the medium may appear undisturbed, three distinct waves are present.

PROOF of the Three-Phase Miracle

The composite wave $\psi_{23,R} = \psi_{2R} + \psi_{3R}$ is readily determined from the two-wave superposition analysis at the start of this chapter. The composite has the same wave number and angular frequency as the input waves. Its phase is equal to the average of the constituent phases, $\varphi_{23} = \frac{1}{2}(2\pi/3 + 4\pi/3) = \pi$. Its amplitude is equal to twice the common input amplitude, multiplied by the `cosine` of one-half the phase difference between the input waves, *viz.*, $A_{23} = 2\,A\,\cos(\pi/3) = A$. Therefore,

$$\psi_{23,R} = A_{23}\sin(k\,x - \omega\,t + \varphi_{23}) = A\,\sin(k\,x - \omega\,t + \pi) = -A\,\sin(k\,x - \omega\,t) = -\psi_{1R}.$$

Thus, the superposition of all three waves cannot fail to vanish.

ASIDE: Actually, this is not a miracle after all. Any set of [equally spaced] phases corresponding to complex roots of unity possesses this property.

RIGHTMOVING HARMONIC WAVES: EQUAL AMPLITUDES, DIFFERENT FREQUENCIES

Two rightmoving harmonic waves sharing common amplitude, ψ_1 and ψ_2, propagate in the same medium. WLOG one may write:

$$\psi_1 = \psi_1(k_1\,x - \omega_1\,t) = A\,\sin(k_1\,x - \omega_1\,t)$$
$$\psi_2 = \psi_2(k_2\,x - \omega_2\,t) = A\,\sin(k_2\,x - \omega_2\,t)\,.$$

ASIDE: **Q:** Aren't phase factors $\{\varphi_1, \varphi_2\}$ needed to describe these waves?

A: Two phases may always be set to zero by felicitous choices of $t = 0$ and $x = 0$. There is just enough latitude to dispense with both phase terms.[1]

The superposition $\psi_1 + \psi_2$ is easily, but not intelligibly, expressed as

$$\psi_{\text{net}}(t, x) = A\Big[\sin(k_1\,x - \omega_1\,t) + \sin(k_2\,x - \omega_2\,t)\Big].$$

The sum of **sines** trigonometric identity, $\sin(a) + \sin(b) = 2\cos\left[\frac{a-b}{2}\right]\sin\left(\frac{a+b}{2}\right)$, enables progress here, too. In this case, one identifies

$$a = k_1\,x - \omega_1\,t \qquad \text{and} \qquad b = k_2\,x - \omega_2\,t,$$

---
[1] While this latitude is inviting, our lassitude makes it compelling.

and then obtains

$$\frac{a-b}{2} = \frac{1}{2}\left(k_1\,x - \omega_1\,t - k_2\,x + \omega_2\,t\right) = \frac{1}{2}\left(k_1 - k_2\right)x - \frac{1}{2}\left(\omega_1 - \omega_2\right)t$$

and

$$\frac{a+b}{2} = \frac{1}{2}\left(k_1\,x - \omega_1\,t + k_2\,x - \omega_2\,t\right) = \frac{1}{2}\left(k_1 + k_2\right)x - \frac{1}{2}\left(\omega_1 + \omega_2\right)t\,.$$

Thus, the superposition of ψ_1 and ψ_2 may be written:

$$\psi_{\text{net}}(t,x) = 2\,A\,\cos\left[\frac{1}{2}\left(k_1 - k_2\right)x - \frac{1}{2}\left(\omega_1 - \omega_2\right)t\right]\,\sin\left(\frac{1}{2}\left(k_1 + k_2\right)x - \frac{1}{2}\left(\omega_1 + \omega_2\right)t\right)\,.$$

It is noteworthy that the argument of the `sine` term involves the average values of the wave numbers and angular frequencies, while the argument of the `cosine` term contains factors which are half the differences in the wave numbers and frequencies.

To visualise this complex waveform, it is helpful to adopt the usual, dual perspectives.

FIXED POINT   From the vantage point of a stationary observer regarding the wave disturbance at fixed X_0, the wave oscillates in time as

$$\psi_{\text{net}}(t,X_0) \simeq 2\,A\,\cos\left[\frac{\Delta\omega}{2}\,t\right] \times \sin\left(\omega_{\text{av}}\,t\right),$$

up to inclusion of irrelevant constant phase factors.

SNAP-SHOT   Suppose instead that a snapshot is taken of the superposed waves at $t = T_0$. Then the wave is seen to vary with position according to

$$\psi_{\text{net}}(T_0,x) \simeq 2\,A\,\cos\left[\frac{\Delta k}{2}\,x\right] \times \sin\left(k_{\text{av}}\,x\right),$$

also up to constant phase factors.

....

EXAMPLE [*Beats Me!*]

Pairs of musicians employ a phenomenon called *beating* to tune their instruments.

First, they agree that a particular instrument produces a true note, say the A at 440 Hz.

> ASIDE: How this initial assessment is made is not our concern here, but a small subset of people possess "perfect pitch," enabling them to accurately identify notes which are played or sung. Most people, having "relative pitch," can reliably order any two notes from lower to higher frequency.   [Batters beware—PK has "wild pitch!"]

Second, both instruments play the same [putative] note with approximately the same *degree of loudness*.[2] Let's suppose for definiteness that one note is exactly 440 Hz, while the other is 436 Hz. The musicians listen carefully[3] to the composite wave, described by

$$\psi_{\text{net}} = 2\,A\,\cos\left[-\frac{1}{2}\Delta\omega\,t + \varphi_c\right]\sin\left(-\omega_{\text{av}}\,t + \varphi_s\right)$$

$$\simeq 2\,A\,\cos\left[2\,(2\,\pi)\,t\right]\sin\left(-438\,(2\,\pi)\,t\right).$$

---

[2] The degree of loudness is correlated with the amplitude of the sound wave.

[3] For instance, they ensure that they remain at rest with respect to each other and their respective instruments, notwithstanding the snarky realisation that [musical] "rest" is a logical impossibility.

In this expression, irrelevant constant phase factors are labelled $\{\varphi_c, \varphi_s\}$, and subsequently ignored. Factors of (2π) are explicitly present to convert the cycle frequencies, ν, to angular frequencies, ω, in the arguments of the trig functions. In addition, the even symmetry of the `cosine` function was exploited to eliminate a minus sign.

The net wave is the product of oscillating `sine` and `cosine` factors. The `sine` term oscillates at the average of the two frequencies produced by the instruments, 438 Hz. Only a person with perfect pitch can tell that it differs from A_{440}. The `cosine` term oscillates at one-half of the frequency difference between the two instruments, *i.e.*, at 2 Hz. This slow oscillation effectively modulates the amplitude of the faster [`sine`] term, as may readily be seen in Figure 24.1.

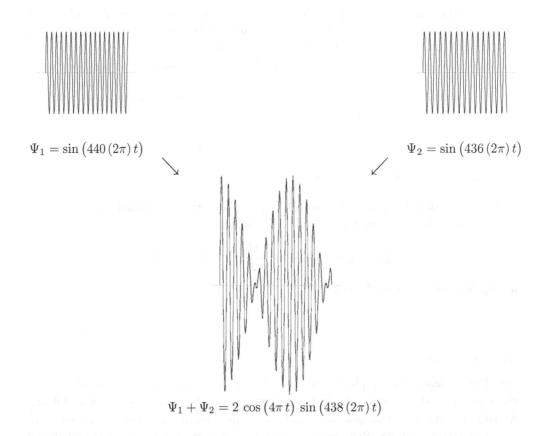

$$\Psi_1 = \sin\left(440\,(2\pi)\,t\right) \qquad\qquad \Psi_2 = \sin\left(436\,(2\pi)\,t\right)$$

$$\Psi_1 + \Psi_2 = 2\,\cos\left(4\pi\,t\right)\,\sin\left(438\,(2\pi)\,t\right)$$

FIGURE 24.1   Production of Beats from Rightmoving Harmonic Waves

The loudness or energy intensity of the net wave depends on the [squared] magnitude of the wave amplitude, and thus the musicians hear the composite note fade in-and-out at twice the slow oscillation frequency. Thus, the **beat frequency** is

$$\nu_{\text{BEAT}} = 2 \times \frac{\Delta\omega}{2} = \Delta\omega\,.$$

In the case of our two musicians, playing at 440 Hz and 436 Hz respectively, the composite note beats at 4 Hz. The person with the second instrument adjusts his instrument and they play the same note again. When the notes are played and the beating is undetectable, the musicians agree that the second instrument is "in tune" with the first.

Chapter 25

Standing Waves

Standing waves may arise from the superposition of left- and rightmoving repeating waves residing in the same medium. A plethora of possibilities exist in that the frequencies [and wavelengths] and/or amplitudes may be the same or differ. Phase effects occur in the situations of greatest interest and give rise to a form of **selection rule**. In this chapter, our attention is given over to analyses of equal frequency/wavelength–equal amplitude cases.

SUPERPOSITION OF LEFT- AND RIGHTMOVING HARMONIC WAVES

A rightmoving harmonic wave, ψ_{1R}, and a leftmoving harmonic wave, ψ_{2L}, both with wave number k, angular frequency ω, and amplitude A, propagate in the same medium with the same speed. The phase offsets of these waves may, by judicious choice of clock and coordinate origins, be set to zero. Having done this, we must be very careful in any further attempt to redefine t or x. The waves themselves, and their superposition, are:

$$\begin{aligned} \psi_{1R} &= A\,\sin(k\,x - \omega\,t) \\ \psi_{2L} &= A\,\sin(k\,x + \omega\,t) \end{aligned} \quad \Longrightarrow \quad \psi_{\text{net}}(t,x) = A\big[\sin(k\,x - \omega\,t) + \sin(k\,x + \omega\,t)\big]\,.$$

The trigonometric identity for combining sums of **sines** that was profitably employed in Chapter 24,

$$\sin(a) + \sin(b) = 2\,\sin\left(\frac{a+b}{2}\right)\cos\left[\frac{a-b}{2}\right]\,,$$

is efficacious here as well. Setting

$$a = k\,x - \omega\,t \qquad \text{and} \qquad b = k\,x + \omega\,t$$

fixes the values of the combinations

$$\frac{a+b}{2} = k\,x \qquad \text{and} \qquad \frac{a-b}{2} = -\omega\,t\,.$$

Upon minor simplification $[\cos(-\theta) = \cos(\theta)]$, the net wave is

$$\psi_{\text{net}}(t,x) = 2\,A\,\sin(k\,x)\,\cos(\omega\,t)\,.$$

The spatial and temporal dependencies have separated. This is a **standing wave**.

Let's carefully distinguish standing harmonic waves from travelling harmonic waves.

TRAVELLING For a harmonic travelling wave, two consecutive snapshots reveal sinusoidal disturbances in space with the same amplitude. As the wave propagates past any fixed point, its periodic disturbances make full-amplitude sinusoidal excursions.

STANDING For the standing wave, two snapshots reveal fixed **nodes** and distinct amplitudes. The medium at location x undergoes sinusoidal oscillation with a position-dependent amplitude.

NODAL POINT   A nodal point, or node, is a place in the medium which remains quiescent [undisturbed] despite the presence of the wave. Fixed endpoints are nodes. Nodes within the medium occur wherever the position-dependent amplitude vanishes.

ANTINODE   An antinode is a point in the medium at which the local wave disturbance is maximal. Free endpoints are antinodes. Antinodes within the medium are those locations at which the amplitude is greatest.

In the above discussion, we have imagined the simultaneous co-existence of leftmoving and rightmoving waves with equal frequencies and amplitudes. That these are the only waves present seems to require that the medium be infinite in extent.

Q: How might the situation of co-existing left- and rightmovers be effected?

A: Encountering a fixed or free endpoint converts an incident leftmover into a reflected rightmover, and *vice versa*. Incident and reflected waves residing in a semi-infinite medium will combine to produce the standing waves discussed above.

Q: No mention has been made of restrictions on the frequency or wavelength of the constituent waves. Are all frequencies allowed?

A: In the limits of infinite and semi-infinite 1-d media, all frequencies are allowed. For media of finite length, all frequencies are *a priori* possible, but multiple reflections from the two endpoints leads to destructive interference for frequencies which are incommensurate.

[This effect is not unlike that exposed in the Three-Phase Miracle example in Chapter 24.]

The essential idea is that equal-amplitude harmonic waves combining with all possible phase shifts will inevitably cancel to zero. Exceptions to this general behaviour occur for particular commensurate frequencies/wavelengths. Despite their relative scarcity, there exist an infinite number of these so-called **harmonics**,[1] and they may be labelled by a natural number, n. An alternate nomenclature, preferred by musicians, speaks of the lowest-order or **fundamental frequency**, $m = 0$, along with an infinite tower of ordinal **overtones**, $m = 1\text{st}, 2\text{nd}, 3\text{rd}, \dots$.

ASIDE: The mapping between harmonics and overtones consists in an identification of the nth harmonic with the $(n-1)$th overtone, and the mth overtone with the $(m+1)$th harmonic.

Q: For standing waves in a homogeneous medium of finite extent, how do the fixed/free boundary conditions matter?

A: The three possible combinations of boundary conditions, FIXED–FIXED, FIXED–FREE, and FREE–FREE, are examined, each in turn, below.

STANDING WAVES: FIXED–FIXED BOUNDARY CONDITIONS

Fixed–fixed boundary conditions imply the presence of nodes at each end of the medium. The four lowest frequency and longest wavelength modes are sketched in Figure 25.1. The black lines show the waves at times at which the local disturbances are maximal, while the grey lines depict the waves at other instants. All these lines cross at the internal nodal points.

For the standing waves to possess nodes at both ends, there must be an integral number of half-wavelengths resident in the medium.

---

[1]This is yet another potential source of lexical confusion and yet we trust that the context will always enable disambiguation of these various usages of "harmonic."

First Harmonic   Second Harmonic   Third Harmonic   Fourth Harmonic

FIGURE 25.1   Standing Harmonic Waves with Fixed–Fixed Boundary Conditions

[The harmonic label, n, is equal to this number of half-wavelengths.]

Cramming additional half-wavelengths into the medium of fixed length L requires that the higher-order harmonics possess diminishing wavelengths and increasing frequencies. The frequency associated with each of these modes is fully determined by the wave speed relation, $c = \lambda \nu$.

Harmonic	Commensuration	Wavelength	Frequency	Overtone
1	$L = \frac{\lambda_1}{2}$	$\lambda_1 = 2L$	$\nu_1 = \frac{c}{2L}$	FUNDAMENTAL
2	$L = \lambda_2$	$\lambda_2 = L$	$\nu_2 = \frac{c}{L}$	1st
3	$L = \frac{3\lambda_3}{2}$	$\lambda_3 = \frac{2}{3}L$	$\nu_3 = \frac{3c}{2L}$	2nd
4	$L = 2\lambda_4$	$\lambda_4 = \frac{1}{2}L$	$\nu_4 = \frac{2c}{L}$	3rd
...	...	...	...	...

The patterns among entries in the above list are captured by expressions of the form

$$\lambda_n = \frac{2L}{n} \qquad \text{and} \qquad \nu_n = \frac{c}{\lambda_n} = \frac{nc}{2L},$$

or

$$\lambda_m = \frac{2L}{m+1} \qquad \text{and} \qquad \nu_m = \frac{(m+1)c}{2L},$$

where n and m label the harmonics and overtones, respectively.

One final comment rounds out this section. The physical realisation of fixed–fixed boundary conditions which springs most readily to mind is that of transverse waves on a stretched string. Precisely the same set of modes occurs for sound waves resident in closed-end organ pipes.

STANDING WAVES: FIXED–FREE BOUNDARY CONDITIONS

Fixed–free boundary conditions imply the existence of a node at one end and an antinode at the other. These constraints entail that the number of half-wavelengths in the medium be augmented by an additional quarter-wavelength. The lowest-order modes appear in Figure 25.2.

First Harmonic   Second Harmonic   Third Harmonic   Fourth Harmonic

FIGURE 25.2 Standing Harmonic Waves with Fixed–Free Boundary Conditions

Harmonic	Commensuration	Wavelength	Frequency	Overtone
1	$L = \frac{\lambda_1}{4}$	$\lambda_1 = 4L$	$\nu_1 = \frac{c}{4L}$	FUNDAMENTAL
2	$L = \frac{3\lambda_2}{4}$	$\lambda_2 = \frac{4}{3}L$	$\nu_2 = \frac{3c}{4L}$	1st
3	$L = \frac{5\lambda_3}{4}$	$\lambda_3 = \frac{4}{5}L$	$\nu_3 = \frac{5c}{4L}$	2nd
4	$L = \frac{7\lambda_4}{4}$	$\lambda_4 = \frac{4}{7}L$	$\nu_4 = \frac{7c}{4L}$	3rd
...	...	...	...	...

The patterns revealed in the table are summarised by the general relations

$$\lambda_n = \frac{4L}{2n-1} \quad \text{and} \quad \nu_n = \frac{(2n-1)c}{4L},$$

or

$$\lambda_m = \frac{4L}{2m+1} \quad \text{and} \quad \nu_m = \frac{(2m+1)c}{4L},$$

where, again, n labels the harmonics and m the overtones.

STANDING WAVES: FREE–FREE BOUNDARY CONDITIONS

Free–free boundary conditions require that antinodes occur at each end. Thus, successive harmonics arise via inclusion of additional half-wavelengths, as illustrated in Figure 25.3.

First Harmonic   Second Harmonic   Third Harmonic   Fourth Harmonic

FIGURE 25.3 Standing Harmonic Waves with Free–Free Boundary Conditions

Harmonic	Commensuration	Wavelength	Frequency	Overtone
1	$L = \frac{\lambda_1}{2}$	$\lambda_1 = 2L$	$\nu_1 = \frac{c}{2L}$	FUNDAMENTAL
2	$L = \lambda_2$	$\lambda_2 = L$	$\nu_2 = \frac{c}{L}$	1st
3	$L = \frac{3\lambda_3}{2}$	$\lambda_3 = \frac{2}{3}L$	$\nu_3 = \frac{3c}{2L}$	2nd
4	$L = 2\lambda_4$	$\lambda_4 = \frac{1}{2}L$	$\nu_4 = \frac{2c}{L}$	3rd
...	...	...	...	...

The wavelengths and frequencies of these modes are the same as those which arise under fixed–fixed boundary conditions, *viz.*,

$$\lambda_n = \frac{2L}{n} \quad \text{and} \quad \nu_n = \frac{nc}{2L},$$

or

$$\lambda_m = \frac{2L}{m+1} \quad \text{and} \quad \nu_m = \frac{(m+1)c}{2L},$$

where n is the harmonic number and m labels the overtones.

ASIDE: Musicians have oft *noted* that the fundamental tone in a pipe which is open at both ends can be lowered by [approximately] one octave by capping one end. Capping both ends restores the original tone.

Two penultimate words follow before we study examples.

⊥ Standing waves are called **resonant waves** by some, on account of the frequency/wavelength selection effects described above. The power spectrum consists of sharp "spikes" at the harmonic frequencies and zero at all others,[2] which is evocative of a collection of [ultra-]high-Q DHOs.

⊥ These same standing wave modes are the Fourier basis functions [studied in Chapter 23] on the interval spanned by the wave medium.

---

[2]In the ideal limit in which the spike is infinitely both tall and narrow, it is called a **delta function**.

EXAMPLE [*Transverse Standing Waves on a String*]

A uniform string of length $L = 3\,\mathrm{m}$ is held straight under tension and clamped firmly at both ends. The wave speed on this particular string[3] happens to be $50\,\mathrm{m/s}$.

Q: What are the wavelength and frequency of the fourth harmonic?

A: For fixed–fixed boundary conditions, the harmonic number corresponds to the number of half-wavelengths accommodated by the medium. Hence,

$$L = \frac{n}{2}\,\lambda_n \qquad \Longrightarrow \qquad \lambda_4 = \frac{L}{4/2} = \frac{3}{2}\,\mathrm{m}\,.$$

The wavelength of the fourth harmonic on the three metre string subject to fixed–fixed boundary conditions is $1.5\,\mathrm{m}$. Given $\nu_n\,\lambda_n = c$, the frequency of the fourth harmonic is

$$\nu_4 = \frac{c}{\lambda_4} = \frac{50\ \mathrm{m/s}}{1.5\ \mathrm{m}} = \frac{100}{3\ \mathrm{s}} = 33\frac{1}{3}\ \mathrm{Hz}\,.$$

This is a very low note, not far from the threshold for normal human hearing, and as such it might be felt as much as heard.

EXAMPLE [*Longitudinal Standing Waves in an Organ Pipe*]

An organ pipe of length $L = 3\,\mathrm{m}$, open at one end and capped at the other, is filled with air. The speed of sound in air[4] may be taken to be $c = 344\,\mathrm{m/s}$.

Q: What are the wavelength and frequency of the fourth overtone?

A: For fixed–free boundary conditions, the mth overtone has one quarter-wavelength plus $(m + 1)$ half-wavelengths completely occupying the medium. Hence,

$$L = \frac{2\,m + 1}{4}\,\lambda_m \qquad \Longrightarrow \qquad \lambda_{m=4} = \frac{4\,L}{9} = \frac{4}{3}\,\mathrm{m}\,.$$

The frequency of the waves is $\nu_m = c/\lambda_m$, and therefore

$$\nu_{m=4} = \frac{344\ \mathrm{m/s}}{4/3\mathrm{m}} = 258\ \mathrm{Hz}\,.$$

This particular note lies a smidgen below **middle C** [261.63 **Hz**].

---

[3]In Chapter 26 a general expression is obtained determining the speed of transverse waves in terms of properties of the medium in which they reside.

[4]The speeds of longitudinal waves in various media, including air, shall be determined in Chapter 27.

Chapter 26

Transverse Waves: Speed and Energetics

Our first task in this chapter is to derive an expression for the propagation speed of transverse waves in terms of physical properties of the medium in which they move. The analysis performed for a solitary pulse generalises to repeated waves. Second, we shall obtain an expression for the amount of mechanical energy associated with the disturbance in the medium throughout one wavelength of a harmonic wave. Granted that waves are propagating disturbances, the conclusion that **transverse waves are able to transfer energy** is inescapable.

DERIVATION OF WAVE SPEED

Consider a differentiable[1] symmetric[2] solitary transverse pulse, such as the one exhibited in Figure 21.1 of Chapter 21 and reproduced here in Figure 26.1, propagating with speed c in a homogeneous 1-d medium. The point of view in these figures is that of an observer who is at rest with respect to the medium.

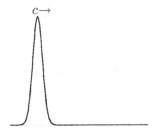

FIGURE 26.1 A Transverse Solitary Wave

That the homogeneous and uniform medium is [conventionally] also considered continuous does not prevent adoption of the viewpoint in which the medium consists of a collection of [identically sized] discrete material elements.

> ASIDE: We remind the gentle reader that this partitioning has nothing to do with the atomic nature of materials. Instead, it is a formal device, applicable on scales which are small compared to the size of the pulse and yet large compared to the size(s) of the microscopic material structures: atoms, molecules, crystal grains, *etc.*, particular to the medium.

Suppose that each material element is a strip of medium with arc length ΔS. The assumed homogeneity [and thus uniformity, too] of the bulk medium ensures that the mass of each element is

$$\Delta M = \mu \, \Delta S \,,$$

---

[1]The insistence on differentiability is to ensure that kinks are not present in the wave form.

[2]Spatial symmetry helps to guarantee that the medium's motion is purely transverse (without any longitudinal back-and-forth sloshing).

where μ is the constant lineal mass density[3] of the medium.

From the perspective of an observer moving alongside the medium with [constant] speed c, the pulse appears to be a static [*i.e.*, unchanging in time] deformation. Furthermore, the medium far from the disturbance is seen to be moving uniformly, toward or away from the observer, at c. In the vicinity of the pulse, the material elements follow a trajectory conforming in shape to the pulse/deformation. This situation is illustrated in Figure 26.2.

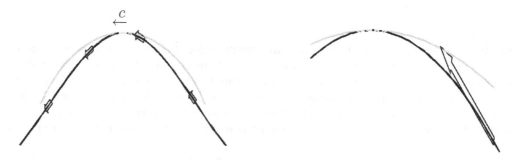

FIGURE 26.2 Two Higher-Resolution Co-moving Frame Views of a Solitary Pulse

Each mass element, as it rounds the crest of the pulse, experiences a brief interval during which its trajectory may be identified with a segment of circular arc[4] of radius R. On account of the [assumed forward–backward] pulse symmetry, the centre of the osculating circle corresponding to the trajectory of the mass element at the peak lies along the straight line joining the equilibrium and disturbed positions of the material element. In other words, the wave disturbance is purely transverse.

The material element at the peak experiences instantaneous uniform circular motion.

A feature specific to uniform circular motion is the expression for centripetal acceleration:

$$\vec{a}_c = \frac{c^2}{R} \, [\text{radially inward}].$$

The centripetal $\Delta M \, \vec{a}_c$ experienced by the material element at the peak of the disturbance is an *effect*. Its associated *cause*, according to N2, is the net external force acting on the material element [treated as a particle of mass ΔM and represented by a thickened line in Figure 26.3] instantaneously at the peak.

Tension in the material comprising the medium provides the dominant[5] force acting on the mass element. A tension force, $\vec{T}_{\text{fwd}}$, acts forward and down from the portion of the medium ahead of the pulse, while another, $\vec{T}_{\text{bkwd}}$, acts backward and down from that part of the medium behind the pulse. The directional subscripts, "fwd" and "bkwd," are with respect to the observed motion of the material elements.

The forces are illustrated *in situ* in Figure 26.3, and in a FBD in Figure 26.4.

---

[3]Lineal mass density was introduced in VOLUME I, Chapter 30 *et seq.*, where it was denoted by "λ." In application to waves, it is conventional to use μ.

[4]Osculating circles were mentioned in the Long and Winding Road digression in Chapter 8 of VOLUME I.

[5]Potential complications are averted by assuming that all forces, aside from the tension in the medium, vanish. In the event that this assumption is not even approximately true, it becomes necessary to carefully consider time scales and judiciously employ the Impulse Approximation, *cf.* VOLUME I, Chapter 29.

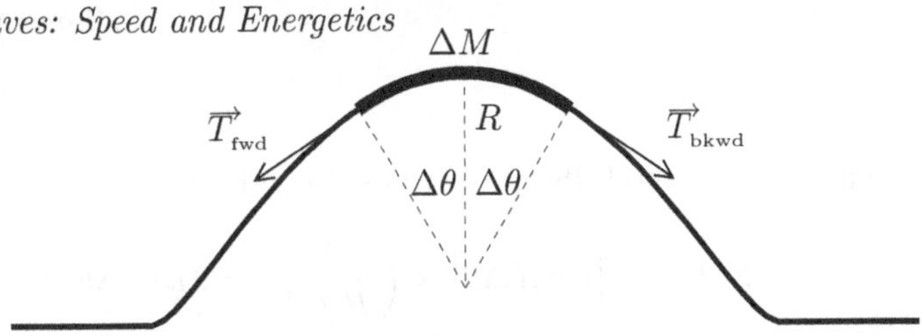

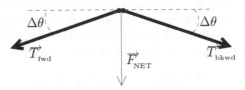

FIGURE 26.3 Forces on the Mass Element (shown by the thick line) at the Pulse Peak

All of the necessary pieces have been assembled for the derivation of the speed of transverse waves in homogeneous media. The argument proceeds as follows.

ΔS   The arc length of the material element at the crest of the pulse is

$$\Delta S = R(2\,\Delta\theta) = 2\,R\,\Delta\theta\,,$$

where it is implicitly assumed that $\Delta\theta$ is a small angle. [The arclength in Figure 26.3 is exaggerated to more clearly illustrate the tension forces.]

ΔM   The inertia of the specified material element is

$$\Delta M = \mu\,\Delta S = \mu\,2\,R\,\Delta\theta\,.$$

$\vec{F}_{\mathrm{NET}}$ The net external force acting on the material element is the vector sum of the forward and backward tension forces,

$$\vec{F}_{\mathrm{NET}} = \vec{T}_{\mathrm{fwd}} + \vec{T}_{\mathrm{bkwd}}\,,$$

as is evident from Figure 26.4.

FIGURE 26.4 The FBD for the Mass Element at the Peak of the Pulse

By symmetry, the forward and backward tensions have the same magnitude, T. Thus, their horizontal components sum to zero and the net force acts purely in the vertical direction, with magnitude

$$F_{\mathrm{NET}} = 2T\,\sin(\Delta\theta) \simeq 2T\,\Delta\theta\,,$$

where the small angle approximation $\sin(\Delta\theta) \sim \Delta\theta$ has been invoked.

N2

LHS   In this instance, the LHS of Newton's Second Law reads

$$\Delta M \, \vec{a}_c = \left[\mu \, 2 \, R \, \Delta\theta \right] \times \left(\frac{c^2}{R} \right) \, [\downarrow] = 2 \, \mu \, c^2 \, \Delta\theta \, [\downarrow] \, .$$

RHS   Meanwhile, in the limit in which the angle becomes infinitesimally small, the RHS of N2 reads

$$\vec{F}_{\text{NET}} = 2 \, T \, \Delta\theta \, [\downarrow] \, .$$

Invoking N2 [by setting LHS = RHS], one obtains

$$2 \, \mu \, c^2 \, \theta = 2 \, T \, \theta \qquad \Longrightarrow \qquad c^2 = \frac{T}{\mu} \, .$$

Thus, the speed of the transverse wave propagating through the medium is completely determined by local material properties of the [unperturbed[6]] medium:

$$c = \sqrt{\frac{T}{\mu}} \, .$$

ENERGETICS OF HARMONIC TRANSVERSE WAVES

A propagating wave is capable of transporting energy. To quantify the amount of energy borne by the wave, we shall

♮   restrict attention to harmonic waves only,[7] and

♮   recognise and exploit the fact that harmonic wave disturbances are, at each and every point in space, of simple harmonic [SHO] form.

The same partitioning of the medium into a collection of material elements [*a.k.a.* chunks] of mass ΔM and small spatial extent ΔS employed in the first part of this chapter is also used here. The assumed smallness of the chunks enables one to associate a unique spatial position, x_i, with the ith chunk. The function describing a leftmoving harmonic wave is

$$\Psi(t,x) = A \, \sin(k \, x + \omega \, t + \phi) \, .$$

When restricted to the ith chunk, this reads

$$\psi(t) = \Psi(t, x_i) = A \, \sin(k \, x_i + \omega \, t) = A \, \sin(\omega \, t + \varphi) \, ,$$

where $\varphi = k \, x_i + \phi$ is a constant phase offset. Thus, when viewed at a fixed location, the transverse motion of the [chunk of] medium associated with the propagation of the harmonic wave is sinusoidal.

---

[6]The presence of a wave slightly perturbs the mass density and internal tension force, providing an avenue through which wave–wave interactions may occur. While our [zeroth order] assumption of linear superposition assiduously eschews such interactions, they do exist and can be modelled.

[7]This is a not-so-onerous restriction in light of Fourier analysis.

Therefore, SHO dynamics governs the [correlated] motions of the chunks.

Hence, at any and every instant, each chunk possesses an admixture of kinetic and potential energies which sum to a constant amount of total mechanical energy, as was discussed in Chapter 15. According to the iconic Figure 15.4, the total mechanical energy is equal to

K   the maximum amount of kinetic energy of the oscillating particle,[8] and

U   the maximum amount of potential energy of the oscillating particle.[9]

METHOD ONE: [*Konsidering Kinetic*]

The maximum kinetic energy is attained as the chunk passes through its equilibrium position. The velocity of the *i*th chunk, at location x_i, is

$$v(t) = \frac{d\psi}{dt} = \omega\, A\, \cos(\omega\, t + \varphi)\,.$$

Therefore, the [time-dependent] kinetic energy of the chunk is

$$\Delta K = \frac{1}{2}\,\Delta M\, v^2 = \frac{1}{2}\,\Delta M\, \omega^2\, A^2\, \cos^2(\omega\, t + \varphi)\,.$$

Inspection reveals that the maximum value of the kinetic energy is

$$\Delta K_{\text{max}} = \frac{1}{2}\,\Delta M\, \omega^2\, A^2\,.$$

The total mechanical energy associated with the material element of the medium through which the wave propagates is equal to ΔK_{max}.

METHOD TWO: [*Pondering Potential*]

The maximum potential energy is attained as the chunk experiences its maximal excursions from equilibrium, *viz.*, $\psi(t) = A$. At these instants, the potential energy of the chunk is

$$\Delta U_{\text{max}} = \frac{1}{2}\, k\, A^2 = \frac{1}{2}\,\Delta M\, \omega^2\, A^2\,.$$

Imposition of the SHO relation, $\omega^2 = k/\Delta M$, led to the final form of the expression for the maximum potential energy of the *i*th chunk. The total mechanical energy of the *i*th chunk is equal to its maximum potential energy.

ASIDE: It is a relief to see that METHOD ONE and METHOD TWO yield results which are in perfect agreement.

$$\Delta E = \frac{1}{2}\,\Delta M\, \omega^2\, A^2\,.$$

---

[8]The particle, in this context, is the chunk of medium disturbed by the wave.
[9]Provided that the potential energy is zero when the chunk is at its equilibrium position, *i.e.*, undisturbed.

Three comments set the stage for our ultimate result.

ENERGY DENSITY   The mechanical energy of the material element can be thought of as being distributed throughout its spatial extent. Since

$$\Delta M = \mu \, \Delta S \,,$$

the amount of energy borne by each mass/material element is

$$\Delta E = \frac{1}{2} \, \mu \, \omega^2 \, A^2 \, \Delta S \,.$$

The average energy density[10] associated with the wave is

$$\frac{\Delta E}{\Delta S} = \frac{1}{2} \, \mu \, \omega^2 \, A^2 \,.$$

With the assumptions that we have been labouring under:

μ homogeneity [constant density],

ω no dispersion [unique fixed value of angular frequency], and

A no dissipation [constant amplitude],

the average energy density of transverse harmonic waves is evidently constant.

Refining the partition, one obtains the local energy density,

$$\frac{dE}{dS} = \frac{1}{2} \, \mu \, \omega^2 \, A^2 \,.$$

NATURAL SCALE   The wavelength, λ, provides the natural spatial scale over which to integrate this energy density. The energy content associated with one complete wavelength of the harmonic wave is

$$E_\lambda = \int dE = \int_0^\lambda \frac{dE}{dS} \, dS = \frac{1}{2} \, \mu \, \omega^2 \, A^2 \int_0^\lambda dS = \frac{1}{2} \, \mu \, \lambda \, \omega^2 \, A^2 \,.$$

POWER   The mechanical energy, computed above, is borne along with the wave. This flow of energy [a.k.a. power] can be quantified by means of the same argument used in Chapter 21 to show that $c = \lambda \nu$.

The wave moves forward one wavelength in every period, and thus the [magnitude of the] average rate at which energy flows is

$$P_{\text{av}} = \frac{E_\lambda}{T} = \nu \, E_\lambda = \frac{1}{2} \, \mu \left(\nu \, \lambda \right) \omega^2 \, A^2 = \frac{1}{2} \, \mu \, \omega^2 \, A^2 \, c \,.$$

This ultimate expression reveals that

$$\text{Average Power} = (\text{Energy Density}) \times (\text{Wave Speed}) \,.$$

---

[10] *I.e.*, the average amount of mechanical energy per unit length in the medium.

Chapter 27

Speed of Longitudinal Waves

Here we shall parallel the perambulations taken in Chapter 26, so as to derive an expression for the propagation speed of longitudinal waves in terms of medium properties. The analysis performed for a compound pulse generalises to repeated waves. The tasks of determining the amount of mechanical energy associated with each wavelength of a harmonic wave and the associated energy flux per unit cross-sectional area are undertaken in Chapter 28.

FIGURE 27.1   Two Views of a Longitudinal Compound Pulse

Consider the longitudinal plane wave compound pulse[1] exhibited in Figures 27.1 and 27.2. The wave is propagating with speed c in a homogeneous medium which fills a channel[2] with constant cross-sectional area $\mathcal{A}$. The point of view in each figure is that of an observer at rest with respect to the medium.

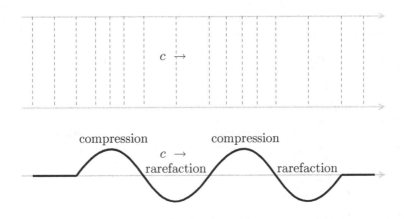

FIGURE 27.2   Medium Compressions and Rarefactions in a Longitudinal Pulse

---

[1] This pulse resembles two cycles of a sinusoid. What appear to be kinks at the leading and trailing edges of the pulse would be rounded in any physical realisation of this wave.

[2] Such channels are often called **waveguides**.

The physical nature of the pulse displayed in Figures 27.1 and 27.2 is left unspecified. Often, Figure 27.1 is interpreted as showing zones of density **enhancement** and **depletion** associated with a 2-d plane wave moving along the channel. This misses the mark if the wave characteristic is pressure, in which case regions of **compression** and **rarefaction** within the medium are indicated.

[The quantity displayed in the plots may even be something else altogether.]

The subtleties of and distinctions between the density and pressure viewpoints of longitudinal waves will be addressed in Chapter 28. Meanwhile, let's accede to the majority opinion and assume that the wave disturbance is being characterised by fluctuations in density.

To an observer moving alongside the medium with constant speed c, the density pulse appears to be **static**, *i.e.*, not changing in time. Far from the disturbance, the medium possesses uniform density ρ_0, and moves toward or away from the observer with speed c. The situation in the midst of the pulse is as illustrated in Figure 27.3. Away from the pulse, the medium may be partitioned into uniform slabs [*a.k.a.* material elements] with face area $\mathcal{A}$ and thickness Δx. The volume and mass of each slab are

$$\Delta V = \mathcal{A}\,\Delta x \qquad \text{and} \qquad \Delta M = \rho_0\,\Delta V = \rho_0\,\mathcal{A}\,\Delta x\,,$$

respectively. Prior to encountering the region in which the pulse is concentrated, all portions of each slab are moving uniformly with speed c.

Upon entry into the pulse zone, the material elements narrow in the **enhanced** regions, and widen in the **depleted** zones. After the material elements pass completely through the disturbance, they move uniformly once again. For the shapes of the slabs to change in this manner, relative motion of their front and rear planes must occur. Concomitantly, the entire slab must undergo an increase or decrease in its speed.

In the midst of $\begin{bmatrix} \text{an enhanced} \\ \text{a depleted} \end{bmatrix}$ zone, the speed of the material is $\begin{bmatrix} \text{decreased} \\ \text{increased} \end{bmatrix}$ to $\begin{bmatrix} c_< \\ c_> \end{bmatrix}$.

ASIDE: Traffic flow on a highway may exhibit similar behaviour. In a rolling traffic jam, cars moving normally come upon a clot of slower-moving vehicles, which are bunched up for no apparent reason. The overtaking cars are slowed and absorbed into the jam as additional cars collect behind them. Meanwhile, cars at the front are able to accelerate and resume travelling at highway speed. By pulling away, they make it possible for the cars behind to gradually make their way through to the front.

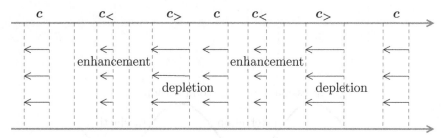

FIGURE 27.3   The Longitudinal Pulse Viewed from the Co-moving Frame

It is not entirely accidental that the compound pulse under study possesses symmetrically enhanced and depleted zones. This is fully consistent with overall conservation of matter in the wave medium, and inspires us to perform the analysis both ways.

METHOD ONE: [*Enhancement*]

The speed of the front edge of the material element is reduced to $c_<$ upon entering into an enhanced region. There is a brief time interval, δt, during which the rear edge continues to move forward at c, until it too is slowed to $c_<$. This is illustrated in Figure 27.4.

For modelling purposes, the manner in which the trailing edge slows is completely ignored,[3] and we suppose that it moves at c for a time interval,

$$\delta t = \frac{\Delta x}{c} \, ,$$

during which a material slab far from the pulse moves forward a distance equal to its length. Meanwhile, the front edge is deemed to move forward by

$$\widetilde{\Delta x} = c_< \, \delta t = \frac{c_<}{c} \, \Delta x \, .$$

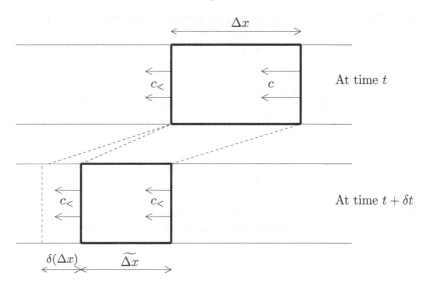

FIGURE 27.4   Narrowing of a Slab as It Slows from c to $c_<$

Therefore, the length of the slab changes by

$$\delta(\Delta x) = \widetilde{\Delta x} - \Delta x = \left(\frac{c_<}{c} - 1 \right) = -\frac{c - c_<}{c} \, \Delta x \, .$$

ASIDE: This negative change in length, $\delta(\Delta x) < 0$, may be regarded as the distance by which the front end falls short of the point where it would have lain in the absence of the denser region.

The fractional change in length of the slab is equal to its fractional change in volume, *i.e.*,

$$\frac{\delta(\Delta x)}{\Delta x} = -\frac{c - c_<}{c} = \frac{\delta(\Delta V)}{\Delta V} \, ,$$

owing to the constancy of the cross-sectional area of the channel.

As the precise dynamical details are unknown to us, this is an instance ripe for application of the Impulse Approximation.[4]

---

[3]This isn't as egregious as it sounds. Corrections to the position are of higher order and vanish in the limit in which the slabs become infinitesimally thin.

[4]The Impulse Approximation was studied in VOLUME I, Chapter 29.

IMPULSE APPROXIMATION: IF the time scale is short, AND a single force, F, dominates the dynamics, THEN the net change in the momentum of a particle or system may be consistently ascribed to F alone, *i.e.*, $F_{\text{NET}} \simeq F$.

Consequently, the average value of F is (approximately) equal to the net change in the momentum divided by the duration of the time interval through which the change occurs.

Prior to its slowing down, the momentum of the slab was $p_i = \Delta M\, c$. Immediately after, it was $p_f = \Delta M\, c_<$. The average net external force acting on the slab is

$$F_{\text{NET}} = \frac{\delta p}{\delta t} = \frac{p_f - p_i}{\delta t} = \frac{\Delta M \left(c_< - c\right)}{\Delta x / c} = -\frac{\Delta M}{\Delta x}\, c \left(c - c_<\right).$$

Invoking $\Delta M = \rho_0\, \mathcal{A}\, \Delta x$ and dividing by the cross-sectional area yield

$$\frac{F_{\text{NET}}}{\mathcal{A}} = -\rho_0\, c \left(c - c_<\right) = \delta P.$$

In the last equality, the average net force per unit area is identified with the **pressure increment** [*cf.* the discussion of Pascal's formula in Chapter 4 and bulk modulus in Chapter 2]. Replacing $\left(c - c_<\right)$ with $c\,\delta(\Delta V)/\Delta V$ and rearranging reveal the presence of a factor identifiable as the bulk modulus,[5] *viz.*,

$$-B = \frac{\delta P}{\delta(\Delta V)/\Delta V} = -\rho_0\, c^2, \qquad \Longrightarrow \qquad c^2 = \frac{B}{\rho_0}.$$

Hence, the speed of a longitudinal wave is determined by the bulk modulus and the [unperturbed] density of the medium through which the wave propagates:

$$c = \sqrt{\frac{B}{\rho_0}}.$$

METHOD TWO: [*Depletion*]

The speed of the front edge of the material element is increased to $c_>$ upon entering into a depleted zone. There is a brief time interval, δt, during which the rear edge continues to move forward at c, until it too is accelerated to $c_>$. This is illustrated in Figure 27.5.

For modelling purposes, the trailing edge is deemed to move at c for time

$$\delta t = \frac{\Delta x}{c},$$

during which a material slab far from the pulse moves forward a distance equal to its length. Meanwhile, the front edge moves forward by

$$\widetilde{\Delta x} = c_>\, \delta t = \frac{c_>}{c}\, \Delta x.$$

---

[5]Recall that the definition of the bulk modulus includes an explicit minus sign, since a positive pressure increment results in a diminution of the volume.

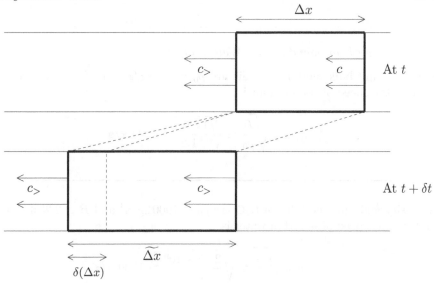

FIGURE 27.5 Elongation of a Slab as It Speeds Up from c to $c_>$

Therefore, the slab is lengthened by

$$\delta(\Delta x) = \widetilde{\Delta x} - \Delta x = \frac{c_> - c}{c}\Delta x\,.$$

One may think of this as the distance by which the front end pulls away from the point where it would lie were there no depleted zone. The fractional change in length of the slab [now positive] is equal to its fractional change in volume, *i.e.*,

$$\frac{\delta(\Delta x)}{\Delta x} = \frac{c_> - c}{c} = \frac{\delta(\Delta V)}{\Delta V}\,,$$

owing to the uniformity of the channel.

Application of the Impulse Approximation in this situation is straightforward. Prior to speeding up, the momentum of the slab was $p_i = \Delta M\,c$. Afterward, $p_f = \Delta M\,c_>$. Thus the average net external force is

$$F_{\text{NET}} = \frac{\delta p}{\delta t} = \frac{p_f - p_i}{\delta t} = \frac{\Delta M\,(c_> - c)}{\Delta x/c} = \frac{\Delta M}{\Delta x}c\,(c_> - c)\,.$$

Once more invoking $\Delta M = \rho_0\,\mathcal{A}\,\Delta x$ and dividing both sides by $\mathcal{A}$ yield

$$\frac{F_{\text{NET}}}{\mathcal{A}} = \rho_0\,c\,(c_> - c) = \delta P\,.$$

The net force per unit area is, again, interpreted as the pressure increment, and replacing $(c_> - c) = c\,\delta(\Delta V)/\Delta V$ leads to an expression in which the bulk modulus appears:[6]

$$B = \frac{\delta P}{\delta(\Delta V)/\Delta V} = \rho_0\,c^2 \qquad \Longrightarrow \qquad c^2 = \frac{B}{\rho_0}\,.$$

Hence,

> the speed of propagation of longitudinal waves is determined by the bulk modulus and density of the material of which the medium is comprised, independently of whether the pulse is an enhancement or a depletion.

---

[6]No explicit minus sign occurs here, as the change in the volume of the slab is positive.

EXAMPLE [*Speeds of Sound in Air, Water, and Steel*]

The density and bulk modulus[7] of air are $\rho_0 \simeq 1.2\,\mathrm{kg/m^3}$ and $B \simeq 142\,\mathrm{kPa}$, respectively. Therefore, the speed of sound in air is

$$c = \sqrt{\frac{B}{\rho_0}} \simeq \sqrt{\frac{142000}{1.2}} \simeq 344\ \mathrm{m/s}\,.$$

The density and bulk modulus of H_2O are $\rho_0 \simeq 1000\,\mathrm{kg/m^3}$ and $B \simeq 2.25\,\mathrm{GPa}$, respectively. Therefore, the speed of sound in water is

$$c = \sqrt{\frac{B}{\rho_0}} \simeq \sqrt{\frac{2.25 \times 10^9}{1000}} \simeq 1500\ \mathrm{m/s}\,.$$

Evidently, sound propagates significantly faster in water than in air.

The density and bulk modulus of a sample of steel are $\rho_0 \simeq 7900\,\mathrm{kg/m^3}$ and $B \simeq 160\,\mathrm{GPa}$, respectively. Therefore, the speed of sound in steel is

$$c = \sqrt{\frac{B}{\rho_0}} \simeq \sqrt{\frac{1.6 \times 10^{11}}{7900}} \simeq 4500\ \mathrm{m/s}\,.$$

This is 4.5 kilometres per second!

---

[7]For this suite of examples, the quoted bulk moduli are of the adiabatic variety.

Chapter 28

Energy Content of Longitudinal Waves

The disturbance characteristic of a repeated longitudinal wave may be described in terms of regions of relative density **enhancement** and **depletion**, or alternatively by **compressions** and **rarefactions**. In conjunction with our analysis of the energetics of longitudinal waves, we shall expose relations between these viewpoints.

ENERGETICS OF HARMONIC LONGITUDINAL WAVES

As discussed in the analysis of transverse waves, no irrevocable loss of generality ensues from limiting attention to a rightmoving harmonic travelling wave. Recalling Chapter 27, we consider a longitudinal wave propagating within a homogeneous medium, confined to a channel with uniform cross-sectional area $\mathcal{A}$. For this analysis, **the quantification of the wave disturbance is provided by the displacement of material elements away from their equilibrium positions.**

[The medium itself does not experience bulk motion.]

A longitudinal harmonic wave is portrayed in both panels of Figure 28.1. On the left, regions of lower/higher density appear with darker/lighter shading. The sketch on the right illustrates the to-and-fro sloshing[1] of thin planar slabs of medium. By squinting carefully at Figure 28.1, it is possible to convince oneself of the two panels' mutual consistency.

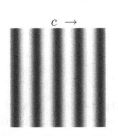
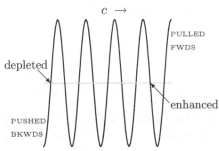

FIGURE 28.1   Two Views of a Longitudinal Harmonic Wave

The harmonic wave disturbance may be presumptively written[2] as

$$S(t,x) = S_0 \cos(k\,x - \omega\,t)\,.$$

The LHS represents the forward–backward displacement of an infinitesimally thin slice of medium which would be at rest at x were the wave disturbance not present. With no wave

---

[1]This is an essential point. The local motions appear to be just so much sloshing around. The WAVE arises through spatially coherent/correlated motions of swathes of the medium.

[2]Phase factors which would otherwise appear may be set to zero by careful and consistent choices of the origins of the spatial and temporal coordinates.

present, a material element [with cross-sectional area $\mathcal{A}$] whose left edge is at x, while its rightmost edge is at $x + \Delta x$, has an [unperturbed] volume of

$$\Delta V = \mathcal{A}\big[(x + \Delta x) - x\big] = \mathcal{A}\,\Delta x.$$

When the harmonic disturbance is present, the left and right edges are shifted to $x + S(t, x)$ and $x + \Delta x + S(t, x + \Delta x)$, respectively. The volume of the slab changes too:

$$\widetilde{\Delta V} = \mathcal{A}\Big[\big(x + \Delta x + S(t, x + \Delta x)\big) - \big(x + S(t, x)\big)\Big]$$

$$= \mathcal{A}\Big[\Delta x + S(t, x + \Delta x) - S(t, x)\Big].$$

The presence of the wave induces a distortion of the material element leading to a change in its volume:

$$\delta(\Delta V) = \widetilde{\Delta V} - \Delta V = \mathcal{A}\Big[S(t, x + \Delta x) - S(t, x)\Big] \to \mathcal{A}\,\frac{\partial S}{\partial x}\,\Delta x = \frac{\partial S}{\partial x}\,\Delta V.$$

The partial derivative in the above expression arises in the limit in which the thickness of each slab approaches zero. Continuing the argument, it follows that

$$\frac{\delta(\Delta V)}{\Delta V} \simeq \frac{\partial S}{\partial x}.$$

Since the wave function is known, its spatial partial derivative can be computed:

$$\frac{\partial S}{\partial x} = \frac{\partial}{\partial x}\big(S_0\,\cos(k\,x - \omega\,t)\big) = -k\,S_0\,\sin(k\,x - \omega\,t).$$

The fractional change in the volume of the material element is the volume strain. The local pressure increment provides the volume stress.

$$\begin{bmatrix} \text{An EXCESS} \\ \text{A DEFICIT} \end{bmatrix} \text{ of pressure leads to } \begin{bmatrix} \text{COMPRESSION} \\ \text{RAREFACTION} \end{bmatrix} \text{ of the medium in the vicinity of the slab.}$$

As the bulk modulus is the ratio of the volume stress and strain,[3] we obtain

$$B = -\frac{\delta P}{\delta(\Delta V)/\Delta V} = \frac{-\delta P}{-k\,S_0\,\sin(k\,x - \omega\,t)} \qquad \Longrightarrow \qquad \delta P = B\,k\,S_0\,\sin(k\,x - \omega\,t).$$

A longitudinal harmonic wave disturbance characterised by the relative displacement of material elements possesses an associated harmonic pressure disturbance with the same frequency and wavelength, amplitude $P_0 = B\,k\,S_0$, and relative phase-shift $\pi/2$.

Two comments on the correspondence between the sloshing and the pressure representations of the longitudinal wave appear below.

UNITS   The product of the bulk modulus of the material comprising the wave medium, the wave number of the harmonic wave, and the sloshing amplitude has the units necessary to describe the pressure amplitude,

$$[\,\mathtt{Pa}\,]\,[\,\mathtt{m}^{-1}\,]\,[\,\mathtt{m}\,] = \mathtt{Pa}.$$

---

[3]The adiabatic bulk modulus is employed when the sloshing time scale is short.

SCALE The displacement and pressure amplitudes, S_0 and P_0, are not unrestricted. Two overlapping constraints apply. First, the displacement amplitude must be [significantly] less than the wavelength. If it were otherwise, then the coherence of the waves could hardly be maintained. Second, the pressure excursions must not be so great as to damage the medium through which the wave passes. We shall insist that $k\,S_0 < 1$, and, equivalently, $P_0 < B$, to guard against these adverse possibilities. A salutary implication of these particular constraints will become apparent at the close of this chapter, where we re-express the wave in terms of local density fluctuations.

It was realised, in Chapter 27, that the speed of the waves is set by the bulk modulus and density of the [unperturbed] medium. This speed relation may be rearranged to yield $B = \rho_0\,c^2$. Also, our original expression for wave speed, $c = \lambda\nu = \omega/k$, reveals that $k = \omega/c$. Combining these two results enables re-expression of $B\,k$ in terms of density, wave speed, and angular frequency:

$$B\,k = \rho_0\,c^2\,\frac{\omega}{c} = \rho_0\,c\,\omega\,.$$

Thus, the pressure amplitude may be recast as

$$P_0 = B\,k\,S_0 = \rho_0\,c\,\omega\,S_0\,.$$

In summary, the pressure disturbance associated with the longitudinal harmonic wave is

$$\delta P(t,x) = \rho_0\,c\,\omega\,S_0\,\sin(k\,x - \omega\,t)\,.$$

ASIDE: In anticipation of an upcoming need, it is remarked that

$$\omega\,S_0 = \frac{P_0}{\rho_0\,c}$$

follows from the expression for the pressure amplitude.

It is assumed that the to-and-fro sloshings of the thin slices of medium exhibit SHO kinematics. Therefore, the dynamics and energetics of the material elements must be presumed to have SHO form as well. Recalling the iconic energetics figure from Chapter 15, one is reminded that the total mechanical energy of the thin slice is equal to its maximum kinetic energy.[4] Therefore,

$$\Delta E = \Delta K_{\text{max}} = \frac{1}{2}\,\Delta M\,v^2_{\text{max}}\,.$$

ASIDE: We shall perform the analysis based upon Kinetic Konsiderations alone, and not revisit the Potential Pondering undertaken in Chapter 26.

The velocity of the thin slice of medium [found at x in the absence of the wave] is

$$v(t,x) = \frac{\partial S}{\partial t} = \omega\,S_0\,\sin(k\,x - \omega\,t)\,,$$

and hence

$$v_{\text{max}} = \omega\,S_0 \qquad \Longrightarrow \qquad \Delta E = \frac{1}{2}\,\Delta M\,\omega^2\,S_0^2\,.$$

---

[4]This is so when the potential energy vanishes at the equilibrium point.

This result is of precisely the same form as that obtained for transverse harmonic waves in Chapter 26. The only distinction is that there the displacement was transverse, while here it is longitudinal. While one might often have a reasonable hope of measuring a transverse amplitude, a longitudinal amplitude is much harder in practice [and in principle] to ascertain. Local pressures are more readily measured, and hence it is more efficacious to adopt the pressure point of view. Invoking $\omega\, S_0 = P_0/(\rho_0\, c)$, we may write

$$\Delta E = \frac{1}{2}\, \Delta M\, \frac{P_0^2}{\rho_0^2\, c^2}\,.$$

Recognising that the mass term is $\Delta M = \rho_0\, \Delta V = \rho_0\, \mathcal{A}\, \Delta x$ and dividing by Δx yield an expression for the average energy [lineal] density within the oscillating thin slab,

$$\frac{\Delta E}{\Delta x} = \frac{\mathcal{A}\, P_0^2}{2\, \rho_0\, c^2}\,.$$

This average is, under the assumptions made about longitudinal wave propagation and the properties of the medium, constant. The infinitesimal limit, $\Delta x \to 0$, may be taken [trivially] to obtain the local lineal energy density:

$$\frac{dE}{dx} = \frac{\mathcal{A}\, P_0^2}{2\, \rho_0\, c^2}\,.$$

As in the transverse case, the wavelength provides the natural scale for the relevant spatial extent of the wave. Thus, the energy per wavelength is

$$E_\lambda = \int_0^\lambda \frac{dE}{dx}\, dx = \frac{\mathcal{A}\, P_0^2}{2\, \rho_0\, c^2} \int_0^\lambda dx = \frac{\mathcal{A}\, P_0^2}{2\, \rho_0\, c^2}\, \lambda\,.$$

Since the travelling wave propagates a distance of one wavelength in each time interval of duration one period, the energy flux [*i.e.,* power] associated with the forward motion of the wave is

$$P_{\text{av}} = \frac{E_\lambda}{T} = E_\lambda\, \nu = \frac{\mathcal{A}\, P_0^2}{2\, \rho_0\, c^2}\, (\lambda\, \nu) = \frac{\mathcal{A}\, P_0^2}{2\, \rho_0\, c}\,.$$

The wave power [energy flux] depends on the cross-sectional area of the channel, and

$$\text{Average Power} = \left(\text{Energy Density}\right) \times \left(\text{Wave Speed}\right),$$

just as was the case for transverse waves.

INTENSITY   Energy intensity is the energy flux per unit cross-sectional area at a point in space at a particular time. The SI units for energy intensity are watts per square metre:

$$[\,I\,] = \frac{\text{W}}{\text{m}^2}\,.$$

The average energy intensity of the 3-d longitudinal plane wave analysed above is

$$I_{\text{av}} = \frac{P_{\text{av}}}{\mathcal{A}} = \frac{P_0^2}{2\,\rho_0\,c}.$$

The units of I_{av} are equivalent to watts per square metre,

$$\frac{[\,\text{Pa}^2\,]}{[\,\text{kg/m}^3\,]\,[\,\text{m/s}\,]} = \frac{\left(\frac{\text{kg}}{\text{m}\cdot\text{s}^2}\right)^2}{\frac{\text{kg}}{\text{m}^2\cdot\text{s}}} = \frac{\text{kg}}{\text{s}^3} = \frac{\text{W}}{\text{m}^2}.$$

ASIDE: A technical consideration further militates for the use of pressure amplitude rather than displacement amplitude in the analysis. The expression for the intensity depends on properties of the medium [density and wave speed] and the [measured] pressure amplitude. Had we persisted with the analysis in terms of S_0, there would have appeared a frequency dependence in the final result. Thus, comparing only the displacement amplitude of waves of differing frequency is not sufficient to determine their relative energy intensity.

The energy intensity is the normalised flux of energy distributed across the face of the waveguide/channel and as such does not have an obvious counterpart in the case of transverse waves. However, there are situations in which transverse waves occupy a 3-d medium, and in these instances it is perfectly meaningful to speak of the energy intensity through [imaginary] surfaces lying perpendicular to the direction of propagation of the wave.

ATTENUATION Attenuation is a diminution of the energy intensity of a particular wavefront in a travelling wave as it propagates.

ABSORPTIVE Absorptive attenuation occurs whenever energy is dissipated within the medium. This can occur via friction, viscosity, or drag. Ultimately, some portion of the mechanical energy borne by the wave is transferred to the medium or its surroundings and is thus lost to the wave. The decrease in the mechanical energy content of the wave is manifest as a corresponding reduction in its amplitude.

GEOMETRIC Geometric attenuation occurs in situations in which the extent of the wave front expands. An increase in the volume through which the mechanical energy of the wave is distributed generally entails a decrease in the energy density.

[Enhancement of energy density, *viz.*, focussing, occurs when the area diminishes.]

Consider an outgoing circular wave in a 2-d medium.[5] The circumference of each wave crest grows at the rate

$$\frac{ds}{dt} = 2\,\pi\,\frac{dr}{dt} = 2\,\pi\,c,$$

where c is the speed of propagation of the wave. In order to maintain constancy of the mechanical energy in the pulse, the amplitude must diminish with increasing distance from the source of the wave.

For outgoing spherical waves, the area of an expanding wave crest increases:

$$\frac{dA}{dt} = 4\,\pi\,\frac{dr^2}{dt} = 8\,\pi\,r\,\frac{dr}{dt} = 8\,\pi\,r\,c.$$

Hence, spherical waves attenuate faster than circular waves, as expected.

---

[5]We were introduced to such waves in Chapter 22.

DISPERSIVE   Waves propagating in inhomogeneous media[6] have variable speeds. Such waves experience attentuation when their speed increases.

[The converse, *viz.*, amplitude enhancement, occurs when waves slow down.]

It has been explicitly verified that, when the planar mass elements slosh back and forth in a simple harmonic manner, the local excess pressure also exhibits SHO with the same frequency and wavelength, but with a phase shift of $\pi/2$. Physically, the regions where the planes are most shifted from their equilibrium positions have little [near zero] excess pressure on account of the fact that they are moving slowly [nearly stopped], whereas the planes which are near to their equilibrium positions are rapidly diverging from or converging toward their near neighbours.

Q: In what manner does the relative density vary in the presence of the sort of longitudinal harmonic wave depicted in this chapter?

A: Let's find out!

The density of the material elements comprising the unperturbed medium is

$$\rho_0 = \frac{\Delta M}{\Delta V}.$$

The longitudinal harmonic wave shifts the boundaries of each slab, leading to a concomitant change in its volume:

$$\Delta V \longrightarrow \widetilde{\Delta V} = \Delta V + \delta(\Delta V) = \Delta V \left(1 + \frac{\delta(\Delta V)}{\Delta V} \right) = \Delta V \left(1 + \frac{\partial S}{\partial x} \right).$$

Thus, the density of each slab changes to

$$\rho_0 \longrightarrow \widetilde{\rho} = \rho_0 + \delta\rho = \rho_0 \left(1 + \frac{\delta\rho}{\rho_0} \right), \quad \text{and} \quad \widetilde{\rho} = \frac{\Delta M}{\widetilde{\Delta V}} = \rho_0 \left(\frac{1}{1 + \frac{\partial S}{\partial x}} \right).$$

Hence, the relative density disturbance,

$$\frac{\delta\rho}{\rho_0} = \frac{1}{1 + \frac{\partial S}{\partial x}} - 1 = \frac{-\frac{\partial S}{\partial x}}{1 + \frac{\partial S}{\partial x}},$$

admits the forms

$$\frac{\delta\rho}{\rho_0} = \frac{k\,S_0 \sin(k\,x - \omega\,t)}{1 - k\,S_0 \sin(k\,x - \omega\,t)} = \frac{\frac{P_0}{B} \sin(k\,x - \omega\,t)}{1 - \frac{P_0}{B} \sin(k\,x - \omega\,t)}$$

when written in terms of wave number and displacement amplitude, or excess pressure amplitude and bulk modulus, respectively. Singular behaviour is avoided by imposing the constraints, already discussed above, limiting $k\,S_0 < 1$ and $P_0 < B$.

It follows that the density disturbance is

O   OSCILLATORY—with the same frequency and wavelength as the displacement disturbance.

H   HARMONIC—the density wave function is infinitely differentiable provided that $k\,S_0 < 1$ and $P_0/B < 1$.

S̸   NOT SIMPLE—the density wave function is not a simple sinusoid when the displacement wave function is.

---

[6]Fused piecewise homogeneous media are studied in the next chapter.

Chapter 29

Inhomogeneous Media

Waves experience partial transmission and reflection upon encountering a discontinuous change in medium. In this chapter, we restrict our focus to transverse waves on a 1-d piecewise homogeneous string or longitudinal waves in a piecewise uniform waveguide.

A ▬ B   The medium consists of two or more sections of abutting homogeneous material. These are labelled by capital Roman letters, $\{A, B, \ldots\}$.

$c_A \neq c_B$ The speed of travelling waves changes abruptly at shared boundaries. Unless otherwise specified, we suppose that $c_A > c_B$.

∩ , ∼   The analyses first consider pulses, then harmonic waves.

PULSE PROPAGATION FROM A FAST MEDIUM INTO A SLOW MEDIUM

A rightmoving pulse, established in medium A, moves toward the boundary with medium B, as illustrated in Figure 29.1. After the pulse reaches the boundary, two smaller pulses are present. A rightmoving transmitted pulse [rather like a continuation of the original pulse] propagates in medium B, while a reflected pulse moves to the left in medium A.

Label	Description, direction in medium	Relative Orientation
I	Incident Pulse, rightmoving in A	⇑
T	Transmitted Pulse, rightmoving in B	⇑
R	Reflected Pulse, leftmoving in A	⇓

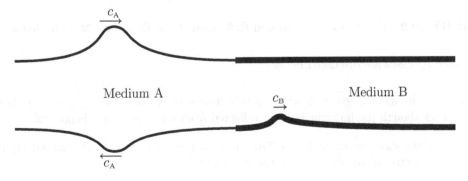

FIGURE 29.1   Pulse Transmission and Reflection at the Boundary of Two Media

There are two important limiting cases.

≫   IF the speed in the slow medium is nearly zero **AND** is much smaller than the speed in the fast medium, *i.e.*, $c_A \gg c_B \simeq 0$, **THEN** the size of the transmitted pulse is negligible, while the reflected [and inverted] pulse is nearly the same size as the incident pulse. It is as though the incident wave is reflected from a fixed end [*cf.* Chapter 22].

=   IF the wave speeds are exactly the same[1] in the two media, **THEN** there is no reflection, only transmission. It is as though there is no boundary present.

PULSE PROPAGATION FROM A SLOW MEDIUM INTO A FAST MEDIUM

A leftmoving pulse in medium B propagates toward medium A as shown in Figure 29.2. After the pulse reaches the boundary, a leftmoving transmitted pulse propagates in medium A, while a reflected pulse moves to the right in medium B.

Label	Description, direction in medium	Relative Orientation
I	Incident Pulse, leftmoving in B	⇑
T	Transmitted Pulse, leftmoving in A	⇑
R	Reflected Pulse, rightmoving in B	⇑

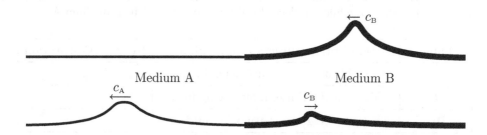

FIGURE 29.2   Pulse Transmission and Reflection at the Boundary of Two Media

Two limiting cases are discussed below.

≫   IF the wave speed in A is very much greater than that in B, $c_A \gg c_B$, **THEN** it is as though the incident pulse is reflected from a free end [*cf.* Chapter 22].

=   IF the wave speeds are equal, **THEN** only transmission occurs, with no reflection, irrespective of the direction of the incident wave.

---

[1] This is certainly the case when the purported boundary is an artifact within a homogeneous medium.

Harmonic Waves Propagating from One Medium into Another

Consider a harmonic wave propagating from medium A into medium B.

[We won't specify which one has the higher wave speed so as to do both cases, formally, together.]

Medium and Wave Interdependent Properties

Wave Speeds	Frequencies	Wave Numbers
$c_A \neq c_B$	$\omega_A = \omega = \omega_B$	$k_A \neq k_B$

The harmonic wave has the same angular frequency everywhere. Its wave number adjusts to accommodate the differing wave speeds. The harmonic wave disturbance will be described by a [piecewise] wave function with support [domain] throughout the regions of different medium. For this investigation, we will employ a complex function representation of the wave, *viz.,*

$$\psi(t, x) = A\, e^{i(k\,x \pm \omega\,t)}\,.$$

ASIDE: Euler's formula,

$$e^{\pm i\theta} \equiv \cos(\theta) \pm i\,\sin(\theta)\,,$$

for all [real-valued] angles, θ, measured in radians, provides the connection between this formal representation of the harmonic wave and the actual disturbance in the medium.

Q: Wait! This wave function is complex-valued. Can we do this?

A: In using the complex exponential form of the wave function, the instruction "Once the calculation of some physical quantity is formally completed, take only its real part" is implicitly understood.

Suppose that the scenario under consideration is a wave initially in medium A, propagating into medium B. We must accept that the steady state will consist of the composition of three distinct waves: the rightmoving incident wave in medium A, a leftmoving reflected wave in medium A, and a transmitted rightmoving wave in medium B.

Directional Wavefunctions

Incident: $\psi_i(t, x)$	Reflected: $\psi_r(t, x)$	Transmitted: $\psi_t(t, x)$
$A_i\, e^{i(k_A\,x - \omega\,t)}$	$A_r\, e^{i(-k_A\,x - \omega\,t)}$	$A_t\, e^{i(k_B\,x - \omega\,t)}$
RIGHTMOVING	LEFTMOVING	RIGHTMOVING

Grouping these disturbances by the medium in which they reside shows most clearly the physical manifestation of the harmonic wave.

In-Medium Wavefunctions

Medium A: $\psi_A(t, x)$ Medium B: $\psi_B(t, x)$

$$\psi_i(t, x) + \psi_r(t, x) = \left[A_i\, e^{i\,k_A\,x} + A_r\, e^{-i\,k_A\,x}\right] e^{-i\,\omega\,t} \qquad \psi_t(t, x) = A_t \left[e^{i\,k_B\,x}\right] e^{-i\,\omega\,t}$$

WLOG it is possible [and desirable] to choose coordinates such that the boundary between the media occurs at $x = 0$. Two conditions on the wave functions exist at the interface.

CONTINUOUS   The wave function must be continuous at $x = 0$, irrespective of the time. *I.e.,*

$$\lim_{x \to 0^-} \psi_A(t, x) = \lim_{x \to 0^+} \psi_B(t, x),$$

where

$$\lim_{x \to 0^-} \psi_A(t, x) = \psi_i(t, 0) + \psi_r(t, 0) = [A_i + A_r] e^{-i \omega t}$$

and

$$\lim_{x \to 0^+} \psi_B(t, x) = \psi_t(t, 0) = A_t e^{-i \omega t}.$$

Therefore,

$$A_i + A_r = A_t$$

is a constraint which must be satisfied by the amplitudes of the incident, reflected, and transmitted waves.

DIFFERENTIABLE   The wave function must not possess a kink at the boundary, *i.e.,* its partial derivative with respect to x must be continuous there. On account of the smoothness of the exponential, no kink will arise if the left and right derivatives agree at the boundary point. The left derivative at the boundary is the $x \to 0^-$ limit of the derivative taken within medium A. Similarly, the right derivative is the $x \to 0^+$ limit of the derivative inside medium B. That is,

$$\lim_{x \to 0^-} \frac{\partial \psi_A}{\partial x} = \lim_{x \to 0^-} \left[i\, k_A\, A_i\, e^{i\, k_A\, x} - i\, k_A\, A_r\, e^{-i\, k_A\, x} \right] e^{-i \omega t} = i\, k_A \left(A_i - A_r \right) e^{-i \omega t},$$

and

$$\lim_{x \to 0^+} \frac{\partial \psi_B}{\partial x} = \lim_{x \to 0^+} i\, k_B\, A_t\, e^{i\, k_B\, x}\, e^{-i \omega t} = i\, k_B\, A_t\, e^{-i \omega t}.$$

Thus, insisting that the wave function not be kinked at the boundary yields

$$k_A(A_i - A_r) = k_B\, A_t,$$

a constraint on the amplitudes of the constituent waves.

These two constraints can be solved to yield expressions for the reflected and transmitted amplitudes in terms of the incident amplitude:

$$A_r = \frac{k_A - k_B}{k_A + k_B}\, A_i \qquad \text{and} \qquad A_t = \frac{2\, k_A}{k_A + k_B}\, A_i.$$

The amounts of reflection and transmission are entirely determined by the wave numbers, up to a trivial scale set by the incident amplitude. Let's think about this a little more. The incident harmonic wave has a well-defined angular frequency which does not change as the medium changes. The wave speed does change, however, since it is fixed by properties of the medium.

$$c = \frac{\omega}{k} \qquad \Longrightarrow \qquad k \downarrow \text{ when } c \uparrow$$

The harmonic wave extending across the boundary has a smaller wave number [larger wavelength] in the region where the speed is larger.

In Figure 29.3, the incident wave resides in the faster medium and impinges upon the slower medium, *i.e.,* $c_A > c_B$. The reflected wave is inverted relative to the incident wave, and thus partial destructive interference ensues in medium A.

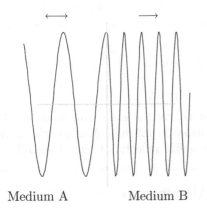
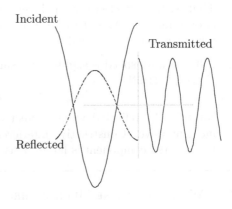

Medium A Medium B

FIGURE 29.3 Harmonic Waves in a Discontinuous Medium with $k_{\mathrm{A}} < k_{\mathrm{B}}$

In Figure 29.4, the incident wave resides in the slower medium and propagates into the faster one, *i.e.*, $c_{\mathrm{A}} < c_{\mathrm{B}}$. The reflected wave is erect, and the wave in medium A experiences partial constructive interference.

To further corroborate these results, let's examine three cases.

= IF the wave speeds are equal, **THEN** $A_r = 0$ and $A_t = 1$. Only transmission occurs; there is no reflection. [This behaviour is a welcome relief, since what we call a boundary may be merely an artifice.]

≫ IF the wave speed is much greater in medium A than in medium B [*i.e.*, $k_{\mathrm{A}} \ll k_{\mathrm{B}}$], **THEN**
$$A_r \to -1^- \quad \text{and} \quad A_t \to 0^+ \, .$$

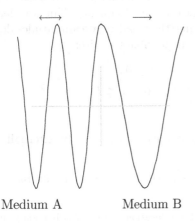
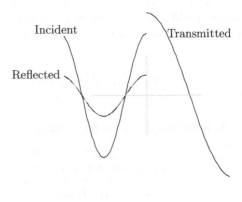

Medium A Medium B

FIGURE 29.4 Harmonic Waves in a Discontinuous Medium with $k_{\mathrm{A}} > k_{\mathrm{B}}$

That is, the reflected wave is inverted, considerable destructive interference occurs in medium A, and the transmitted component is small. This is not unlike reflection from a fixed end.

≪   IF the wave speed is significantly smaller in medium A than in medium B [*i.e.*, $k_A \gg k_B$], THEN
$$A_r \to 1^- \qquad \text{and} \qquad A_t \to 2^- \,.$$

That is, the reflected wave is erect and of nearly the same amplitude as the incident wave. Constructive interference ensues in medium A. Meanwhile, the transmitted component is large. This is not unlike reflection from a free end.

WAVES FROM ONE MEDIUM THROUGH ANOTHER AND INTO A THIRD

Three media, A, B, and C, are arranged as shown in Figure 29.5. A rightmoving harmonic wave with angular frequency ω is incident from the left.

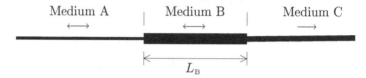

FIGURE 29.5   Three Media with Two Separate Boundary Layers

Let's think iteratively: the incident, rightmoving, wave produces a leftmoving (A) reflected wave and a transmitted (B) rightmoving wave. The transmitted wave propagates to the border with C, whereupon it too is partially reflected (B) and transmitted (C). This reflected wave impinges upon the border with medium A, whereupon there occurs transmission into (A) and re-reflection back into (B). Thus, the net wave in A is formed by the superposition of the incident wave, the first reflection, and an infinite series of waves arising from multiple scatterings between the (AB) and (BC) boundaries.[2] The net waves in B and C are superpositions of infinite numbers of waves, also.

Thinking iteratively, in this particular instance, does tend to make one's head hurt. Thinking like a fisherman[3] invites one to apply the boundary conditions directly to the net or steady state wave. Since the ultimate source of the waves is harmonic, with fixed angular frequency ω, the waves in each region are harmonic at the same frequency. The wavelengths depend on specific aspects of the respective media, while the steady state amplitudes depend on material and geometric factors. All of this makes it possible to write:[4]

$$\psi_A = \left[A_i e^{i k_A x} + A_r e^{-i k_A x} \right] e^{-i\omega t}, \quad x \le 0,$$
$$\psi_B = \left[B_R e^{i k_B x} + B_L e^{-i k_B x} \right] e^{-i\omega t}, \quad 0 \le x \le L_B,$$
$$\psi_C = C_t e^{i(k_A x - \omega t)}, \quad x \ge L_B,$$

where A_i is the amplitude of the rightmoving incident wave and is usually externally prescribed.

---

[2] Energy conservation suggests that the amplitudes of these waves fall off sufficiently rapidly that the net amplitude of their linear superposition is finite everywhere.

[3] "Net" behaviour is of paramount importance to fishermen.

[4] Here the origin of coordinates is chosen to coincide with the location of the (AB) boundary, just as was done in the previous [two-medium] analysis. With this choice, the (BC) boundary is at $x = L_B$. An alternative, more symmetric, approach is to set the origin of coordinates at the centre of the middle section. In this case the (AB) and (BC) boundaries occur at $\mp L_B/2$ respectively.

[Remember that the observed amplitude includes superposition effects.]

All five amplitude factors are enumerated in the table below.

A_i	Rightmoving in A, incident
A_r	Leftmoving in A, overall reflected
B_R	Rightmoving in B, self-consistent sum
B_L	Leftmoving in B, self-consistent sum
C_t	Rightmoving in C, overall transmitted

The four unknown amplitude factors may be solved for (in terms of A_i) by demanding continuity of the waveform and its derivative at each of the two material boundaries.

> ASIDE: In other circumstances, one might measure the transmitted amplitude and then infer the amplitude of the incident wave.

ADDENDUM: Waves in the Deep and at the Beach

Water waves are complicated [they are neither purely transverse, nor purely longitudinal] and dispersive. However, it is chiefly from our aquatic experiences that we acquire experiential knowledge of wave behaviours.

An accurate model for the dispersion of water waves was finally worked out in the mid-nineteenth century by George Airy,[5] after many false starts by other scholars. In Airy's formulation, the wave speed satisfies

$$c^2 = \frac{g}{k} \tanh(h\,k) \,,$$

where g is the local acceleration due to gravity, k is the wave number, $k = 2\pi/\lambda$, and h is the depth of the water beneath the point where the wave is propagating. In relatively deep water, *i.e.*, $h > \lambda/2$, the hyperbolic tangent term approaches 1, and the speed of the waves is very nearly

$$c \simeq \sqrt{\frac{g}{k}} = \sqrt{\frac{g\,\lambda}{2\pi}} \,.$$

In shallow water, $h/\lambda \to 0$, and the hyperbolic tangent term is well-approximated by its series expansion:

$$\tanh(\epsilon) \sim \epsilon - \frac{1}{3}\,\epsilon^3 + \frac{2}{15}\,\epsilon^5 - \cdots .$$

Taking just the linear term to get the dominant behaviour,

$$c^2 \to \frac{g}{k}\,(h\,k) = g\,h \to 0 \,.$$

Thus, according to this model, ocean waves slow down as they approach the beach.

---

[5]Airy (1801–1892) was an English mathematical physicist and astronomer. He briefly held the Lucasian Professorship at Cambridge (previously occupied by Newton and other luminaries). As Astronomer Royal during an exciting age of discovery and tremendous gains in precision, Airy helped to establish the convention in which the **Prime Meridian**, passing through Greenwich (England), serves as the global reference for measurements of longitude (and thus also local time).

Now, let's consider how the dispersive model comports with our observations of how water waves pile up on the shore.

Ξ   The wavelength diminishes.

The frequency remains constant while the wave speed is reduced. Since $c = \lambda \nu$, it must be that the wavelength shortens.

Ξ   The waves turn toward the beach.

IF the waves impinge obliquely, THEN for a specified crest the edge near the beach is slowed, while the parts farther out remain moving at the deep-water speed. Thus, the faraway portion gets ahead of the nearer parts, making it appear that the wave is turning toward the beach.   [This is **refraction**, *cf.* Chapter 35.]

Ξ   The seas get choppier.

The wavelength shrinks in proportion to the wave speed, since the frequency is fixed. This alone would make the waves steeper, were the amplitude to remain constant. However, the amplitude of the waves increases with the diminished wave speed, further increasing the choppiness.

> ASIDE: When the wavelength decreases, the energy borne by the wave becomes more concentrated. According to the energetic analyses of Chapters 26 and 28, the lineal energy density of a travelling wave is proportional to the square of its amplitude. Hence, in order that energy might be conserved, the amplitude grows.

Tsunami

Consider an offshore earthquake causing a portion of the seabed that is 20 km long, lying beneath water which is 3 km deep, to heave 1 m upward. A surface wave with amplitude $A_0 \sim 1$ m will result from the upthrust. The dominant Fourier mode of this disturbance will have wavelength equal to twice the length of the moving portion of seabed, *i.e.*, $\lambda_0 \sim 4 \times 10^4$ m. According to Airy's dispersion relation [with $g \sim 10 \, \mathrm{m/s^2}$], the wave speed is roughly estimated to be

$$c = \sqrt{\frac{g \lambda_0}{2\pi} \tanh\left(\frac{2\pi h}{\lambda_0}\right)} \sim \sqrt{\frac{2 \times 10^5}{\pi} \tanh\left(\frac{3\pi}{20}\right)} \sim 167 \, \mathrm{m/s} = 600 \, \mathrm{km/h} \, .$$

> ASIDE: Such enormous propagation speeds have led to many past instances in which a tsunami struck faraway coasts before word of the earthquake (and warning of tsunami danger) could be broadcast. Presently, many nations' geophysical observatories make available real-time seismic data to other agencies and non-governmental organisations. When these data are combined with telemetric and wave pattern information obtained from networks of ocean buoys, more timely tsunami-risk assessments are possible.

As the tsunami nears the shore, its speed decreases until, crudely speaking, it is reduced to about 1 m/s. Recalling that the energy content of a harmonic wave scales with the product of the wave speed and with the square of the amplitude,

$$c A^2 \sim 167 \times (1)^2 \sim 1 \times (13)^2 \, ,$$

and thus conservation of energy impels the fast-moving small-amplitude deep-water ocean wave to become a slow-moving towering wall of water, ~ 13 m high, as it reaches the shore.

Chapter 30

Doppler Shifts

In all of our discussions of waves thus far, we have glossed over the production of waves [emission] and their subsequent detection [reception]. Furthermore, in the dual representations of waves provided by snapshots at instants in time and observers attentive to particular points in space, there was an unrealistic element of omniscience.[1] Here we investigate situations in which emitters or receivers are in motion with respect to the medium through which the wave propagates.

Both Emitter and Observer at Rest wrt Medium

An emitter, E, sends a regular pulse train of narrow disturbances outward from a point lying in a homogeneous 2-d medium at a time which he and the observer, O, agree to call $t = 0$. By prior measurement, E and O have ascertained that they are at mutual rest a distance D apart.

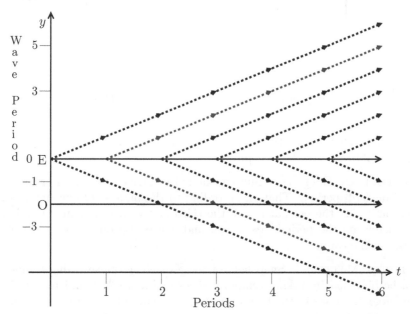

FIGURE 30.1   A Pulse Train with Emitter and Observer at Rest

Already there exist the complicating factors of units for time and space[2] and the particular value of the wave speed in terms of these units. This *Gordian Knot* is cleaved by adopting

---

[1]Not quite omniscience *per se*, but data distilled from a myriad of observers.
[2]This is written ironically, and yet there is no denying the point which follows.

the convention of measuring time, t, in periods: $1\ \mathcal{T} =$ the time interval between pulses, and employing "wave-periods" for determinations of distance,[3] *i.e.*,

$$1\ \text{w-p} =\ \text{the distance travelled by the wave in one period}$$
$$=\ \text{the spatial distance between pulses} = \lambda.$$

Thus, the speed of the wave is $c \equiv 1$ [wave-period per period].

The emitter, observer, and pulse train are illustrated in the spacetime diagram displayed in Figure 30.1. That the actors are at rest is indicated by their horizontal worldlines. The emitter lies at the origin of spatial coordinates, $(x, y) = (0, 0)$, while in this instance the observer happens to be 2 w-p away on the negative side, *i.e.*, at $(0, -2)$.

Below, we elaborate upon salient features of Figures 30.1 and 30.2.

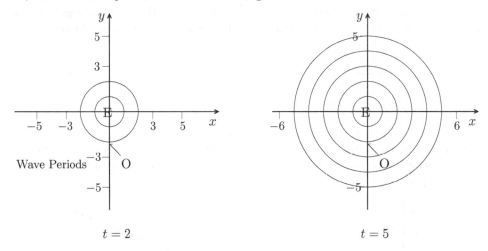

FIGURE 30.2   Two Snapshot Views of the Outgoing Two Dimensional Circular Pulses

⊙   RETARDATION EFFECTS   While E commences emission at $t = 0$, O must wait until $t = 2$ to receive the first pulse.

⊙   FREQUENCY EFFECTS   Subsequent pulses arrive at time intervals equal to one [standard] period. That is, according to the figures, O encounters the first pulse at $t = 2$, the next at $t = 3$, the third at $t = 4$, *etc.* Thus O and E agree completely on the period [frequency], and hence also on the wavelength of the pulses in the train.

⊙   TIMELY MATTERS   Snapshots of the [expanding] pulse train at times $t > 0$ can be obtained by taking temporal slices of Figure 30.1 and recognising that the outgoing pulses in the homogeneous 2-d medium have circular symmetry. In Figure 30.2, this is done for $t = 2$, the instant at which the leading pulse reaches O, and for $t = 5$.

⊙   GEOMETRY   The pulses form concentric circles about the stationary emitter. At $t = 2$, the leading edge of the first pulse has just reached O, whereas at $t = 5$, three of the five emitted pulses have already passed by O and the fourth is impinging upon him.

---

[3] A wave-period is akin to the "lightyear" favoured by many astronomers.

That was the easy case, because the emitter and receiver of the pulses were both at rest with respect to the medium. When one, the other, or both are in motion, **Doppler**[4] **shifting** occurs.

DOPPLER SHIFTS   Doppler shifts are correlated changes in the apparent frequency and wavelength of travelling waves owing to the motion of the source, and/or the receiver, with respect to the medium within which the wave propagates.

EMITTER IN MOTION AND OBSERVER AT REST WRT MEDIUM

Here we posit that the emitter, E, moves at constant speed[5] and thus $E_t = E(t) = E_0 - v\,t$ in the direction of decreasing y, as indicated by his constant-slope worldline in Figure 30.3. For purposes of dramatic illustration, the speed of the emitter is taken to be $v = \frac{1}{2}$, $i.e.$, E moves a distance of one wave-period in a time interval of two periods.

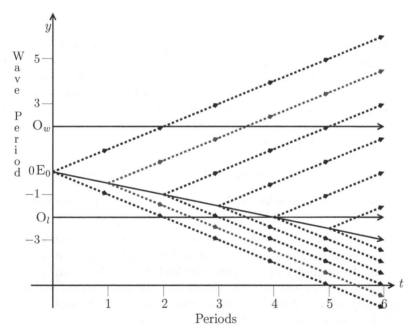

FIGURE 30.3   A Pulse Train with Emitter in Motion and Observers at Rest

Two observers, LEEWARD and WINDWARD, O_l and O_w, receive the pulses.

> ASIDE: The observers could have been consistently labelled LEFT and RIGHT or UP and DOWN or TOP and BOTTOM or STRANGE and CHARM or what have you. LEEWARD, in the direction of the prevailing wind, and WINDWARD, opposed to the wind, has the advantage of distinguishing the receivers by the direction of motion of the source.

---

[4]Christian Doppler (1803-1853) was an Austrian mathematical physicist.

[5]The speed with which the emitter moves is typically much less than the wave speed of the pulse train. The curious situation which ensues when the emitter moves at the wave speed is briefly discussed in the Addendum to this chapter.

Both observers are at rest with respect to the medium, and consequently their worldlines in Figure 30.3 are horizontal and parallel.

To facilitate comparison with the previous case, E commences emission at $t = 0$, when he is exactly 2 w-p away from both his LEEWARD and WINDWARD observer friends. Finally, it must be noted that the worldline trajectories of E and O_l cross over at time $t = 4$. After this instant, both observers are WINDWARD of the wave source.

⊙   RETARDATION EFFECTS

    l   The LEEWARD observer must wait until $t = 2$ to receive the first pulse. Even though E is in forward motion toward O_l when the first pulse is emitted, the wave does not reach him any sooner than it would have if E were at rest.

    w   The WINDWARD observer likewise receives the first pulse at time $t = 2$, despite the motion of the emitter.

> **The speed of the wave with respect to the medium is not affected by the motion of the source.**

⊙   FREQUENCY EFFECTS

    l   The first batch of pulses arrive in compressed temporal intervals as seen in Figure 30.3. Careful inspection reveals that O_l determines the period to be $1/2$ and the frequency to be 2, since two pulses pass by in each unit time interval [for $2 \leq t \leq 4$, *i.e., post* retardation effect and *ante* crossover]. The wavelength ascribed to the pulse train by O_l is shorter than that established by E. This must be so for the two friends to agree on the speed of the wave with respect to the medium:

$$\nu_{\text{emitter}}\,\lambda_{\text{emitter}} = c = \nu_{\text{observer}}\,\lambda_{\text{observer}}\,.$$

As the clock and coordinate choices force $c = 1$, O_l infers a wavelength of $1/2$ for the train of pulses that he initially receives.

    w   The pulses which impinge on the windward observer occur at regular intervals for as long as the pulse train and the uniform motion of the emitter persist. Examination of the figure reveals that two pulses arrive in the period from $t = 2 \rightarrow 5$, and thus O_w determines the pulse period and associated frequency to be $3/2$ and $2/3$, respectively. Consistent determination of wave speed requires that the wavelength be $3/2$.

    X   After crossover, O_l's observations conform to those of O_w.

⊙   TIMELY MATTERS   Snapshots of the expanding pulse train from the moving source in Figure 30.3 are shown below. The initial, $t = 0$, position of the emitter at the origin of coordinates is noted in both figures. The first pulse maintains circular symmetry about the origin as time advances.

t = 2   At $t = 2$, the leading edge of the first pulse has just reached the two observers. The second pulse is centred on the point $E_1 = (0, -1/2)$, at which the source was located at the instant the pulse was emitted. At the instant $t = 2$, captured in the first figure, the emitter, located at $(0, -1)$ is just about to release the third pulse.

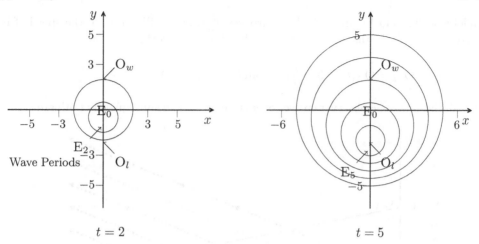

FIGURE 30.4 Two Snapshots of Outgoing 2-D Circular Pulses: Moving Source

t = 5   At $t = 5$, the leading edge of the first pulse lies a distance of 5 from the origin. The subsequent pulses are increasingly shifted down the y-axis. The fifth pulse is centred on the leeward observer, because crossover occurred just as it was emitted [at $t = 4$]. At the instant shown in the figure, the sixth pulse is being generated at the point tagged E_5.

⊙   GEOMETRY   As a result of the motion of the emitter, later concentric pulses are offset from the centres of earlier pulses. At time $t = 2$, the leading edge of the first pulse simultaneously reaches the two observers. At $t = 5$, all five of the previously-emitted pulses have passed by O_l [formally including the one generated at LEEWARD's position], while O_w has only encountered three of them.

⊙   GENERALISED   Extending the above analysis to the general case, in which the speed of the emitter is $v_E < c$, the apparent frequencies and wavelengths perceived by windward and leeward observers are determined to be

$$\nu_w = \frac{1}{1 + v_E/c} = \frac{c}{c + v_E} \nu_E \qquad \text{and} \qquad \nu_l = \frac{1}{1 - v_E/c} = \frac{c}{c - v_E} \nu_E$$
$$\lambda_w = \left(1 + \frac{v_E}{c}\right) = \frac{c + v_E}{c} \lambda_E \qquad\qquad \lambda_l = \left(1 - \frac{v_E}{c}\right) = \frac{c - v_E}{c} \lambda_E .$$

A very common instance of the source-moving Doppler effect occurs when an emergency vehicle passes by with its siren blaring. As it approaches, its apparent pitch is higher, and as it recedes, its pitch seems lower.

EMITTER AT REST AND OBSERVERS IN MOTION WRT MEDIUM

We shall again endeavour to facilitate comparison with the previous cases. The stationary source, E, commences emission at $t = 0$, at which time he lies 1 w-p above RECEDING, O_R, and 3 w-p below ADVANCING, O_A. The observers move with the same velocity and thus have parallel worldlines in Figure 30.5. For purposes of dramatic illustration, the speed of both observers is $v = \frac{1}{2}$. In this particular case, the observers' worldlines are

$$O_{At} = O_A(t) = 3 - \frac{1}{2}t \quad \text{and} \quad O_{Rt} = O_R(t) = -1 - \frac{1}{2}t.$$

At $t = 6$, the ADVANCING observer crosses over the emitter and subsequently both are RECEDING.

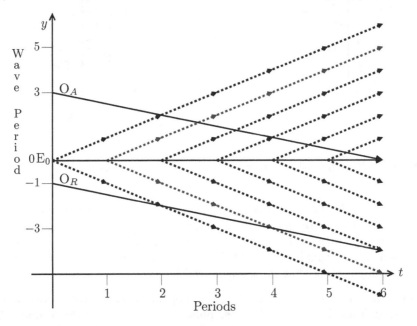

FIGURE 30.5   A Pulse Train with Emitter at Rest and Observers in Motion

⊙   RETARDATION EFFECTS

A,R   The ADVANCING and RECEDING observers each receive their first pulse at $t = 2$.

The speed of the wave with respect to the medium is not affected by the motion of the observer.

⊙   FREQUENCY EFFECTS

A   The ADVANCING observer sees compressed pulses throughout the time interval from first reception to crossover. Careful inspection of Figure 30.5 reveals that O_A receives three pulses in a time equal to two [emitted, standard] periods, and thus measures the period, frequency, and wavelength to be: 2/3, 3/2, and 2/3, respectively.

R   The pulses which impinge on the RECEDING observer occur at regular intervals for as long as the pulse train and the uniform motion of the observer persist. Examination reveals that one pulse arrives in each two-period time interval, and thus O_R determines the pulse period, frequency, and wavelength to be 2, 1/2, and 2, respectively.

X   After crossover, O_A's observations conform to those of O_R.

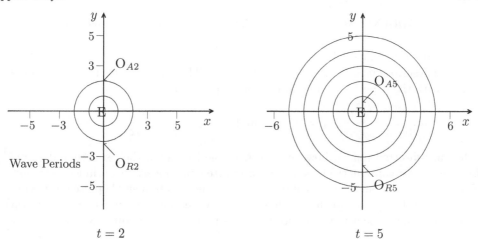

FIGURE 30.6 Two Snapshots of Outgoing 2-D Circular Pulses: Moving Receivers

⊙   TIMELY MATTERS   Snapshots of the expanding pulse train from the stationary source in the presence of the moving observers are illustrated in Figure 30.6.

t = 2   At $t = 2$, the leading edge of the first pulse reaches the two observers.

t = 5   At $t = 5$, the ADVANCING observer has encountered the five previously emitted pulses and is soon to meet up with the sixth. Meanwhile, RECEDING has had only two pulses overtake him; pulse three will reach him at $t = 6$.

⊙   GEOMETRY   The pulses form concentric circles about the origin.

⊙   GENERALISED   Extending the above analysis to the general case, in which the speeds of the advancing and receding observers are $v_{A,R} < c$, the wave properties perceived by the observers are determined to be

$$\nu_A = \left(1 + \frac{v_A}{c}\right) = \frac{c + v_A}{c}\,\nu_E \qquad \text{and} \qquad \nu_R = \left(1 - \frac{v_R}{c}\right) = \frac{c - v_R}{c}\,\nu_E$$

$$\lambda_A = \frac{1}{1 + v_A/c} = \frac{c}{c + v_A}\,\lambda_E \qquad\qquad\qquad \lambda_R = \frac{1}{1 - v_R/c} = \frac{c}{c - v_R}\,\lambda_E \;.$$

When one drives by pounding jackhammers while passing through a road construction zone one can often discern a frequency shift.

—————————

BOTH EMITTER AND OBSERVER IN MOTION WRT MEDIUM

Composition of the results derived above leads to

$$\nu_O = \frac{c \pm v_O}{c \mp v_E}\,\nu_E \qquad \text{and} \qquad \lambda_O = \frac{c \mp v_E}{c \pm v_O}\,\lambda_E \;,$$

where the signs must be chosen according to the directions (towards/away) of the motions of the emitter (E) and observer (O).

ADDENDUM: Emitter in Motion at Wave Speed WRT Medium

In Figure 30.7 the emitter, E, moving with speed c, follows a worldline which admits the parametric description $E_t = E(t) = E_0 - ct = 3 - t$. The pulses emitted rearward propagate normally with the expected source-in-motion Doppler shift:

$$\nu_w\Big|_{v_E = c} = \frac{c}{c+c}\,\nu_E = \frac{\nu_E}{2} \quad \text{and} \quad \lambda_w\Big|_{v_E = c} = \frac{c+c}{c}\,\lambda_E = 2\,\lambda_E .$$

The forward-directed pulses, however, are not able to outrun their source and subsequent pulses superpose constructively with those emitted previously. Continued emission augments the amplitude and energy content of what amounts to a single pulse co-moving with the source. Such a pulse is called a shock wave. When an aircraft moves at [or surpasses] the speed of sound, the attendant shock wave is perceived as a **sonic boom**.

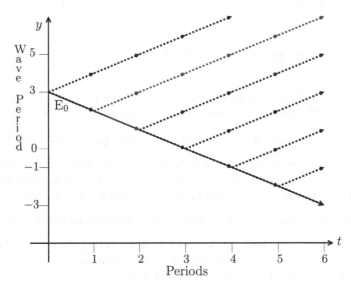

FIGURE 30.7   A Pulse Train with Emitter in Motion at the Wave Speed

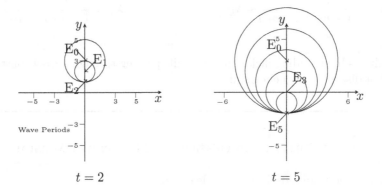

FIGURE 30.8   Two Snapshots of Outgoing 2-D Circular Pulses: Sonic Boom

Chapter 31

Huygens' Principle, Interference, and Diffraction

Christiaan Huygens (1629–1695) was yet another brilliant and prolific[1] contemporary of Newton. Huygens studied waves and optics and made numerous discoveries, including those which bear his name.

HUYGENS–FRESNEL PRINCIPLE   The Huygens–Fresnel[2] Principle[3] states that at any instant, every point on a wavefront belonging to a wave with frequency ν acts as a point source of spherical wavelets [circular in 2-d] with the same frequency. At a later time, a new wavefront, at a distance $d = c\,\Delta t$ ahead[4] of the old one, will have formed from the superposition of the wavelets.

The essential idea is that wave propagation, a non-local effect, arises through the coherent local superposition of local processes.

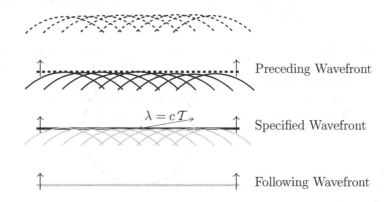

FIGURE 31.1   Each Point on Plane Wavefront Acts as a Source of Wavelets

A powerful example of Huygens' Principle is evident in the propagation of monochromatic[5] plane waves through a [2-d] homogeneous medium, as illustrated in Figure 31.1. In the figure, a particular wavefront is denoted by a dark solid line. The dotted black and solid

---

[1]One of Huygens' achievements was the design of escapement mechanisms for pendular clocks, so as to eliminate the amplitude dependence in the period noted in Chapter 13. This invention had profound effects on **horology**, the science dedicated to the mensuration of time, and led to advances in navigation and industry. Some of his other accomplishments were in diverse fields of acoustics, observational astronomy, calculus, and probability theory.

[2]Augustin-Jean Fresnel (1788–1827) refined Huygens' ideas on the propagation of waves.

[3]Recall that a principle is an axiomatic statement about a property of nature.

[4]It would appear that symmetry demands that a new wave front be formed behind the old one also. This inconvenient result is avoided by assumption that the intensity of the spherical waves is non-zero in the forward direction only, or it is ignored.

[5]Literally, this means "single colour." Our usage specifies a harmonic wave with a particular frequency and wavelength.

grey lines represent the wavefronts which are directly ahead of and behind the one which is singled out. Wavelets emitted by a sample of points on each respective wavefront, having propagated for a time interval equal in duration to one period of the original wave, are shown as segments of circular arc.

The wavelets from the following wavefront have superposed to *regenerate* a new wavefront at the specified location. The preceding wavefront is *reconstituted* by the superposition of wavelets with sources on the specified front. This happens throughout the region of space occupied by the plane wave. In a sense, via the Huygens–Fresnel principle, the wave *begets* a phase-advanced copy of itself.

One may have a qualm or two about the mathematical convergence properties of the sums of interfering wavelets spawned according to the Huygens–Fresnel Principle.

$\overset{?}{\infty}$ Here is a line of false reasoning involving a single wavefront:

> IF there are an infinite number of Huygens sources on each wavefront,
> AND each source emits waves (forward and/or backward), THEN surely
> these waves must yield an infinite superposition somewhere.

We shall [mostly[6]] ignore this conundrum by issuing the disclaimer that our usage of Huygens' Principle is intended to be heuristic.

> ASIDE: Attention to limits reveals that the Huygens superposition converges.
> Surprisingly, the situation in 2- or 3-d is less pathological than in 1-d owing
> to the effects of geometric attenuation.

$\overset{?}{\infty}$ Having an infinite series of past wavefronts reproducing themselves into the future is a situation ripe for blow-up. Here, the convergence problems are solved either by insisting that each present wavefront produces only its preceding wavefront one period later, or by accepting that the putative wave train extending from the infinite past to the present is a mathematical abstraction wholly unable to be physically realised.

WAVE INTERFERENCE   Interference is a form of correlated multi-wave linear superposition.

PLANE WAVES IMPINGING UPON TWO THIN SLITS: QUALITATIVE

Plane waves with angular frequency ω and wavelength λ, incident from the left, encounter an opaque partial barrier. The waves may pass only through two thin slits. Lying at distance L from the barrier is an absorbing screen[7] on which fixed point observations of the intensity of the wave disturbance are made. In Figures 31.2 and 31.3, the barrier, slits, screen, and wavefronts extend perpendicularly out of the page. The region to the left of the barrier contains rightmoving incident and leftmoving reflected waves. Here, we are concerned with what happens to the right of the barrier.

---

[6]Careful analysis of diffraction, in Chapter 32, requires superposition of an infinite set of Huygens emitters.

[7]It is assumed that waves impinging on the screen do not reflect. Otherwise, the reflected waves would [partially] re-reflect from the partial barrier and impinge again on the screen (with some additional phase shift). Multiple scatterings would disrupt the simple two-slit effect which we are attempting to discern.

Perhaps inspired by geometric optics[8] [light moves in straight lines, shadows are sharp], one might expect to see on the screen two bright images of the slits with darkness everywhere else. This naive view is not unreasonable in situations in which the slit widths and separations are large [on the scale set by the wavelength].

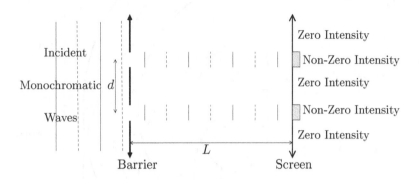

FIGURE 31.2   Two Slit [Non-]Interference: Naive Model

The naive model is fallacious.

To formulate a better model, we'll invoke Huygens' Principle. To wit, each slit acts as a line source[9] for waves of frequency ω and wavelength λ. These waves superpose and interfere, as illustrated in Figure 31.3.

Salient features of this interference pattern include the following.

1. CENTRAL MAXIMUM

   A zone of maximal intensity occurs on the screen at the point directly across from the midpoint of the barrier between the slits. The paths from the two slits to dead centre are precisely the same length. Thus, the wavelets from each slit arrive with perfect phase alignment and interfere constructively. Ironically, this intensity maximum occurs in what would be, according to the naive model, a region of shadow lying between the pair of slit images.

2. FADE TO BLACK

   Moving away from the central maximum, the path length difference increases, and thus so too does the phase difference between the waves arriving from the slits. Just this situation was encountered in the example of the superposition of two rightmoving harmonic waves with identical frequency (wavelength) and monotonically increasing relative phase found in Chapter 24. What begins as incomplete constructive interference grows[10] into increasingly distructive interference.

---

[8]Coming soon in Chapters 33–36.

[9]These sources are lineal because the slits extend perpendicularly out of the page. This geometry makes for easier visualisation on the printed page. It is possible to generalise this analysis to study the interference pattern produced by small circular holes with centres separated by d.

[10]In the present context, a snarky person might rather say *diminishes*.

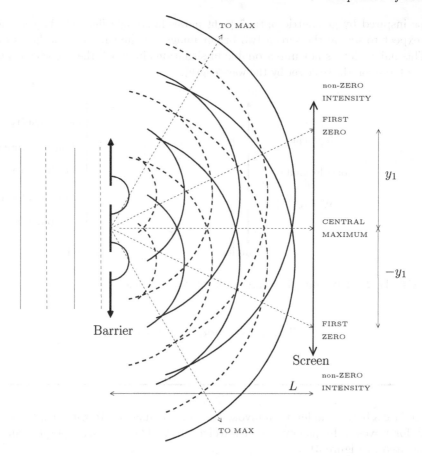

FIGURE 31.3 Two Slit Interference: According to Huygens' Principle

3. FIRST MINIMA

    There are points [actually lines extending out of the page] shown on the screen in Figure 31.3 where the net intensity is zero owing to destructive interference of the waves from the two slits. The first pair lie at distance y_1 from the central maximum when the phase difference is π, which happens when the path lengths differ by one-half wavelength.

4. SECONDARY MAXIMA/MINIMA

    Proceeding along the screen away from the centre beyond the first minimum, the path difference increases and so too does the phase shift. The destructive interference diminishes and eventually constructive interference occurs once again. As the phase shift approaches 2π, waves from the two slits enjoy complete constructive interference at the first of the secondary maxima. Further increasing the path difference leads to diminished constructive interference, then partial destructive, and complete destructive at the first pair of secondary minima.

5. TERTIARY and HIGHER MAXIMA/MINIMA

    This pattern of alternating maxima/minima continues *ad infinitum* ... or until a cutoff is reached beyond which the intensity is too small to measure.

ASIDE: The intensity decreases owing to the expansion of the wavefronts with distance from the slits, and their oblique incidence on the flat screen.

PLANE WAVES IMPINGING UPON TWO THIN SLITS: QUANTITATIVE

Plane waves with wavelength λ impinge at normal incidence upon two thin uniform parallel slits[11] in an otherwise opaque barrier. The distance between the slits, d, is taken to be much larger than their width, w. The most dramatic effects are achieved when w is comparable to the wavelength of the incident waves, ensuring that the phase differences between Huygens wavelets emitted from the midpoint of the slit and those from other points elsewhere in the same slit are relatively small. Thus, the wave emerging from a single narrow slit may be modelled as a coherent, single-phase, forward-propagating, [half-] circular wave centred on the middle of the slit.

Each slit acts this way and thus, in the region past the barrier, the resultant wave is the linear superposition of these two travelling waves, exactly as illustrated in the preceding qualitative analysis. Since the incident plane waves have constant phase fronts, the waves from the centres of the two slits possess identical phases.

Further geometric assumptions are that the screen/detector on which the interference pattern is observed lies parallel to the barrier at a distance which is much greater than the distance between the slits, $L \gg d$. The point on the screen directly opposite the midpoint between the slits is equidistant from the centres of both slits and hence, recalling the superposition analysis of equal amplitude harmonic waves undertaken in Chapter 24, the superposition gives rise to [complete] constructive interference.

The separation of the slit and screen distance scales, $L \gg d$, justifies a set of further approximations. Consider a point, $\mathcal{P}$, on the screen at distance y from the middle of the central maximum. The path lengths from the upper and lower slits to $\mathcal{P}$ are readily determined to be

$$L_u = \sqrt{L^2 + \left(y - \frac{d}{2}\right)^2} \qquad \text{and} \qquad L_l = \sqrt{L^2 + \left(y + \frac{d}{2}\right)^2} \,.$$

For $L \gg y \gg d$, the approximate difference between these lengths may be determined via binomial expansion:

$$\Delta L = L_l - L_u = L \left[\sqrt{1 + \left(\frac{y - d/2}{L}\right)^2} - \sqrt{1 + \left(\frac{y + d/2}{L}\right)^2} \right]$$

$$\simeq L \left[1 + \frac{1}{2}\left(\frac{y + d/2}{L}\right)^2 - 1 - \frac{1}{2}\left(\frac{y - d/2}{L}\right)^2 \right] = \frac{1}{2L}\left[2\,y\,d \right] = \frac{y}{L}\,d \,.$$

In the approximation, higher-order terms in the binomial expansion were dropped. To reach the penultimate expression, the terms in the square bracket were identified as the difference of squares and simplified. In the final equality, the factor $\frac{y}{L}$ admits geometrical interpretation as the **tangent** of the angle lying between the normal and the ray directed

---

[11]These assumptions effectively "flatten" the resulting interference pattern into 1-d. One can obtain more complicated interference effects with circular holes.

from the midpoint of the slits to $\mathcal{P}$. Under the assumptions made above, it follows that the angle is small, and hence the approximation $\tan(\theta_{\mathcal{P}}) \simeq \sin(\theta_{\mathcal{P}})$ is valid. Thus, the path difference is well-approximated by the simplified expression

$$\Delta L \simeq d \, \sin(\theta_{\mathcal{P}})\,.$$

In Figure 31.4, the paths from the slits to the point on the screen are drawn as though they were parallel.[12] By similar triangles, it is evident that the difference in length between the two paths is (approximately) $d \, \sin(\theta_{\mathcal{P}})$.

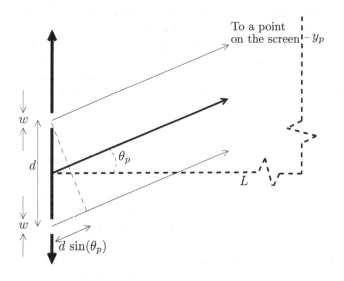

FIGURE 31.4   Approximate Analysis of Two-Slit Interference

The difference in path lengths gives rise to a phase difference in the waves from the two slits as they combine at $\mathcal{P}$. As we move up the screen from the centre, the phase shift increases from zero, exactly as was inferred qualitatively. Being able to determine HOW the path lengths differ equips us to ascertain the precise form of relations governing the occurrences of interference minima and maxima, in terms of $\theta_{\mathcal{P}}$, where

$$\tan\left(\theta_{\mathcal{P}}\right) = \frac{y_{\mathcal{P}}}{L}\,.$$

MIN   IF the ray is directed to the nth minimum, THEN

$$d \, \sin(\theta_n) = \left(n - \tfrac{1}{2}\right)\lambda\,, \quad \text{for } n = 1, 2, 3, \ldots\,.$$

MAX   IF the ray is directed to the mth maximum, where, by convention, the central maximum is labelled the 0th, THEN

$$d \, \sin(\theta_m) = m \, \lambda\,, \quad \text{for } m = 0, 1, 2, \ldots\,.$$

---

[12]This is, of course, not possible, since truly parallel lines do not converge. However, given the scales, these lines are nearly parallel.

ASIDE: Here we have elaborated upon the conventions introduced in Chapter 25, where

$$n \in \{1, 2, 3, \ldots\} \quad \text{for harmonics and minima,}$$
$$m \in \{0, 1, 2, \ldots\} \quad \text{for overtones and maxima.}$$

EXAMPLE [*Employing Interference to Determine Wavelength*]

Waves pass through two very thin slits with separation $d = 100\,\mu\text{m} = 0.1\,\text{mm}$, and fall upon a screen lying $L = 1.6\,\text{m}$ away. For a small neighbourhood on the screen, directly across from the slits,[13] the interference pattern and the variable local intensity of the waves are shown in Figure 31.5. [The scale on the left edge of the pattern measures distance in cm from the central point.]

Q: From the interference pattern and the geometry of the slits and screen, might one infer the wavelength of the incident waves?

A: Certainly! Also, as the wavelength can be inferred from the locations of either the minima or maxima, we shall compute both and compare the results.

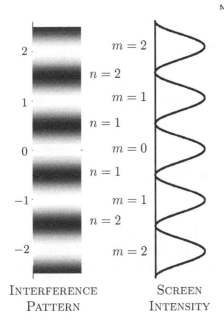

MIN The nth intensity troughs occur $n - \frac{1}{2}$ cm from the centre of the pattern. Hence, for L expressed in cm,

$$d\,\frac{n - \frac{1}{2}}{L} = \left(n - \tfrac{1}{2}\right)\lambda\,,$$

implying that

$$\lambda = \frac{d}{L}\,,$$

where d bears units, while L is the pure number expressing the distance from the slits to the screen in cm. Here,

$$\lambda = \frac{0.1\,\text{mm}}{160} = 6.25 \times 10^{-7}\,\text{m}$$
$$= 625\,\text{nm}\,.$$

INTERFERENCE PATTERN   SCREEN INTENSITY

FIGURE 31.5 Employing Interference to Determine Wavelength

MAX The mth non-central intensity peak is at m cm from the central maximum. Hence, when L is the slit–screen distance in centimetres,

$$d\,\frac{m}{L} = m\,\lambda\,, \quad \text{and thus} \quad \lambda = \frac{d}{L}\,,$$

---

[13]Under these conditions, the amplitudes of the interfering waves are taken to be equal despite the fact that, since the sources are distinct and separated, they experience different amounts of geometrical attenuation.

leading to the conclusion that the waves in this instance have wavelength $\lambda = 625\,\mathrm{nm}$, in exact agreement with the result from analysis of the minima.

DIFFRACTION   Diffraction is a remarkable instance of self-interference. Suppose that waves are normally incident from the left upon the opaque barrier with a slit of width w shown in Figure 31.6. Each and every point in the gap acts as a source of Huygens wavelets propagating to the right. In contrast to the situation prevailing in Figure 31.1, where the incident plane wave begets another complete copy of itself, the narrow slit has the effect of cutting off what would have been the contributions from adjacent [neighbouring] wavelets. The disruption that this produces is more pronounced near the edges than near the centre.

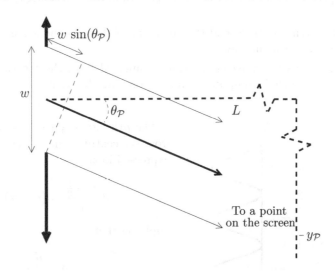

FIGURE 31.6   Approximate Analysis of Single Slit Diffraction

Consider in particular, the superposition of wavefronts appearing at the point $\mathcal{P}$, on an absorbing screen[14] far from the slit. Provided that the screen distance, L, is sufficiently large, $L \gg w$, the rays from each end of the slit to $\mathcal{P}$ are effectively parallel, at angle $\theta_{\mathcal{P}}$ with respect to the forward direction.[15] The path length difference for the Huygens wavelets from opposite ends of the slit is then well approximated by

$$\Delta L = w\, \sin(\theta_{\mathcal{P}})\,.$$

Since all points in the gap coherently emit wavelets, this path length difference yields a concomitant [continuous] phase difference.

WIDE   For wide gaps and angles $\theta_{\mathcal{P}} \simeq \tan(\theta_{\mathcal{P}}) > \frac{w}{L}$, the phase difference between the endpoints is typically very large, corresponding to many wavelengths. Superposition of waves with a wide range of phase angles is almost always [nearly completely] destructive.

A similar, incoherent superposition effect leading to broad cancellation was encountered in the analysis of standing waves in Chapter 25.

---

[14]By absorbing the waves, backwards reflections onto the barrier are avoided.
[15]The angle is defined by the ray from the midpoint of the slit to the point $\mathcal{P}$, as shown in Figure 31.6.

NARROW   When the gap is narrow, the phase difference may be small, even at large angles. This mostly coherent superposition leads to a wave with an appreciable fraction of its initial intensity. We confine our attention to the narrow case in the rest of this discussion.

The phase differences of the wavelets emitted from points in the slit are minimised at the central maximum, $y = 0$. As $\mathcal{P}$ migrates away from dead centre, the phase shift grows with increasingly destructive effect. IF the angle to $\mathcal{P}$ is such that the path length difference corresponds to n complete wavelengths, *i.e.*,

$$w \sin(\theta_{\mathcal{P}}) = n\lambda, \quad \text{for } n = 1, 2, 3, \ldots,$$

THEN the superposition is completely destructive and the intensity of the wave vanishes. Making the same sorts of approximations here as were made earlier for interference,

$$\sin(\theta_{\mathcal{P}}) \simeq \tan(\theta_{\mathcal{P}}) = \frac{y_{\mathcal{P}}}{L} \quad \Longrightarrow \quad \frac{w\, y_n}{L} = n\lambda,$$

where y_n denotes the position of the nth diffraction minimum.

Careful analysis[16] reveals a systematic reduction of amplitude with increasing departure from centrality:

$$\psi(y) \propto \frac{\sin\left(\frac{w\,y}{\lambda L}\pi\right)}{\frac{w\,y}{\lambda L}\pi}.$$

As the intensity of the diffracted wave depends on the square of the amplitude, it oscillates within an envelope which decays as y^{-2}.

Q: Shouldn't both interference and diffraction effects occur simultaneously when waves pass through two or more slits?

A: You betcha!

EXAMPLE [*Interference and Diffraction Together*]

Light waves, $\lambda = 625\,\text{nm}$, pass through two very thin slits separated by $d = 100\,\mu\text{m}$, and fall upon an absorbing screen 160 cm away. The banded pattern shown in Figure 31.7 is centred upon the point directly across from the pair of slits. The scale is marked in centimetres. The interference pattern in the previous example is reproduced in the region $y \in [-2, 2]$ in this figure. The novelty here is that the interference pattern stripes do not all have the same intensity. Instead, the intensity is modulated by the effects of diffraction [from each slit].

Q: What is the width of the pair of slits?

A: Invoking the various approximations discussed in the derivation of the diffraction effect, one may write

$$n\lambda = w\frac{y_n}{L},$$

---
[16]See the next chapter for details.

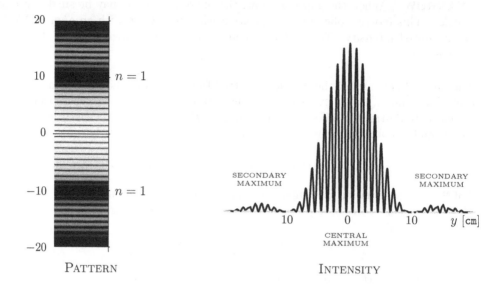

FIGURE 31.7   Interference and Diffraction Together

relating the positions of the nth diffraction minima to the wavelength, screen distance, and slit width. From the figure, the first minima lie 10 cm from the centre of the pattern. Thus,

$$w = \frac{\lambda L}{y_1} = \frac{10^{-6} \text{ m}^2}{0.1 \text{ m}} = 10^{-5} \text{ m} = 10 \ \mu\text{m}.$$

The two narrow slits each have width 10 μm, so

$$\lambda \ll w \ll d \ll L,$$

justifying all of the approximations that have been made in these examples.

Diffraction effects are present in a great many common situations. When you peer carefully at a shadow cast by an object and realise that its edges are not perfectly sharp, you may be seeing diffraction effects. When you see waves in a harbour gently rocking boats in their mooring bays, think diffraction. When you hear around corners and thereby avoid collisions with people and things, thank diffraction.

Chapter 32

Say Hello, Wave Goodbye

Before advancing to the proper subject of this chapter, we shall mop up a topic spilling out of Chapter 31. Girding ourselves for action, we recall Captain Kirk's exhortations[1] to "Set phasors on stun."

In Chapter 23, the superposition of two equal amplitude rightmoving harmonic waves with differing phases was effected directly by invocation of a trigonometric identity:

$$A \sin(k\,x - \omega\,t) + A \sin(k\,x - \omega\,t + \varphi) = 2\,A \cos\left[\frac{\varphi}{2}\right] \sin\left(k\,x - \omega\,t + \frac{\varphi}{2}\right).$$

Phasor diagrams afford a more general approach to the superposition of equal frequency harmonic functions. The starting point is to represent the function as an arrow from the origin in a 2-d space, with length equal to the amplitude. The angle that it makes with respect to a reference direction [conventionally the x-axis] is equal to the phase argument. The phasor rotates about the origin with the angular frequency, ω, of the harmonic function. At each instant, its projection onto the y- [or x-] axis is equal to the value assumed by the harmonic function at that time.

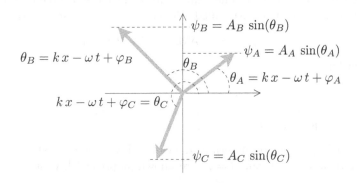

FIGURE 32.1 Three Phasors

Multiple comments are in order.

AMPLITUDE The amplitude of the harmonic function, and thus the length of the phasor, need not be constant in time [*cf.* the underdamped DHO].

PHASE Phases expressed in absolute terms, $\{\theta_A, \theta_B, \theta_C, \dots\}$, are readily converted into relative phases: $\varphi_{BA} = \theta_B - \theta_A = (k\,x - \omega\,t + \varphi_B) - (k\,x - \omega\,t + \varphi_A) = \varphi_B - \varphi_A$.

$k\,x - \omega\,t$ The uniform rotation of all phasors can be self-consistently ignored. Also, one may choose to align the reference direction with one of the phasors and thenceforth deal with relative phases.

---

[1] James T. Kirk commanded the Starship Enterprise in the original *Star Trek* TV series and films.

CLAIM: Linear superposition of harmonic functions is effected by vector addition[2] of the associated phasors. Two examples of phasor addition appear just below.

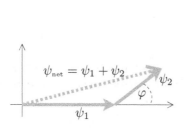
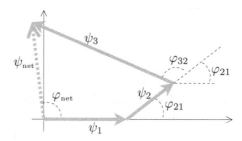

FIGURE 32.2   Vector Addition of Phasors

Rather than prove the equivalence[3] of phasor sums and superpositions of harmonic functions, we shall instead

∧   verify the two equal-amplitude rightmoving harmonic waves result, and

△   revisit the Three-Phase Miracle example from Chapter 24.

∧   Two equal-length phasors [with relative phase φ], ψ_1, ψ_2, and their sum, $\psi_{\text{net}} = \psi_1 + \psi_2$, appear on the left in Figure 32.3. The triangle formed by these three phasors is isosceles, with its common angle, α, left unspecified. Planar geometry constrains the angles so that

$$\bar{\varphi} + \varphi = \pi \qquad \text{AND} \qquad \bar{\varphi} + 2\,\alpha = \pi\,,$$

which together fix

$$\alpha \equiv \frac{\varphi}{2}\,.$$

The phase of the superposition lies exactly half-way between the phases of the constituents [as was shown to be the case in Chapter 24].

The amplitude of ψ_{net} may be determined via the LAW OF COSINES: $c^2 = a^2 + b^2 - 2\,a\,b\,\cos(\theta)$. Here, $a = b = A$, and $\theta = \bar{\varphi}$, are known [in principle], while $c = |\psi_{\text{net}}|$. Hence,

$$\left|\psi_{\text{net}}\right|^2 = 2\,A^2\left(1 - \cos(\bar{\varphi})\right)\,.$$

Realising that $-\cos(\bar{\varphi}) = -\cos(\pi - \varphi) = \cos(\varphi) = \cos(2\,\alpha)$ and $\left(1 + \cos(2\,\alpha)\right) = 2\,\cos^2(\alpha)$ allows us to write

$$\left|\psi_{\text{net}}\right|^2 = 4\,A^2\,\cos^2(\alpha) = 4\,A^2\,\cos^2\left[\frac{\varphi}{2}\right] = \left(2\,A\,\cos\left[\frac{\varphi}{2}\right]\right)^2\,.$$

[Not surprisingly, the amplitude factor matches that which was derived in Chapter 24.]

△   Hearkening back to Chapter 24, the superposition of three equal-amplitude waves with relative phase shifts of $2\,\pi/3$ is cast in terms of phasors in the right hand panel of Figure 32.3. The three phasors form an equilateral triangle, and hence this superposition necessarily vanishes.

---

[2] *Cf.* VOLUME I, Chapter 5.
[3] It's just linear superposition and effectively tautological.

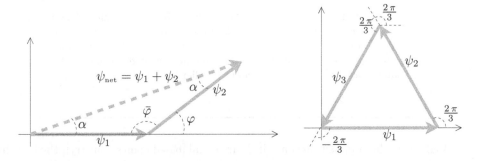

FIGURE 32.3 Particular Sums of Equal-Length Phasors

While all of this has been great fun, our current concern is the superposition of waves with a continuous distribution of phase shifts considered in the analysis of single-slit diffraction.

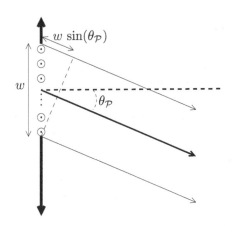

Recall the previous analysis and notation from Chapter 31.

The path length difference for waves emitted from the opposite edges of the slit and impinging upon the same distant point is

$$\Delta L \simeq w \sin(\theta_{\mathcal{P}}) \simeq \frac{w\,y}{L}\,.$$

The maximal phase shift is $\Delta\varphi$, where

$$\frac{\Delta\varphi}{2\pi} = \frac{\Delta L}{\lambda} \simeq \frac{w\,y}{\lambda\,L}\,.$$

FIGURE 32.4 Diffraction Analysis

When confronted with a continuum, we call upon the PARTITION, COMPUTE, SUM, and REFINE strategy.

P   Partition the gap into $\mathcal{N}$ equal-sized intervals. Each of the $\mathcal{N}+1$ boundary points [including the slit edges] may be considered a source of Huygens wavelets. A phasor is associated with each wavelet source: $\{\psi_i,\ i=0\ldots\mathcal{N}\}$.

C   The ith phasor possesses magnitude A_i and phase φ_i.

The geometry of the diffraction scenario ensures that the phase shifts of the individual phasor constituents are computable as fractions of $\Delta\varphi$, *i.e.*,

$$\varphi_i = \frac{i}{\mathcal{N}}\Delta\varphi\,,\quad i\in[\,0\,,\mathcal{N}\,]\,.$$

That all the nearest-neighbour relative phase differences are $\Delta\varphi/\mathcal{N}$ stems from the regularity of the partition.

ASIDE: Formally, choosing particular [rational fraction] values of $\Delta\varphi/(2\pi)$ and allowing $i = 0 \to \mathcal{N} = 2\pi/(\Delta\varphi)$ affords a way of generating regular $\mathcal{N}$-gons in the plane. A theorem states that the sum of the exterior angles of an $\mathcal{N}$-gon is 2π, while the sum of the interior angles is $(\mathcal{N} - 2)\pi$. The triangle case, $\mathcal{N} = 3$, was considered earlier in this chapter.

Aside from the supposition that all A_i are equal [based upon symmetry], the lengths of the phasors are *a priori* unknown. In the following analysis, we shall resort to a "reverse engineering" procedure to express the length in physical terms.

At the central maximum point on the observation screen, far from the slit, the diffracted wave reaches its maximum intensity, and thus attains its maximum amplitude, $A_{\max}$. The wavelets all arrive at that point with minimal [*i.e.,* zero] relative phase shift. In the language of phasors, all of the $\mathcal{N} + 1$ constituents are parallel, and therefore the magnitude of the sum is equal to the sum of the magnitudes. This is illustrated in Figure 32.5.

$$\psi_0 \quad \psi_1 \quad \psi_2 \quad \dots \quad \psi_{\mathcal{N}-1} \; \psi_{\mathcal{N}} \quad \equiv \quad \psi_{\text{net}} = (\mathcal{N} + 1)\,\psi_i$$

FIGURE 32.5   Phasors Combine Simply at the Central Maximum

Thus, each of the $\mathcal{N} + 1$ phasors in the gap may be assigned an amplitude of

$$A_i \equiv \frac{A_{\max}}{\mathcal{N} + 1}.$$

Therefore an expression for the ith phasor is

$$\psi_i = \frac{A_{\max}}{\mathcal{N} + 1}\left(\cos\left(\frac{i}{\mathcal{N}}\,2\pi\,\frac{w\,y}{\lambda\,L}\right),\ \sin\left(\frac{i}{\mathcal{N}}\,2\pi\,\frac{w\,y}{\lambda\,L}\right)\right),\quad \text{for } i = 0\dots\mathcal{N}.$$

S   The phasor sum over the $\mathcal{N}$-partitioned interval is equivalent to a partial sum of $\mathcal{N} + 1$ sides of a regular $\frac{2\mathcal{N}\pi}{\Delta\varphi}$-gon. The geometry is illustrated in Figure 32.6.

Computation of the phasor sum, shown in grey in the figure, is straightforward [albeit tedious] for any finite $\mathcal{N}$.

R   To obtain analytic results, we refine the partition by taking $\mathcal{N} \to \infty$. We are aided in this endeavour by the realisation that **regular plane polygons with increasing number of edges tend toward circles**. Thus, the continuous superposition of phasors across the slit yields a segment of circular arc.

Parsing Figure 32.7 requires care. Consideration of finite $\mathcal{N}$ cases reveals that the arc length of the segment must be equal to the sum of the magnitudes of the phasors, which we have identified with the maximal intensity, $A_{\max}$. Also, the angle through which the tangent to the arc has turned is equal to the overall

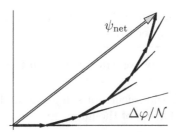

FIGURE 32.6   Sum of $\mathcal{N}$ Phasors across a Finite Width Slit

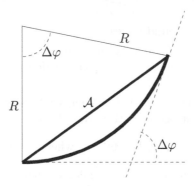

FIGURE 32.7   Continuous Sum of Phasors across a Finite Width Slit

phase shift, $\Delta\varphi$, and this in turn is the open angle for the arc segment. The usual circular arc length relation then determines the radius of the segment of arc to be

$$R = \frac{A_{\max}}{\Delta\varphi}.$$

The magnitude of the net phasor, $\mathcal{A}$, is equal to the length of the secant forming the base of an isosceles triangle with sides R set at angle $\Delta\varphi$. The most straightforward way to determine $\mathcal{A}$ is to bisect the base [and the angle] by means of an altitude.[4] Doing so,

$$\frac{\mathcal{A}}{2} = R \sin\left(\frac{\Delta\varphi}{2}\right) = \frac{A_{\max}}{\Delta\varphi} \sin\left(\frac{\Delta\varphi}{2}\right)$$

$$\implies \quad \frac{\mathcal{A}}{A_{\max}} = \frac{2}{\Delta\varphi} \sin\left(\frac{\Delta\varphi}{2}\right) = \frac{\sin\left(\frac{\Delta\varphi}{2}\right)}{\frac{\Delta\varphi}{2}}.$$

The overall phase shift may be expressed in terms of the diffraction geometry, *i.e.*, $\Delta\varphi = 2\frac{w\,y}{\lambda\,L}\pi$, and thus

$$\psi_{\mathrm{net}}(y) \propto \frac{\sin\left(\frac{w\,y}{\lambda\,L}\pi\right)}{\frac{w\,y}{\lambda\,L}\pi}.$$

---

[4] An altitude of a triangle is a line perpendicular to a side and passing through the opposite vertex. In the present case, the altitude is the perpendicular bisector.

This is the expression for the amplitude of the diffracted wave, essential for the phenomenology of diffraction, quoted in Chapter 31.

With this discursion out of the way, we can get on with this chapter's intended purpose.

ACOUSTICS   Acoustics is the science of sound. Sound consists of longitudinal density/pressure/sloshing disturbances of the air.

Several branches of acoustics are concerned with human factors. Some of these include: the **production** of sound [*e.g.,* in the vocal tract, by musical instruments and devices, *etc.*]; the **propagation** of sound [*e.g.,* in concert halls, through heterogeneous media (sonography), *etc.*]; and the **response** to, or **perception** of, sound [*e.g.,* in the ear and brain, by microphones and other devices].

Q: How do humans respond to sound waves?

Aqualitative**:** One's perceptions of intensity and frequency are best understood as being roughly logarithmic, thereby enabling one to hear a wide range of intensities throughout a large band of frequencies.

Aquantitative**:** Suppose that a sound wave has an energy intensity of I watts per square metre. An assignment of loudness that people with normal hearing can agree upon is

$$\text{Loudness} = 10 \log_{10}\left(I/I_0\right) \text{ dB}, \qquad \text{where} \qquad I_0 = 10^{-12} \text{ W/m}^2.$$

The decibel unit [dB] is so-named after Alexander Graham Bell (1847–1922). This scale [10 dB constitute 1 Bel(l)] is chosen so as to make the sound intensities found in normal human experience fall into the [convenient] range: $0 \to 100$, or so.

[The same rationale applies to the Fahrenheit temperature scale, to be discussed in Chapter 37.]

The reference value, I_0, corresponds to the faintest sound that can be reliably detected by the human ear. It is assigned a loudness of 0 dB. The loudest sound that can be tolerated, at the threshold of pain, is roughly 120 dB, or 1 W/m^2.

Sound	ENERGY INTENSITY $[\text{W/m}^2]$	PRESSURE AMPLITUDE $[\text{Pa}]$	Loudness $[\text{dB}]$
Faintest Audible Sound	10^{-12}	2×10^{-5}	0
Leaves Rustling	10^{-11}	$6\frac{1}{3} \times 10^{-5}$	10
Quiet Conversation	10^{-8}	2×10^{-3}	40
Professor Gassing	10^{-6}	2×10^{-2}	60
Highway Noise	10^{-4}	2×10^{-1}	80
Jackhammer	10^{-2}	2	100
Pain Threshold	1	20	120
Nearby Jet Engine	100	200	140

We make the following observations about intensity before considering frequency.

- These energy intensities span [in excess of] twelve orders of magnitude.

- Intensities smaller than I_0 are consistently assigned negative values of loudness. *E.g.,* $I = 10^{-14}$ W/m^2 corresponds to -20 dB.

- The ear's non-linear response throughout the intensity range leads to compression, or loss of intensity resolution, at higher intensities. For example, sound intensities of 10^{-11} W/m$^2 = 10$ dB and 10^{-10} W/m$^2 = 20$ dB are distinguishable. The latter is twice as loud as the former. However, intensities of $10^{-6} + 10^{-11}$ W/m^2 $\simeq 60.00004$ dB and $10^{-6} + 10^{-10}$ W/m$^2 \simeq 60.0004$ dB are indistinguishable.

  > ASIDE: From the standpoint of evolutionary biology, an individual's ability to survive is enhanced by having a large dynamic [intensity] range, along with with acute sensitivity at low intensities, because the presence of predators, or prey, may be detected while they are still at some distance.
  >
  > **Q:** Why don't we have equal sensitivity at all intensities throughout this tremendous range? **A:** Any potential gain associated with having such discriminating power in the entire hearing range would be overwhelmed by the neurological/computational overhead required.

- Some sources quote 130 dB, or more, for the pain threshold.

  [PK opts for the value 120 dB.]

- Whether the sound of a professor gassing should be distinct from, and below, the pain threshold is subjective and context-dependent.

Typical young people are able to hear sounds with frequencies ranging from 20 Hz to 20 kHz. Outside this band, spanning three orders of magnitude, our ears have very little or no sensitivity. With age and exposure to loud or persistent sound, one's frequency response may diminish at its extrema or in specific intermediate ranges.

Sound	Frequency [Hz]
Deepest Organ Note	< 20
Lowest Piano C	32.7
AC Electrical Hum	60
Piano Middle C	261.63
A 440	440
Highest Piano C	4186
Annoying CRT Hum	16000 − 17000
Ultrasound	> 20000

♯ One's perception of frequency is also logarithmic. Musical scales,

$$\{ \text{ do, re, mi, fa, so, la, ti, do } \},$$

involve a[n approximate] doubling of frequency from one *do* to the next.

ASIDE: In the early days of music synthesisers, the frequencies of the notes were assigned by a crude 'logarithmically equidistant' model ($\nu_i = \nu_0 \, 2^{i/7}$, where $i \in [\,0\,,7\,]$ and ν_0 is the frequency of the first/lowest note), and the waveforms were purely sinusoidal. Listeners were underwhelmed and musicians were appalled by the abject sterility of tone produced by these instruments. In response to vociferous criticism, the chastened manufacturers tweaked the frequency spacings and modified the shapes of the waveforms. Nowadays, many modern digital instruments reproduce sound from a library of samples of analogue instruments.

♯   Concomitant with the logarithmic nature of our perception is enhanced sensitivity at low frequencies and compression at high frequencies. Distinct notes at the low end of the piano may be 10 Hz or so apart, while at the upper end, the same 10 Hz difference will barely register (if at all). Above 10^4 Hz, a 10 Hz discrepancy is imperceptible.

ASIDE: Evolutionary biology again provides a paradigm within which one can appreciate how it might be that our frequency response is logarithmic, covers the observed range of frequencies that it does, and has enhanced awareness at the lower end.

♯   Looking at the chart and realising that the highest piano C is quite high-pitched, one might wonder about the seemingly empty expanse of sensitivity lying above 5000 Hz. Far from being superfluous, sensitivity to these higher frequencies affords us better perception of the "colour," or "timbre," of the sounds we hear.

Although our formal study of waves has focussed nearly exclusively on simple sinusoidal travelling waves, our brief [and joyful!] glimpse of Fourier Analysis in Chapter 23 has shown that more complicated repeating waveforms can be constructed [to any prescribed degree of approximation] by the superposition of higher-order harmonics.

Thus, it is not so much to hear pure tones at high frequencies that our frequency response extends as high as it does, but rather to perceive details pertinent to lower-frequency sounds which are encoded in the shape of the waveform.

ASIDE: A familiar example of this effect is the manner in which people's voices are altered over the telephone. This is because the telephone frequency bandwidth is roughly 300–3000 Hz. Essentially all of the important speech information needed for intelligibility is found in this range[5] of frequencies. Careful listening leads one to notice *missing* fricatives, like '*s*' and '*th*' (which have the bulk of their energy intensity concentrated at higher frequencies). People compensate by interpolation, the filling in by context, of those parts of the signal that are dropped.

At this juncture we wave goodbye to sound and aim to "see the light" [waves] as we begin the study of [classical] **Geometric Optics**.

---

[5] Excising the rest of the auditory spectrum reduces the amount of information transmitted, enables more efficient digitisation and compression of the signal, and allows use of cruder signal gathering (microphone) and output (speaker) technologies.

Chapter 33

Optics

Optics is the study of visible light and [some of] its interactions with matter. In specialising to classical optics, we shall make two essential assumptions, and several observations having to do with man's perception of light.

WAVE Light is a transverse electromagnetic wave.[1] We eschew discussion of the precise details of these waves and of the mechanics of the media in which they propagate.

> ASIDE: Einstein's careful consideration of properties of light led to Special Relativity and Quantum Mechanics [via wave–particle duality].

SCALE The wavelengths of visible light are very small on the scale of the optical effects we consider.

HUMAN FACTORS The eye has logarithmic sensitivity to the intensity of light in the visible part of the spectrum. As in acoustics [discussed in the previous chapter] this feature enables perception over a wide range of intensities, with particular sensitivity at low intensity.

The visible frequencies and wavelengths constitute a very small part of the electromagnetic spectrum. For all intents and purposes, one's perception of colour is essentially linear.

The frequency [wavelength] spectrum of electromagnetic waves spans [more than] sixteen orders of magnitude. Names are given to certain ranges as described in the table below.

Spectral Range	Frequency [Hz]	Wavelength [m]
Long Waves	$\nu < 10^4$	$\lambda > 3 \times 10^4$
Radio	$10^4 < \nu < 10^{11}$	$3 \times 10^4 > \lambda > 3 \times 10^{-3}$
Infrared	$10^{11} < \nu < 4 \times 10^{14}$	$3 \times 10^{-3} > \lambda > 7 \times 10^{-7}$
Visible	$4 \times 10^{14} < \nu < 7.5 \times 10^{14}$	$7 \times 10^{-7} > \lambda > 4 \times 10^{-7}$
Ultraviolet	$7.5 \times 10^{14} < \nu < 10^{17}$	$4 \times 10^{-7} > \lambda > 3 \times 10^{-9}$
X-rays	$10^{17} < \nu < 10^{20}$	$3 \times 10^{-9} > \lambda > 3 \times 10^{-12}$
γ-rays	$10^{20} < \nu$	$3 \times 10^{-12} > \lambda$

Examination of the data in the table reveals that the product of wavelength and frequency,

$$c = \lambda \nu \simeq 3.0 \times 10^8 \text{ m/s},$$

appears to be [approximately] constant throughout the spectrum.

---

[1]Read VOLUME III of these notes to better appreciate this.

INDEX OF REFRACTION   The speed of light in a medium is proportionally reduced from its speed in vacuum, $c \simeq 3.0 \times 10^8$, by the refractive index, n, of the medium:

$$c_n = \frac{c}{n}.$$

Consistency demands that $n = 1$ for the vacuum and $n > 1$ for all materials found in nature. Typical values of n range from 1.0003 for air to 2.4 for diamond.

At least four matters merit mention.

DISPERSION   Dispersion is modelled by allowing the refraction index to be wavelength-dependent, *i.e.*, $n(\lambda)$. Pure harmonic waves retain their shape, but Fourier compositions of such waves do not, since differing propagation speeds inevitably result in loss of phase correlation.

IMAGINARY PART   Indices of refraction are sometimes made complex, by addition of an imaginary part, to account for absorptive attenuation.

EXOTICA   Diverse meta-materials, some with enormous indices (*i.e.*, greater than 10) and others with negative [effective] indices, have been developed.

NORMALS   Light passing from one medium to another at normal incidence is transmitted and partially reflected, as were the mechanical waves propagating in 1-d discontinuous media studied in Chapter 29. The three-dimensional situation, **refraction**, is studied in Chapter 35.

LIGHT INTERACTIONS WITH MATTER: POLARISATION

POLARISATION   Each wavelet comprising a plane electromagnetic wave has a polarisation vector. These lie in the plane of the wave and are normal to the wave velocity vector. The collection of these polarisation vectors may be utterly uncorrelated [in which case the light is said to be **unpolarised**], nearly completely coincident [*a.k.a.* **highly polarised**], or anywhere on the continuum between these two extremes.

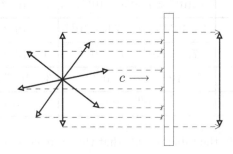

FIGURE 33.1   Highly Polarised Light Obtained by Selective Filtering

Materials can selectively filter light. As illustrated in Figure 33.1, when plane waves possessing an admixture of polarisations encounter a polarising filter, only those constituents having polarisation within a narrow range of angles are able to pass through. The rest are absorbed or reflected. As a necessary consequence, the transmitted light has reduced intensity.

Highly polarised light may be produced and observed in everyday settings.

↔ The harsh glare from a surface [*e.g.*, a roadway or cars on a bright summer's day] is mostly polarised in the horizontal direction.

[*Sunglasses with vertical polarising filters diminish the glare appreciably.*]

↕ Light from electronic displays employing Liquid Crystal Device [LCD] technology [*e.g.*, gas pumps and telephones] is often strongly polarised.

[*The display may appear to fade to black if you wear sunglasses and tilt your head.*]

LIGHT INTERACTIONS WITH MATTER: REFLECTION

COHERENT PLANAR REFLECTION   Light rays reflect coherently from a planar surface, with the angle of reflection equal to the angle of incidence. [Convention dictates that both angles are measured WRT the normal to the surface].

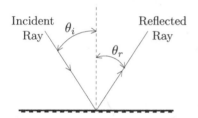

In NORMAL INCIDENCE, $\theta_i = 0 = \theta_r$, the ray is reflected directly backwards along the path from whence it came.

Some refer to this equal-angle property as THE LAW OF REFLECTION.

HUYGENS' DERIVATION OF THE LAW OF REFLECTION

Monochromatic[2] plane waves strike a plane reflecting surface, at angle θ_i, in Figure 33.2.

[In this first figure, we have not attempted to illustrate the reflected waves.]

The incident waves move with speed $c_n = c/n$, and the frequency, wavelength, period, and speed are related via

$$c_n = \nu \lambda \qquad \Longleftrightarrow \qquad \lambda = c_n T.$$

We pay careful attention to the shaded parallelogram comprised of parts of two successive wavefronts and the displacement vectors at the endpoints of the trailing wavefront. These are denoted by the heavier lines in the figure.

The reflected waves obey exactly the same relations, since they, too, propagate in the common medium. Figure 33.3 shows the same portion of reflecting surface as Figure 33.2,

---

[2]Monochromatic waves are harmonic and simple in that they possess a single unique frequency–wavelength combination.

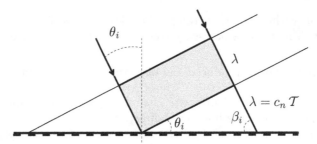

FIGURE 33.2   Huygens' Derivation of the Law of Reflection—Part One

one wave period later. In the course of this time interval two things will have occurred. The trailing wavefront, shown in grey, will have moved forward to the position previously occupied by the leading wavefront, while the highlighted parts of the leading wavefront will have reflected from the surface. The leftmost edge of the leading wavefront was impinging on the surface at the start of the period. A Huygens wavelet propagating radially outward from this point at that instant will have assumed the circular form indicated in Figure 33.3 after the elapse of one period. The rightmost edge will have just reached the surface at the end of the interval and will be the source of a zero-radius Huygens wavelet.

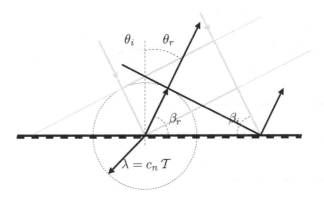

FIGURE 33.3   Huygens' Derivation of the Law of Reflection—Part Two

Coherence of the reflected wave can only arise if the reflected wave is also a plane wave, and this further requires that the reflected wavefront lie tangent to the Huygens circle and through the surface point on the right.

This fixes the locations of all of the points on the reflected wavefront.

The wavefront meeting these conditions is shown, along with its perpendicular, which indicates the direction in which the reflected wave propagates, in Figure 33.3. All of the pieces necessary to prove the Law of Reflection have been assembled.

The right triangle formed by the marked portion of the leading wavefront [altitude], the propagation vector to the surface [base], and a segment lying along the reflecting surface [hypotenuse] is congruent[3] to the right triangle whose sides consist of the reflected wavefront [altitude], its propagation vector [base], and the common hypotenuse.

---

[3]Congruence of two triangles means that they are exactly the same size and shape, *i.e.*, isomorphic.

Congruence of the triangles forces the angles marked β_i and β_r to be equal, which, owing to complementary relations:

$$\theta_i + \beta_i = \frac{\pi}{2} \qquad \text{AND} \qquad \theta_r + \beta_r = \frac{\pi}{2},$$

in turn requires that $\theta_r = \theta_i$.

MINIMUM TIME DERIVATION OF THE LAW OF REFLECTION

The **Principle**[4] **of Minimal Time** avers that plane waves passing from an initial point in space, $\vec{r}_1$, to a final point, $\vec{r}_2$, follow the [unique[5]] path which minimises the transit time of the wave.

> ASIDE: The Minimal Time Principle is operative whenever a lifeguard must both run along the beach and swim in the water to rescue a foundering swimmer.

In a homogeneous medium, the shortest path [*viz.*, the straight line] is the fastest.

In Figure 33.4, the initial and final positions are $\vec{r}_1 = (0\,, y_1)$ and $\vec{r}_2 = (a\,, y_2)$, respectively. The x-axis is aligned with the planar reflecting surface. *A priori*, the path of the wave is not specified. This is alluded to in Figure 33.4 by there being several arrows emanating from $\vec{r}_1$ and converging on $\vec{r}_2$. The putative actual path consists of two straight line segments which meet at an intermediate point, $\vec{X} = (x\,, 0)$.

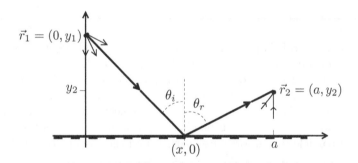

FIGURE 33.4   Minimal Time Derivation of the Law of Reflection

The incident and reflected path lengths are $\sqrt{x^2 + y_1^2}$ and $\sqrt{(a-x)^2 + y_2^2}$, respectively. As the wave speed is $c_n = c/n$ for both, the times taken for the two legs of the journey are:

$$t_1(x) = \frac{n\sqrt{x^2 + y_1^2}}{c} \qquad \text{and} \qquad t_2(x) = \frac{n\sqrt{(a-x)^2 + y_2^2}}{c},$$

respectively, leading to a total elapsed time of

$$t_{12}(x) = t_1(x) + t_2(x) = \frac{n\sqrt{x^2 + y_1^2}}{c} + \frac{n\sqrt{(a-x)^2 + y_2^2}}{c}.$$

---

[4]Recall that physical principles are axiomatic statements.
[5]The lawyers insist that we add the proviso "in most cases."

To ascertain the path leading to the minimum elapsed time, one determines the condition(s) on x which set the derivative of $t_{12}(x)$ with respect to x equal to zero. That is,

$$\frac{dt_{12}}{dx} = \frac{n}{c}\frac{x}{\sqrt{x^2 + y_1^2}} - \frac{n}{c}\frac{(a - x)}{\sqrt{(a - x)^2 + y_2^2}},$$

and therefore, at the extremum,

$$\frac{dt_{12}}{dx} = 0 \quad \Longleftrightarrow \quad \frac{x}{\sqrt{x^2 + y_1^2}} = \frac{(a - x)}{\sqrt{(a - x)^2 + y_2^2}}.$$

It is apparent from Figure 33.4 that

$$\frac{x}{\sqrt{x^2 + y_1^2}} = \sin(\theta_i) \quad \text{and} \quad \frac{(a - x)}{\sqrt{(a - x)^2 + y_2^2}} = \sin(\theta_r),$$

and hence $\sin(\theta_i) = \sin(\theta_r)$, which holds IFF $\theta_r \equiv \theta_i$.

ASIDE: The second derivative of t_{12} with respect to x is

$$\frac{d^2 t_{12}}{dx^2} = \frac{d}{dt}\left[\frac{n}{c}\frac{x}{\sqrt{x^2 + y_1^2}} - \frac{n}{c}\frac{(a - x)}{\sqrt{(a - x)^2 + y_2^2}}\right]$$

$$= \frac{n}{c}\frac{y_1^2}{\left(x^2 + y_1^2\right)^{3/2}} + \frac{n}{c}\frac{y_2^2}{\left((a - x)^2 + y_2^2\right)^{3/2}} > 0\,, \ \forall\, x\,,$$

ensuring that the extremum is indeed a minimum.

Our tasks in the next several chapters are to apply the Law of Coherent Planar Reflection to derive properties of mirrors [Chapter 34], to develop Snell's Law of Refraction from the refractive index effect on the speed of light [Chapter 35], and to determine the optical properties of simple lenses [Chapter 36]. Before doing all of this, however, we pause to develop a classical model of the energetics for the photoelectric effect. The classical model is falsified by experimental evidence. Einstein's explication of the photoelectric effect in 1905—employing a particle model for light—earned him the Nobel Prize in 1921.

ADDENDUM: The Classical Model of the Photoelectric Effect

Recall that it was mentioned in Chapter 28 that the energy intensity of a system of transverse waves is not readily calculable, except, in macroscopic terms, by dividing the total average power by the cross-sectional area of the medium through which the waves propagate.

To render this point explicitly, suppose that we have two sources of harmonic transverse electromagnetic waves [a.k.a. monochromatic light, see Chapter 48 in VOLUME III] with wavelengths λ_1 and $\lambda_2 > \lambda_1$.

ASIDE: LASERs are laboratory sources of intense uniform beams of monochromatic light.

We assume that each light source delivers a uniform beam of power P_0 distributed across an area, A_{Total}. The energy intensity of both beams is the same:

$$I_0 = \frac{P_0}{A_{\text{Total}}}.$$

Further suppose that the beams are made to shine on [ultra-clean] solid pure metallic surfaces in the presence of a device which registers the appearance of electrons emitted from the metal and counts them.

The light waves incident upon the surface of the metal bear energy. [This was shown for transverse mechanical waves in Chapter 26. The case of classical light waves is explicitly dealt with in Chapter 49 of VOLUME III.] Suppose that all of this energy is absorbed into the surface layer of metallic atoms.

[This is unrealistic in that it ignores reflection from the surface and transmission into the block.]

The metallic atoms are arranged in a crystalline lattice pattern.[6] As illustrated in Figure 33.5, each atom on the surface presents an effective cross-sectional area for absorption of the incident waves, a_0.

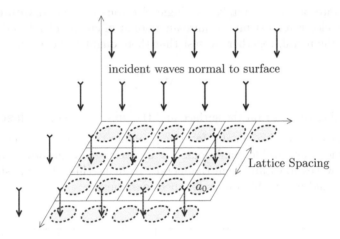

incident waves normal to surface

Lattice Spacing

a_0

FIGURE 33.5 Cross-Section for Absorption of Incident Waves

Let's further assume [again, rather unrealistically] that each surface atom aggregates all of the energy incident upon it and bestows it upon one nearby electron. Under this collection of assumptions, one electron in the vicinity of each surface atom is accruing energy at a rate of

$$P_\text{e} \simeq I_0 \, a_0 = \frac{a_0}{A_\text{Total}} \, P_0 \, .$$

Although this is expected to be a very small power [owing to the micro/macro scales of a_0/A_Total], it can still produce dramatic effects because the electron has so very little inertia.

A metal has the property that a [substance-dependent] fraction of its electrons are free to move about throughout its bulk, but strongly constrained to remain within its confines. This behaviour may be simply modelled by positing a square-well potential energy function for electrons in the material, as pictured in Figure 33.6.

The potential energy function is flat and smooth locally everywhere outside and inside the material. The electron is *free* [*i.e.,* unforced, experiencing NEUTRAL EQUILIBRIUM] whether it be found inside or outside of the metal. However, an electron on the inside must climb out of the energetic well to appear on the outside [with positive—physical—kinetic energy]. Alternatively, those steep well sides imply a strong impulsive force acting to prevent the

---

[6]This is discussed in the context of the microscopic model for electric current developed in Lecture 17 of VOLUME III.

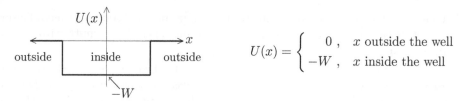

FIGURE 33.6   Electron Potential Energy Function for a Metallic Solid

electron from leaving the interior of the metal. The [relative] depth of the potential well, W, is called the **work function** of the metal, because this is the amount of work which must be absorbed by an electron to liberate it.

A constant rate of energy uptake by single electrons associated with each surface atom, and that any electron must first acquire an amount of energy equal to the work function to escape from the metal, together suggest that there ought to be a time delay,

$$\Delta t = \frac{W}{P_{\text{e}}},$$

between shining the laser on the surface and the appearance of the first photoelectrons.

According to this classical model for the photoelectric effect, it ought to take roughly this long, under favourable conditions, for an electron to absorb sufficient energy from the incident waves to escape from the metal. To make this all less abstract, let's consider some explicit and reasonable values for these parameters.

EXAMPLE [*Classical Photoelectric Effect: Theory*]

The laser sources have wavelengths $\lambda_1 = 250\,\text{nm}$ [near ultraviolet] and $\lambda_2 = 500\,\text{nm}$ [a pleasant shade of green], respectively. Both have power $P_0 = 0.2\,\text{mW}$ and beam area $A_{\text{Total}} = 10\,\text{mm}^2 = 10^{-5}\,\text{m}^2$. Suppose that the metal is copper, with a lattice spacing of roughly $3.6\,\text{Å}$, where one Ångstrom [Å] is equal to 10^{-10} m. Taking the atomic diameter of copper to be approximately $2.5\,\text{Å}$ suggests that the cross-sectional area for absorption of the light waves is $a_0 = \frac{\pi}{4}\,(2.5 \times 10^{-10})^2 \simeq 5 \times 10^{-20}\,\text{m}^2$. Using these parameters, the power input to an electron associated with each atom is approximately

$$P_{\text{e}} \simeq \frac{5 \times 10^{-20}}{10^{-5}} \times 0.2 \times 10^{-3} = 10^{-18}\,\text{W}.$$

Energies associated with atomic phenomena are better described using the electron volt: $1\,\text{eV} = 1.6 \times 10^{-19}$ J. The power input to the electron from the incident waves [modulo all of the crude assumptions] is roughly $6.25\,\text{eV/s}$.

Copper's work function is $W_{\text{Cu}} \simeq 4.7\,\text{eV}$. Hence, one would expect that there should be a delay of about

$$\Delta t \simeq \frac{4.7\ \text{eV}}{6.25\ \text{eV/s}} \sim 0.75\ \text{s}$$

before any electrons are emitted, and then there should be emission at a constant rate, independent of the source of the light. This expected behaviour is plotted in Figure 33.7.

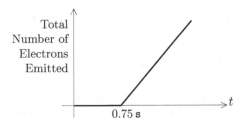

FIGURE 33.7 Classical Prediction of Total Number of Emitted Photoelectrons *vs.* Time

ASIDE: Hold on! We have completely neglected to account for thermal effects. Since the metallic substance is not at ABSOLUTE ZERO, the electrons within must have [a range of] thermal energies dictated by the EQUIPARTITION PRINCIPLE [see Chapter 45 and 49]. Should these be large, then electrons may be emitted sooner than predicted by this model, or may not require any additional energy from the light to break free of the metal. The thermal energy at room temperature, roughly 300 K, is

$$\langle E \rangle = \frac{3}{2} k_B T \simeq \frac{3}{2} 1.38 \times 10^{-23} \, 300 = 6.21 \times 10^{-21} \text{ J} \simeq 3.9 \times 10^{-2} \text{ eV}.$$

Thus, thermal effects are not significant in the consideration of the photoelectric effect for copper at laboratory temperatures.

Q: So much for the theoretical model. What does the experimental data look like?

A: Experimental counts of emitted electrons *vs.* time are shown in Figure 33.8. These results clearly falsify the classical model that we have constructed.

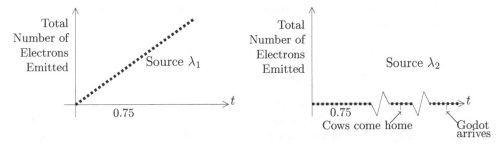

FIGURE 33.8 Number of Emitted Photoelectrons *vs.* Time: Experimental Observations

♯   For the shorter wavelength source, λ_1, there are prompt electrons—those which appear without the expected time delay. [This phenomenon defies classical explanation.]

♯   For the longer wavelength source, λ_2, the expected photoelectrons fail to materialise irrespective of how long one waits.

> ASIDE: A potential explanation for this null result is that the unreasonable assumptions in the model [*e.g.,* that all the light wave energy is absorbed on the surface and that it is then directed to a single electron] fail to hold. Were this the case, however, one would expect no photoelectrons to appear for both shorter and longer wavelengths, in clear contradiction of the λ_1 results.

These photoelectric observations and the abject explanatory failure of the wave model of light caused consternation among physicists in the late nineteenth and early twentieth centuries. All was resolved by Einstein in 1905 with the proposal that the energy carried by light is not dispersed as in the model of electromagnetic waves, but is instead borne by quantised particles of light [PHOTONS].

> ASIDE: A recurrent theme throughout the history of physics (and more generally, the philosophy of nature) has been the rise and fall of discrete and continuous model descriptions of fundamental aspects and constituents of nature. Newton advocated a particle theory of light [based on his studies of geometric optics—light travelling in straight lines within uniform media acts not unlike the inertial motion of particles in the absence of force]. By 1800 or so, the particle model was superseded by the wave model as diffractive phenomena and interference became better understood through careful study. Einstein, and others, reintroduced the notion of particulate constituents of light (wave–particle duality). This formed the scientific consensus for most of the twentieth century. Modern [super]string theory views all so-called particles as wavelike excitations of 1-d string-like objects.
>
> [*Plus ça change*, eh?]

EXAMPLE [*Photoelectric Effect: Einstein*]

Einstein's resolution of the photoelectric effect is that electromagnetic energy comes in [frequency- and wavelength-dependent] packets [QUANTA]: $E = h\nu = \frac{hc}{\lambda}$. The product of Planck's constant and the speed of light [in vacuum] is $hc \simeq 1240\,\text{eV} \cdot \text{nm}$, in the units we have used throughout this analysis. For the lasers employed in the previous example,

$$E_1 \simeq \frac{1240}{250} = 4.96\,\text{eV} \qquad \text{and} \qquad E_2 \simeq \frac{1240}{500} = 2.48\,\text{eV}.$$

This amount of energy [or less] may be transferred to a single electron in the metal via a collision-like interaction.

λ_1 A single quantum of the ultraviolet light is sufficient to [more than] compensate for the work function of copper. Thus, electrons may be ejected without delay once the light shines on the surface.

λ_2 A single quantum of the green light is unable to provide an electron with enough energy to extract it from the potential well. Two or more of these quanta must be absorbed by a single electron for it to acquire sufficient energy to be liberated. This is exceedingly unlikely because there are a vast number of electrons which can absorb the light quanta, and interactions [scattering events] with other electrons and the lattice of metallic ions will tend to cause those with a single excitation to quickly lose part, or all, of their absorbed energy.

The photoelectric effect is a paradigmatic instance of the breakdown of CLASSICAL PHYSICS. Two more comments are germaine.

o   For the two light sources to have the same energy flux [intensity, too], the flux of photons in the 500 nm wavelength beam must be twice that of the 250 nm source.

o   Photoelectric emission is an improbable event for a variety of reasons. The struck electron is most likely to be propelled deeper into the block. The photon may bestow its energy upon two or more electrons. An excited electron may collide with [scatter from, interact with] other electrons or the metallic ions and transfer some of its energy to them.

Chapter 34

Mirror Mirror

The unabridged title of this chapter is: *Mirror mirror, On the wall, In makeup kit, And at the mall.* We shall investigate each clause separately below.

MIRROR MIRROR

We shall examine coherent reflection of light from planar and uniformly-curved surfaces. The most important [local] aspect of reflection is summarised in the Law of Reflection:

The angle of reflection is equal to the angle of incidence.

The situation here is distinct from the [mere] reflection of plane waves, because we are assuming the presence of a particular **object** [assumed to be small and localisable in space] which acts as the source of the light eventually reflected by the mirror.

[We assume that the object is illuminated or luminous.]

The light from the object interacts with the mirror and is scattered. We presume the existence of an observer whose eye and brain collect the scattered light and reconstruct [*i.e.,* synthesise] an **image**. The image appears[1] to reside at a point in space other than that where the object is located.

[Image reconstruction at finite distance requires that the light gathered by the eye be diverging.]

Images are classified according to various properties, all of which are based upon the relative positionings of image and object. The first distinction is between **real** and **virtual** images.

R   A real image resides in a region of space accessible to the observer, and is formed by light which travels directly from the image location to the observer.

V   A virtual image appears to lie in an inaccessible region of space. Such images are synthesised from light which does not emanate directly from the image.

To quantify the object–image relations, first let

$$p = \text{the [minimum, perpendicular] object distance [from mirror], and}$$
$$i = \text{the [minimum, perpendicular] image distance [from mirror]}.$$

---

[1]There are many artful manners in which a *trompe l'oeil* (trick of the eye) may be contrived.

OPTICS MASTER EQUATIONS   There exist two generic MASTER EQUATIONS relating p and i for mirrors of all kinds. The **Object-Image-Focal-Length Relation** reads

$$\frac{1}{p} + \frac{1}{i} = \frac{1}{f}\,,$$

where the focal length, f, is derivable from the [uniform] curvature of the reflecting surface. IF $i > 0$, the image is **real**, WHEREAS IF $i < 0$, it is **virtual**.

The **magnification** associated with the image is determined by $M = \dfrac{-i}{p}$.

The sign of the magnification distinguishes between images which are **erect**, $M > 0$, and **inverted**, $M < 0$. IF $|M| > 1$, the image is said to be **enlarged**, WHEREAS IF $|M| < 1$, it is **reduced** or **diminished**.

ON THE WALL[2]

Consider the reflection of an object in a plane mirror, as illustrated in Figure 34.1. Of the infinitude of optical rays leaving the object and striking the mirror, two have been drawn to illustrate [in an exaggerated manner] the angular range of the rays which may be collected by an observer's eye. The eye and brain trace these diverging rays back to where they appear to originate and thus construct the image [behind the mirror, in the present case].

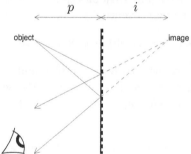

FIGURE 34.1   Reflection of an Object in a Plane Mirror

In quantitative terms, the local curvature of a plane mirror is everywhere zero, and hence the focal length diverges, $f \to \infty$ and $\frac{1}{f} \to 0$. For plane mirrors, then, the object-image-focal-length relation reduces to

$$0 = \frac{1}{p} + \frac{1}{i} \qquad \Longrightarrow \qquad i = -p \quad \text{AND} \quad M = +1\,.$$

This shows that the image resides within/behind the mirror at the same distance that the object lies in front. As the image is in the unphysical region, it is virtual. The magnification formula shows that the image is erect and appears neither enlarged nor reduced, but at its proper size, given the distance that the reflected light has travelled.

Our model of the plane mirror behaves just as expected.

---

[2]The snarks among us might quip that we are *off the wall*.

IN MAKEUP KIT

Consider a concave spherical section, with radius of curvature equal to R, coated with a reflective material. To see how such a mirror produces an image of a particular object, we must build up an inventory of rays whose reflections we can readily trace. In this endeavour we are guided by the principles that (1) in planar reflection, the angle of incidence is equal to the angle of reflection, and (2) a differentiable surface, however much it may be curved, is everywhere locally flat.

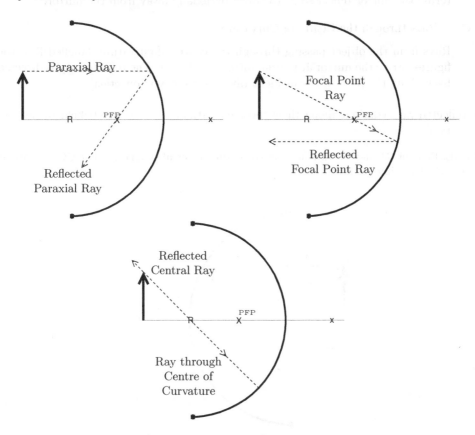

FIGURE 34.2 Rays through the Focal Point and the Centre of a Concave Mirror

The three important classes of rays are illustrated in Figure 34.2. The particular point on the object from which the light rays stream is indicated by the tip of an arrow. In our analysis, the base of the arrow lies on the axis of symmetry of the section of spherical shell. This axis is termed the OPTIC[AL] AXIS. By the rules for image construction, developed just below, object points on the axis are mapped to image points also on the axis. Hence, for objects which extend from the axis, it is only necessary to find the image of a single significant point on the object to infer the location of the entire image.

ASIDE: Images of objects which do not have an extremity on or near the axis may be ascertained by the mapping of two or more points, treated as arrows extending from the axis.

PA   Paraxial Rays

These rays initially travel parallel to the axis until they are reflected by the mirror. The angle of incidence of each ray depends on its distance from the axis and the curvature of the mirror. Astoundingly, all paraxial rays are diverted toward and through the **focal point**, which lies along the optical axis, exactly half-way between the mirror and the centre of curvature.[3]

[The principal focal point is labelled PFP and marked with an X in the figures.]

FP   Rays through the Focal Point

Rays from the object which pass through the focal point on their way to encountering the mirror reflect so as to travel paraxially away from the mirror.

CC   Rays through the Centre of Curvature

Rays from the object passing through the centre of curvature, labelled R in the figures, strike the mirror at normal incidence, $\theta_i = 0$. The reflected ray is directed back along the path of the incident ray [in the opposite direction].

The reflected rays continue along their specified paths until they are beheld by the eye of an observer.

ASIDE: Pairs of PA and FP rays are related by time-reversal symmetry. The CC rays are left invariant by time-reversal.

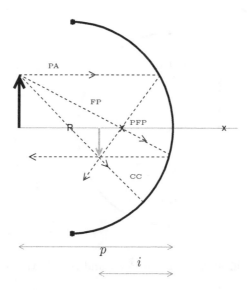

FIGURE 34.3   Formation of an Image by a Concave Mirror

The spherical concave mirror with radius of curvature R has focal length $f = +R/2$.

[The explicit plus sign emphasises that the focal point lies in front of the mirror.]

---

[3]Yes, this is indeed too good to be completely true! For objects far from the optical axis, the rays do not quite meet at a single point. This effect is called **spherical aberration** [of the mirror]. We presume that our objects lie near enough to the axis to justify the unique focal point approximation.

Hence, the object-image-focal-length master equation reads

$$\frac{1}{p} + \frac{1}{i} = \frac{2}{R} \qquad \Longrightarrow \qquad i = \frac{pR}{2p - R},$$

while the magnification is $M = \dfrac{-i}{p} = -\dfrac{R}{2p - R}$.

Two rather different behaviours are possible here, separated by a critical case.

$\boldsymbol{p > f}$ IF $p > R/2$ [*i.e.*, the object distance is greater than the focal length], THEN $i > 0$. The image is real and inverted.
IF $p > R$, THEN the image is reduced. IF $R > p > R/2$, THEN it is enlarged.

$\boldsymbol{p = f}$ IF $p = R/2$ [*i.e.*, the object resides at the focal point], THEN $i \to \infty$. No image is formed in this scenario.

$\boldsymbol{p < f}$ IF $p < R/2$ [*i.e.*, the object is nearer to the mirror than the focal length], THEN $i < 0$. The image is virtual [located behind the mirror], erect, and enlarged.

AND AT THE MALL

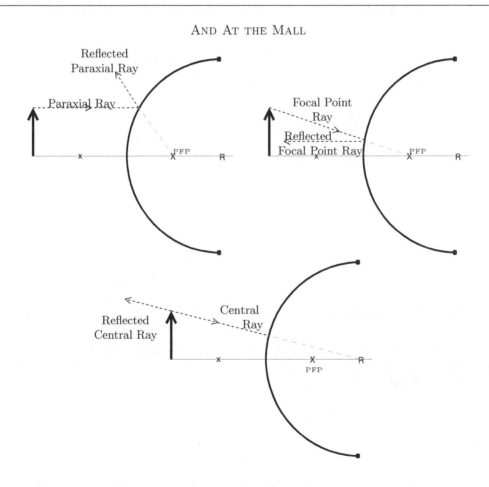

FIGURE 34.4   Rays through the Focal Point and the Centre of a Convex Mirror

Consider a convex spherical reflector with radius of curvature R. In this case also, three principal rays are sufficient to determine the placement and properties of the image of a generic object. The considerations underlying the analysis of the concave mirror are operative here as well.[4]

PA   Paraxial Rays

Rays from the object which are directed parallel to the axis encounter the mirror and are redirected as though they originated at the focal point of the convex mirror. The geometry of uniform curvature places the focal point inside the mirror, one-half of the distance to the centre of curvature.

[The principal focal point is labelled PFP and marked with an X in the figures.]

FP   Rays Toward the Focal Point

Rays initially directed toward the focal point wind up parallel to the axis.

CC   Rays Toward the Centre of Curvature [R]

These rays strike the mirror at normal incidence and are reflected directly back along their initial paths.

ASIDE: Pairs of these PA and FP rays exchange under time reversal, while the CC rays are unaffected.

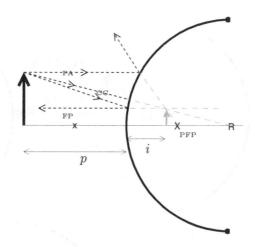

FIGURE 34.5   Formation of an Image by a Convex Mirror

As the focal point lies within the convex mirror, *i.e.*, in the physically inaccessible region, the focal length acquires a negative sign, $f = -R/2$. Thus, the object-image-focal-length master equation now reads

$$\frac{1}{p} + \frac{1}{i} = -\frac{2}{R} \qquad \Longrightarrow \qquad -i = \frac{pR}{2p+R},$$

while the magnification is $M = \dfrac{-i}{p} = \dfrac{R}{2p+R}$.

In the convex mirror, there is only one reflection regime on account of the inaccessibility of the focal point. All images are virtual [appearing inside the mirror], erect, and reduced.

---

[4]In particular, we assume that the object is sufficiently close to the optical axis that the effects of spherical aberration may be neglected.

Chapter 35

Refraction

Refraction effects arise due to changes in the speed of light as it crosses medium boundaries. In each medium, the speed is determined by the index of refraction, n, and the speed of light *in vacuo*, c, *i.e.*,

$$c_n = \frac{c}{n}.$$

Here we neglect dispersion effects and posit a single [wavelength and frequency independent] value of the index of refraction for each material through which light waves travel.

Representative Indices of Refraction						
Substance	Vacuum	Air	Water	Glass	Heavy Glass	Diamond
Refractive Index	1	1.0003	1.33	1.5	1.8	2.4

SNELL'S LAW   The angles of incidence, θ_1, and refraction, θ_2, at a [locally] planar boundary between two media with refractive indices n_1 and n_2, respectively, satisfy

$$n_1 \sin(\theta_1) = n_2 \sin(\theta_2).$$

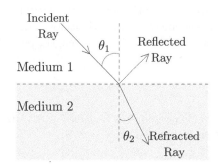

NORMAL INCIDENCE, $\theta_1 = 0 = \theta_2$, corresponds to the 1-d case(s) studied in Chapter 29.

FIGURE 35.1   Snell's Law

Snell's Law is sometimes called THE LAW OF REFRACTION.

HUYGENS' DERIVATION OF SNELL'S LAW

Figure 35.2 shows plane monochromatic waves passing from Medium 1, where they propagate with speed $c_1 = c/n_1$, into Medium 2, in which the wave speed is $c_2 = c/n_2$.

[Here we have assumed that $n_1 < n_2$ and $c_1 > c_2$.]

For the explication of Snell's Law, we pay careful attention to the quadrilateral straddling the boundary between the media, with sides consisting of: portions of two successive wave fronts [each of which resides entirely in a single medium], and their perpendiculars [showing the respective direction of travel of each end of the trailing wavefront].

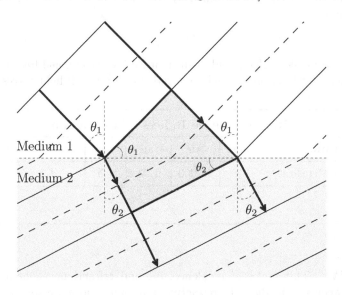

FIGURE 35.2   Huygens' Derivation of Snell's Law

In the time interval of one period, the trailing wave crest, originally lying entirely in Medium 1, advances through the interface between the media. The rightmost edge of the crest moves only in Medium 1 and ends up at the boundary. The wavelength, period, and speed of the wave in Medium 1 are related via

$$\lambda_1 = c_1\, \mathcal{T}\,.$$

Meanwhile, the leftmost edge of the trailing crest [impinging upon the boundary at the start of the interval] propagates forward within Medium 2, and the relation

$$\lambda_2 = c_2\, \mathcal{T}$$

holds. **The frequency and period are not affected by the passage of the waves from one medium into the other.**

The quadrilateral shown in Figure 35.2 is divided into two right triangles by the media boundary. Letting L represent the length of the boundary segment enclosed within the quadrilateral, it is apparent that

$$\lambda_1 = L\,\sin(\theta_1) \qquad \text{and} \qquad \lambda_2 = L\,\sin(\theta_2)\,.$$

Combining the above relations to form the ratio of the wavelengths [in the different media], one obtains

$$\frac{\lambda_1}{\lambda_2} = \frac{c_1}{c_2} = \frac{n_2}{n_1} \quad \text{AND} \quad \frac{\lambda_1}{\lambda_2} = \frac{\sin(\theta_1)}{\sin(\theta_2)} \qquad \Longrightarrow \qquad n_1 \sin(\theta_1) = n_2 \sin(\theta_2) \,.$$

MINIMUM TIME DERIVATION OF SNELL'S LAW

The PRINCIPLE OF MINIMAL TIME[1] avers that a plane wave passing from an initial point in space, $\vec{r}_1$, to a final point, $\vec{r}_2$, follows the path which minimises its transit time. Once more, we claim as self-evident that, in a homogeneous medium, the shortest path [*viz.*, the straight line from one site to the other] is the fastest.

In Figure 35.3, the initial and final positions have coordinates $\vec{r}_1 = (0\,,\,y_1)$ and $\vec{r}_2 = (a\,,\,y_2)$, respectively. The x-axis is aligned with the planar boundary between the media. *A priori*, the path of the wave is not specified. This indeterminacy is noted by allowing several arrows to emanate from $\vec{r}_1$ and converge on $\vec{r}_2$. The putative actual path consists of two straight-line segments joined at the crossing-point $\vec{X} = (x\,,\,0)$, found on the boundary.

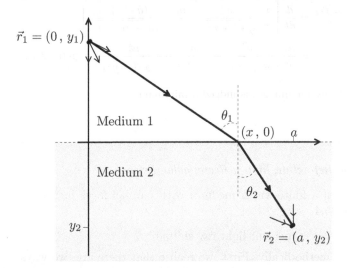

FIGURE 35.3 Minimal Time Derivation of Snell's Law

The path lengths in Medium 1 and Medium 2 are $\sqrt{x^2 + y_1^2}$ and $\sqrt{(a - x)^2 + y_2^2}$. As the wave speeds are $c_1 = c/n_1$ and $c_2 = c/n_2$, the respective propagation times are:

$$t_1(x) = \frac{n_1 \sqrt{x^2 + y_1^2}}{c} \qquad \text{and} \qquad t_2(x) = \frac{n_2 \sqrt{(a - x)^2 + y_2^2}}{c} \,.$$

Thus, the total time taken for the light to pass from $\vec{r}_1$ to $\vec{r}_2$ is

$$t_{12}(x) = t_1(x) + t_2(x) = \frac{n_1 \sqrt{x^2 + y_1^2}}{c} + \frac{n_2 \sqrt{(a - x)^2 + y_2^2}}{c} \,.$$

---

[1] *Cf.* the discussion found in Chapter 33.

To determine the minimal path, we set the derivative of $t_{12}(x)$ with respect to x equal to zero, obtaining a constraint which fixes the critical value of the crossing point.[2] The derivative of $t_{12}(x)$ with respect to x is

$$\frac{dt_{12}}{dx} = \frac{n_1}{c}\frac{x}{\sqrt{x^2 + y_1^2}} - \frac{n_2}{c}\frac{(a - x)}{\sqrt{(a - x)^2 + y_2^2}}.$$

Therefore, at the extremum, $x = x_c$,

$$\left.\frac{dt_{12}}{dx}\right|_{x=x_c} = 0 \iff n_1\frac{x_c}{\sqrt{x_c^2 + y_1^2}} = n_2\frac{(a - x_c)}{\sqrt{(a - x_c)^2 + y_2^2}}.$$

The geometry of the triangles appearing in Figure 35.3 makes it evident that

$$\frac{x_c}{\sqrt{x_c^2 + y_1^2}} = \sin(\theta_1) \quad \text{and} \quad \frac{(a - x_c)}{\sqrt{(a - x_c)^2 + y_2^2}} = \sin(\theta_2),$$

and hence $n_1 \sin(\theta_1) = n_2 \sin(\theta_2)$.

ASIDE: The second derivative of t_{12} with respect to x is

$$\frac{d^2 t_{12}}{dx^2} = \frac{d}{dt}\left[\frac{n_1}{c}\frac{x}{\sqrt{x^2 + y_1^2}} - \frac{n_2}{c}\frac{(a - x)}{\sqrt{(a - x)^2 + y_2^2}}\right]$$

$$= \frac{n_1}{c}\frac{y_1^2}{(x^2 + y_1^2)^{3/2}} + \frac{n_2}{c}\frac{y_2^2}{((a - x)^2 + y_2^2)^{3/2}} > 0, \forall x,$$

ensuring that the extremum, x_c, is indeed a minimum.

EXAMPLE [*Double Refraction from a Rectangular Prism*]

Light passing through a finite-width medium with parallel faces incurs a lateral shift, as exhibited in Figure 35.4.

Q: By what amount is the incident light ray shifted?

A: Let's work it out methodically. First, we realise that there are two ways of quantifying the displacement of the beam: the sideways displacement [*i.e.*, the horizontal distance from the point at which the extension of the initial ray would strike the bottom edge of the medium to that at which the actual ray emerges] and the perpendicular distance between the extended initial ray and the actual ray [*i.e.*, lateral shift].

If the medium were non-refracting [$n_r = n_i$], then the incident ray would emerge (from the base of the block) a distance of $L \tan(\theta_i)$ away from the point directly beneath its entry point. The point at which the refracted ray emerges is $L \tan(\theta_r)$. Thus, the sideways displacement of the beam is $\Delta X = L \tan(\theta_i) - L \tan(\theta_r)$. Trigonometrical definitions and identities allow the expression of **tangent** in terms of **sine**:

$$\tan(\theta) = \frac{\sin(\theta)}{\sqrt{1 - \sin^2(\theta)}}.$$

---

[2]It is critical in the sense that any other crossing point corresponds to an increased transit time.

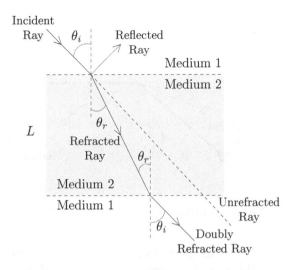

FIGURE 35.4   Double Refraction is Tantamount to a Lateral Translation or Shift

Snell's Law, $\sin(\theta_r) = \frac{n_i}{n_r} \sin(\theta_i)$, allows for expression of ΔX entirely in terms of the incident angle and the indices of refraction:

$$\Delta X = L \left[\frac{\sin(\theta_i)}{\sqrt{1 - \sin^2(\theta_i)}} - \frac{\sin(\theta_r)}{\sqrt{1 - \sin^2(\theta_r)}} \right] = L \frac{\sin(\theta_i)}{\sqrt{1 - \sin^2(\theta_i)}} \left[1 - \frac{\frac{n_i}{n_r} \sqrt{1 - \sin^2(\theta_i)}}{\sqrt{1 - \frac{n_i^2}{n_r^2} \sin^2(\theta_i)}} \right] .$$

Hence, the sideways displacement of the beam is a fixed fraction of the undeviated emerging distance, *i.e.*,

$$\Delta X = L \tan(\theta_i) \left[1 - \frac{\frac{n_i}{n_r} \sqrt{1 - \sin^2(\theta_i)}}{\sqrt{1 - \frac{n_i^2}{n_r^2} \sin^2(\theta_i)}} \right] .$$

The lateral shift of the beam is the projection of the sideways displacement perpendicular to the incident direction, *i.e.*, $\Delta d = \Delta X \cos(\theta_i)$. Therefore,

$$\Delta d = L \sin(\theta_i) \left[1 - \frac{\frac{n_i}{n_r} \sqrt{1 - \sin^2(\theta_i)}}{\sqrt{1 - \frac{n_i^2}{n_r^2} \sin^2(\theta_i)}} \right] .$$

Light passing from a slower medium into a faster one is bent away from the normal. Careful analysis of this effect reveals the phenomenon of **total internal reflection**.

$\theta_i = 0$ At normal incidence, the refracted ray is also normal, $\theta_r = 0$.

$0 < \theta_i < \theta_c$ For small values of the incident angle, refraction occurs, and $\theta_r > \theta_i$. This situation is illustrated by the dotted rays in Figure 35.5.

> ASIDE: Such behaviour is expected. According to Snell's Law, light is bent toward the
> normal when it passes into a medium of greater refractive index. The time-reversed
> beam (passing into the region of lesser refractive index) must then bend away from
> the normal.

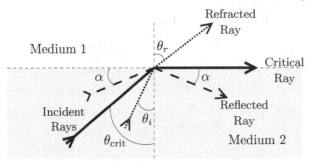

FIGURE 35.5   Total Internal Reflection

$\theta_i = \theta_c$ For a critical value of the incident angle, $\theta_i = \theta_c \in \left(0, \frac{\pi}{2}\right)$, $\sin(\theta_c) = \frac{n_i}{n_r}$, and therefore $\theta_r = \pi/2$. In this case, the refracted ray is directed along the interface between the media, as shown by the solid ray in Figure 35.5.

$\theta_i > \theta_c$ Once the incident angle exceeds its critical value, the refraction angle becomes imaginary-valued. The light ray in this instance cannot escape the medium within which it is travelling, but instead **reflects** from the interface between the media as though it were a plane mirror. The dashed line in Figure 35.5 evinces this behaviour.

EXAMPLE [*Total Internal Reflection in a Swimming Pool*]

Imagine that you are lying on your back on the bottom of a swimming pool which is 1.5 m deep and 6 m wide. Take the index of refraction of the water in the pool to be 4/3.

Q: How high do the walls of the pool appear to be?

A: First we shall construct a crude model. [In the Addendum, we'll consider a better version.]

The ray which runs along the top surface of the water and refracts to your eye [shown by the heavier dashed line in Figure 35.6] satisfies $\frac{4}{3} \sin(\theta_c) = 1$. Therefore, $\theta_c = \sin^{-1}(3/4) \simeq 0.84806$ rad $\simeq 48.59°$. Thus, the apparent height of the point on the wall corresponding to the water-level in the pool is h, satisfying:

$$\tan(\theta_c) = \frac{3}{h} \quad \text{and} \quad \tan(\theta_c) \simeq 1.1339 \qquad \Longrightarrow \qquad h \simeq 2.65\,\text{m}.$$

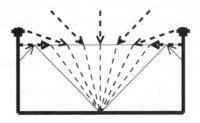

FIGURE 35.6   The View from the Bottom of a Swimming Pool

EXAMPLE [*Measuring the Index of Refraction of a Material*]

Q: How might one measure the index of refraction of a transparent material?

A: Here is one reliable way to deduce indices of refraction.

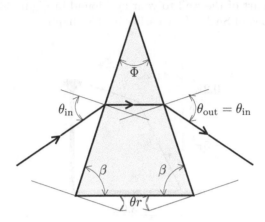

FIGURE 35.7   Measuring Index of Refraction with a Trigonal Prism

(1)   Fabricate a trigonal prism comprised of the material, like that shown in Figure 35.7. Measure the angle, Φ, at one of the corners.

(2)   Shine a narrow beam of light [approximating a single ray] into one side of the prism and observe how it emerges from the other.

(3)   Rotate the prism [or the light source] until the angle at which the light emerges matches precisely the angle at which the light enters the prism, *viz.*, $\theta_{\text{out}} = \theta_{\text{in}}$. This situation is illustrated in Figure 35.7.

(4)   Theorems pertaining to the geometry of triangles [the sum of the interior angles is exactly π, and the sum of complementary angles is precisely $\pi/2$] provide relations among the angles in the figure. In the context of the trigonal prism, these read:

$$\pi = 2\,\beta + \Phi \qquad \text{and} \qquad \beta + \theta_r = \tfrac{\pi}{2}\,.$$

Together, these relations determine that

$$\theta_r = \frac{\Phi}{2}\,.$$

(5)   Approximating the index of refraction of the medium [air, usually] surrounding the sample by 1 and applying Snell's Law yield an expression for the unknown index in terms of measured quantities:

$$\begin{aligned} n_i\,\sin(\theta_i) &= n_r\,\sin(\theta_r) \\ \sin(\theta_{\text{in}}) &= n\,\sin(\Phi/2) \end{aligned} \qquad \Longrightarrow \qquad n = \frac{\sin(\theta_{\text{in}})}{\sin(\Phi/2)}\,.$$

Knowledge of the geometry of the prism, specified by the angle Φ, and experimental determination of the unique angle at which the light both enters and emerges are sufficient to determine the index of refraction of a homogeneous material.

ADDENDUM: A Better Model for the Apparent Height of the Pool Walls

The pool walls are unlikely to end at the top surface of the water [because if they did, then the water would flow too readily out of the pool]. A more accurate model will account for the portion of the wall, taken to be 0.2 m, extending above the waterline. The optical path of the ray from the top part of the wall to your eye, found in Figure 35.8, is not unlike those shown in the derivations of Snell's Law earlier in this chapter.

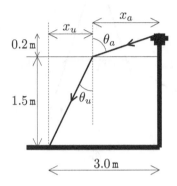

FIGURE 35.8   A Closer Look at the Apparent Height of the Pool Walls

A ray originates from the point $(3.0, 1.7)$ m, strikes the surface at $(x_u, 1.5)$, and thence proceeds to your eye. The horizontal distances traversed under and above the water, x_u and x_a, sum to 3 m. The incident and refraction angles satisfy

$$\tan(\theta_a) = \frac{x_a}{0.2} \quad\text{and}\quad \tan(\theta_u) = \frac{x_a}{1.5}.$$

Finally, the `tangent` function can be re-expressed in terms of `sine` via the identity

$$\tan(\alpha) = \frac{\sin(\alpha)}{\sqrt{1 - \sin^2(\alpha)}}.$$

Therefore,

$$3 = 0.2\,\frac{\sin(\theta_a)}{\sqrt{1 - \sin^2(\theta_a)}} + 1.5\,\frac{\sin(\theta_u)}{\sqrt{1 - \sin^2(\theta_u)}}.$$

Snell's Law relates the angles above and under the water's surface: $\sin(\theta_a) = \frac{4}{3}\sin(\theta_u)$. With all of these pieces, the value of θ_u is constrained to satisfy

$$3 = \frac{4}{5}\,\frac{\sin(\theta_u)}{\sqrt{9 - 16\sin^2(\theta_u)}} + \frac{3}{2}\,\frac{\sin(\theta_u)}{\sqrt{1 - \sin^2(\theta_u)}}.$$

This equation is not easily solved. The standard technique is to contrive to eliminate all the square root terms by carefully squaring twice. This leads to an expression which is quartic in $\sin^2(\theta_u)$, and is both tedious and unenlightening. A more direct path to the solution, $\sin(\theta_u) \simeq 0.7417932\ldots$, is afforded by numerical or graphical techniques. With this result in hand, $\theta_u \simeq 47.88° \simeq 0.83574\,\mathrm{rad}$.

At this angle, the top of the 1.7 m wall, whose base is 3 m away, appears to be at height $h = 3/\tan(\theta_u) \simeq 2.71\,\mathrm{m}$!

Chapter 36

Through a Glass Darkly

In this chapter we concern ourselves with the double refraction of light passing through a lens, a slab of transparent material with non-parallel sides [medium boundaries]. Our analysis is predicated upon, and guided by, the fact that any smooth boundary is locally flat, and hence

Snell's Law applies locally.

Although all shapes are allowed, we shall restrict our study to uniformly-curved surfaces [spherical sections] sharing a single axis of symmetry. Many possible combinations arise, of which a few illustrative examples are found in the table just below.

‖		**plano-plano** This is the double refraction through a parallelepiped prism situation encountered in Chapter 35. Horizontal light rays impinging from either the left or the right meet the boundary at normal incidence and thus refractive effects do not occur.
D		**plano-convex** Here one side is flat and the other bulges outward. Light rays entering horizontally from the left pass through the flat surface at normal incidence and meet the curved surface at an angle which increases with distance from the optical axis.
)(		**plano-concave** One aspect of the lens is flat, while the other dimples inward. Paraxial rays from the left strike the planar side at normal incidence and proceed to the curved side, where they experience position-dependent scattering.
()		**double-convex** [*a.k.a. biconvex*] In this chapter, we'll examine equal curvature thin double-convex lenses in some detail.
)(		**double-concave** [*a.k.a. biconcave*] We will also consider equal curvature thin double-concave lenses.

The curvatures and the thickness of lenses may differ [as illustrated]. In addition, lenses may be made of non-homogeneous composite materials. These added complications can be modelled. Here, we specialise to symmetric curvature and thin lenses.

SYMMETRIC THIN LENSES

The double refraction occurring at the medium boundaries of [thin] lenses enables light emitted from an object to undergo focussing [redirection], which leads to formation of images.

The MASTER EQUATIONS developed for mirrors perfectly accommodate lenses also. Defining object and image distances as in the mirror case,

  $p = $ [minimum, perpendicular] distance from object to the [centre of the] lens and

  $i = $ [minimum, perpendicular] distance from image to the [centre of the] lens,

the master equations read:

$$\frac{1}{p} + \frac{1}{i} = \frac{1}{f} \quad \text{and} \quad M = \frac{-i}{p}.$$

Next, we shall investigate the optical effects of thin double-convex/concave lenses.

THIN DOUBLE-CONVEX LENS

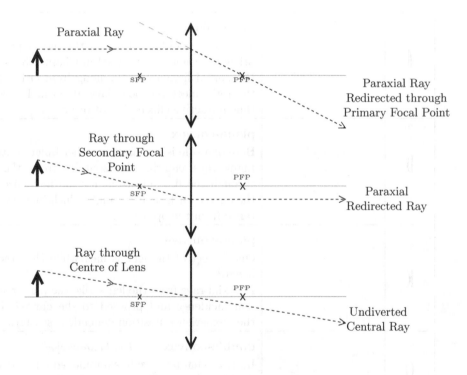

FIGURE 36.1 Rays through the Foci and the Centre of a Double-Convex Lens

Three particular rays emanating from the object suffice to determine the optical properties of the image produced by a thin symmetric double-convex lens. These three rays are directed paraxially, through the secondary focus, and through the centre of the lens, respectively. Each ray is illustrated in Figure 36.1 and discussed below. The biconvex shape of the lens [*i.e.*, thin at its extremities and thick in the middle[1]] is denoted by the arrowheads ↕.

---

[1]Not unlike PK when he fails to exercise adequately and consumes too many sweets.

PA   Paraxial Rays

Rays which are parallel to the optical axis are diverted through the **primary focal point** [PFP] by the action of the lens.[2]

FP   Rays through the Focal Point

Rays which pass through the **secondary focal point** [SFP] *en route* to the lens are redirected parallel to the optical axis.

CL   Rays through the Centre of the Lens

Rays striking the lens at low angle near the optical axis behave like those experiencing double refraction through a parallelepiped. The direction of travel of the ray is minimally affected, and, provided that the lens is thin enough, the finite shift in the path of the ray is also inconsequential. Thus, under the approximations that are in force, these rays pass undeflected through the lens.

That each of these three types of rays emanate from [a reference point on] an object may be used to construct the image of the object. This is illustrated in Figure 36.2.

ASIDE: The PA and FP rays occur in time-reversed pairs. The CL rays are time-reversal symmetric.

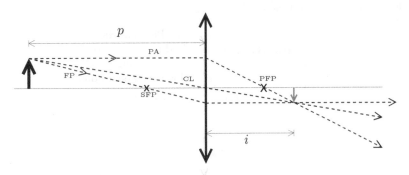

FIGURE 36.2   Formation of an Image by a Double-Convex Lens

In applying the master relations to the double-convex lens, we note that its focal length is positive, *i.e.*, $f > 0$.

ASIDE: How the focal length might be predicted from the shape of the lens, rather than simply measured, is addressed in the Addendum to this chapter.

According to the object-image-focal-length relation, the image appears at

$$i = \frac{p\,f}{p - f} \qquad \text{with magnification} \qquad M = -\frac{f}{p - f}.$$

In this instance, there are two image regimes separated by a critical case.

$p > f$ IF $p > f$, THEN $i > 0$, AND the image is real and inverted.

For an object residing beyond the focal distance, an inverted real image is formed on the other side of the lens.

$$\text{IF} \quad \left\{ \begin{array}{c} p > 2f \\ f < p < 2f \end{array} \right\}, \quad \text{THEN the image is} \quad \left\{ \begin{array}{c} \text{REDUCED} \\ \text{ENLARGED} \end{array} \right\}.$$

---

[2]This is approximately the case for rays near the optical axis. Rays lying further away are increasingly affected by **spherical aberration** [of the lens] and do not pass through a common focal point.

$p = f$ IF $p = f$, THEN $i \rightarrow \pm\infty$, AND no image is formed [at finite distance].

The PA and CL rays from an object located at the focal point lie parallel to one another. The FP ray is ill-defined when the object resides at the focal point.

$p < f$ IF $p < f$, THEN $i < 0$, AND the image is virtual and erect.

An object lying within the focal distance yields an image whose apparent location is on the object side of the lens, and thus it is virtual. Such images are always erect and enlarged. [Biconvex lenses are commonly used as magnifying glasses.]

THIN DOUBLE-CONCAVE LENS

Let's investigate the image-forming properties of a thin symmetric double-concave lens. Again, we start with the behaviours of three canonical rays.[3] [These thin lenses are represented schematically by inverted arrowheads.]

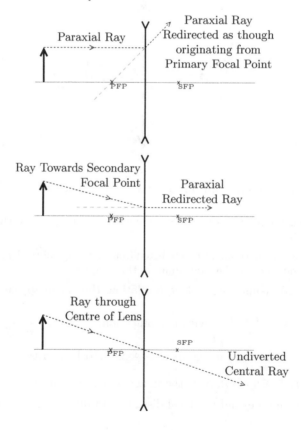

FIGURE 36.3   Particular Rays Passing through a Double-Concave Lens

---

[3] Neglecting to consider spherical aberration.

PA   Paraxial Rays

Rays which are parallel to the optic axis are made to diverge in such manner that they appear to come from the primary focal point [PFP].

FP   Rays Directed Toward the Secondary Focal Point

Rays which would pass through the secondary focal point, in the absence of the lens, are redirected parallel to the optic axis.

CL   Rays through the Centre of the Lens

These rays are undeflected [for the reasons invoked in the double-convex case].

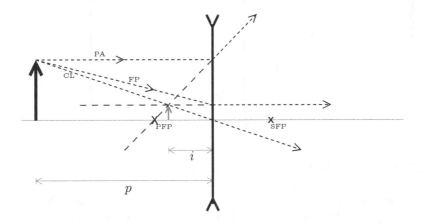

FIGURE 36.4   Formation of an Image by a Concave Lens

Figure 36.4 employs the three canonical rays to relate an object to its image produced by a double-concave lens. The focal distance is deemed negative for the double-concave lens, *i.e.*, $f < 0$, and thus [under normal circumstances, *i.e.*, $p > 0$] i is always negative.

$$-i = \frac{p\,|f|}{p + |f|}, \qquad \Longrightarrow \qquad M = \frac{|f|}{p + |f|}.$$

Typical images formed by double-concave lenses are virtual, erect, and diminished in size.

LENSES IN SERIES AND PARALLEL

P   Parallel arrangements of lenses occur in instances of binocular or multi-ocular [fly's eye] vision. The two or more channels function independently.

S   Series combinations of two or more lenses exhibit a wide range of behaviours. Nonetheless, two fundamental principles govern all cases and inform our analyses.

The lenses act independently of one another. Each takes an object and produces an image.

The image formed by a lens serves as the object for the next lens in the series.

We shall examine a small set of cases in the sequence of examples which follow below.

EXAMPLE [*Two Thin Double-Convex Lenses in Series—1*]

Two thin double-convex lenses have focal lengths $f_1 = 20\,\text{cm}$ and $f_2 = 15\,\text{cm}$. The lenses are arranged in series on a common optical axis and their centres are separated by 55 cm. An object extending from the axis is located 60 cm to the left of the first lens. This configuration of lenses is illustrated in Figure 36.5.

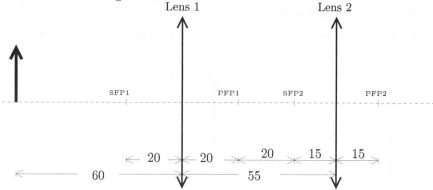

FIGURE 36.5   Two Biconvex Lenses in Series—1

Light from the object passes through the first lens and forms an image. The optical properties of this image are determined by application of the master equations and corroborated by analysis of the three principal rays. This is shown in Figure 36.6.

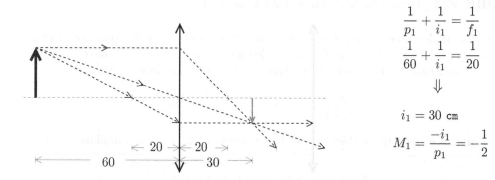

$$\frac{1}{p_1} + \frac{1}{i_1} = \frac{1}{f_1}$$

$$\frac{1}{60} + \frac{1}{i_1} = \frac{1}{20}$$

$$\Downarrow$$

$$i_1 = 30 \ \text{cm}$$

$$M_1 = \frac{-i_1}{p_1} = -\frac{1}{2}$$

FIGURE 36.6   The Action of the First Lens in Series

The image produced by Lens 1 [with all of its attendant properties, *e.g.*, that it is located 30 cm to the right of Lens 1, and that it is inverted and one-half the size of the original object] becomes the object for Lens 2. The object distance, p_2, then is 25 cm. The effects of the second lens are ascertained by applying the master equations and tracing the three principal rays as is explicitly shown in Figure 36.7.

The effect of these two lenses in series has been to take the original object and produce a real and erect image at a distance of 37.5 cm to the right of Lens 2. The overall magnification of the image is

$$M_{21} = M_2\, M_1 = +\frac{3}{4}\,.$$

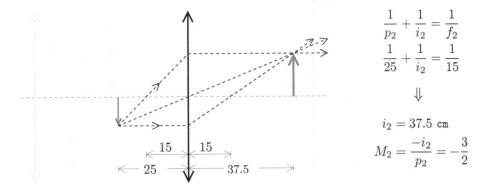

$$\frac{1}{p_2} + \frac{1}{i_2} = \frac{1}{f_2}$$

$$\frac{1}{25} + \frac{1}{i_2} = \frac{1}{15}$$

⇓

$$i_2 = 37.5 \text{ cm}$$

$$M_2 = \frac{-i_2}{p_2} = -\frac{3}{2}$$

FIGURE 36.7 The Action of the Second Lens in Series—1

This last result is consistent with the final image's being erect and somewhat diminished.

EXAMPLE [*Two Thin Double-Convex Lenses in Series—2*]

Consider the same lenses as in the previous example, placed 40 cm apart, as illustrated in Figure 36.8.

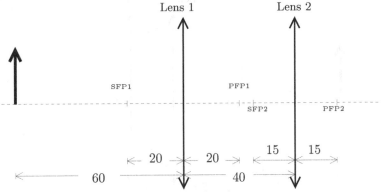

FIGURE 36.8 Two Biconvex Lenses in Series—2

Everything about the passage of light through the first lens remains exactly as it was in the preceding example. Therefore, the object for the second lens is 10 cm to the left of the lens, is inverted, and is diminished in height by a factor of two. This is illustrated in Figure 36.9, where the location and properties of the image formed by the second lens are determined. On account of the object's being found within the focal distance of the lens, the ray tracing had to be done with some additional care.

The effect of these two lenses in series has been to take the original object and produce a real and erect image appearing 30 cm to the left of Lens 2. The overall magnification of the image is

$$M_{21} = M_2 \, M_1 = -\frac{3}{2}.$$

This last result is consistent with the final image's being inverted and somewhat magnified.

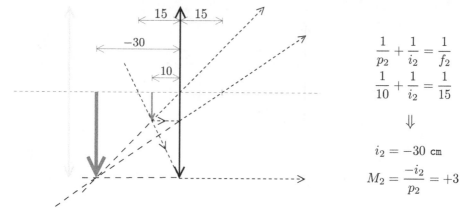

$$\frac{1}{p_2} + \frac{1}{i_2} = \frac{1}{f_2}$$

$$\frac{1}{10} + \frac{1}{i_2} = \frac{1}{15}$$

$$\Downarrow$$

$$i_2 = -30 \text{ cm}$$

$$M_2 = \frac{-i_2}{p_2} = +3$$

FIGURE 36.9   The Action of the Second Lens in Series—2

EXAMPLE [*Two Thin Double-Convex Lenses in Series—3*]

Consider the same lenses as in the previous examples, this time placed 15 cm apart, as illustrated in Figure 36.10.

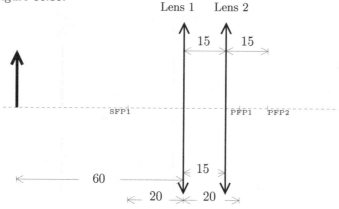

FIGURE 36.10   Two Biconvex Lenses in Series—3

Everything about the passage of light through the first lens remains exactly as before. So, the object for the second lens is 15 cm to the right of the lens, is inverted, and is diminished in height by a factor of two. At first glance, this situation seems contrary to reason, since the VIRTUAL OBJECT lies beyond [to the right of] the second lens. Not to worry, though: with proper care one can reliably ascertain the location and properties of the final image [as is done in Figure 36.11].

The proper procedure to deal with a virtual object goes as follows.

    PA   There is a paraxial ray coming from the left of the second lens directed toward the position of the virtual object [PA → (PA)]. Upon its encounter with the lens this ray is diverted through the principal focal point and continues to the right [PA].

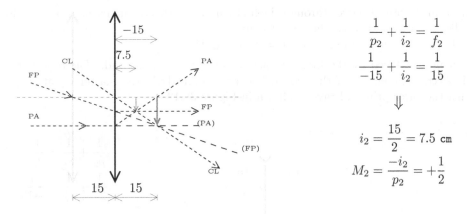

$$\frac{1}{p_2} + \frac{1}{i_2} = \frac{1}{f_2}$$

$$\frac{1}{-15} + \frac{1}{i_2} = \frac{1}{15}$$

$$\Downarrow$$

$$i_2 = \frac{15}{2} = 7.5 \text{ cm}$$

$$M_2 = \frac{-i_2}{p_2} = +\frac{1}{2}$$

FIGURE 36.11   The Action of the Second Lens in Series—3

CL   The ray directed toward the virtual object through the centre of the second lens [CL] is not affected by its passage.

FP   There is a ray coming from the left which passes through the secondary focal point of lens 2 on its way toward the virtual object [FP → (FP)]. [It is an accident of this arrangement of lenses that this ray also passed through the centre of lens 1.] Upon reaching the second lens this ray is redirected parallel to the optical axis.

These three [modified] rays intersect at a point located $i = 7.5 \text{ cm}$ to the right of lens 2. Hence, the overall magnification of the ultimate image is $M_{21} = M_2 M_1 = -\frac{1}{4}$. The final image is real, inverted, and diminished by a factor of four.

EXAMPLE   [*Double-Convex–Double-Concave Lenses in Series—1*]

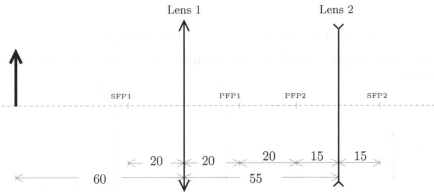

FIGURE 36.12   Biconvex–Biconcave Lenses in Series—1

A thin double-convex lens having focal length $f_1 = 20 \text{ cm}$ and a thin double-concave lens with focal length $f_2 = -15 \text{ cm}$ are placed in series along a common optical axis. The centres of the lenses are separated by 55 cm. An object extending from the axis is located 60 cm to the left of the double-convex lens. This configuration is illustrated in Figure 36.12.

Light from the object passes through the first lens and forms the image shown in Figure 36.6 and discussed as part of the first multi-lens example. We shall not reproduce the analysis here, but merely quote the results: $i = 30\,\text{cm}$ and $M_1 = -\frac{1}{2}$.

The image produced by the first lens acts as object for the second. The object distance, p_2, is $25\,\text{cm}$. The effects of the second lens are obtained via the master equations and by tracing the three principal rays as shown in Figure 36.13.

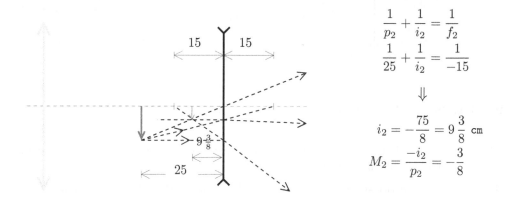

$$\frac{1}{p_2} + \frac{1}{i_2} = \frac{1}{f_2}$$

$$\frac{1}{25} + \frac{1}{i_2} = \frac{1}{-15}$$

$$\Downarrow$$

$$i_2 = -\frac{75}{8} = 9\frac{3}{8}\ \text{cm}$$

$$M_2 = \frac{-i_2}{p_2} = -\frac{3}{8}$$

FIGURE 36.13   The Action of the Second Lens in Series—1

The effect of these two lenses in series has been to take the original object and produce a virtual and inverted image appearing $9.375\,\text{cm}$ to the left of Lens 2. The overall magnification of the image is

$$M_{21} = M_2\,M_1 = -\frac{3}{16}\,.$$

This is consistent with the final image's being inverted and much diminished in size.

EXAMPLE   [*Double-Convex–Double-Concave Lenses in Series—2*]

Consider the same lenses as in the previous example, placed $40\,\text{cm}$ apart, as illustrated in Figure 36.14.

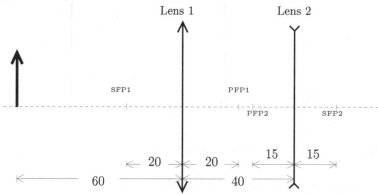

FIGURE 36.14   Biconvex–Biconcave Lenses in Series—2

Everything about the passage of light through the first lens remains exactly as it was in the preceding example. Therefore, the object for the second lens is 10 cm to the left of the lens, is inverted, and is diminished in height by a factor of two. This is illustrated in Figure 36.15, where the location and properties of the image formed by the second lens are determined. On account of the object's being found within the focal distance of the lens, the ray tracing had to be done with a wee bit of additional care.

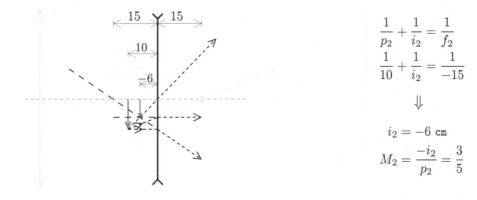

$$\frac{1}{p_2} + \frac{1}{i_2} = \frac{1}{f_2}$$

$$\frac{1}{10} + \frac{1}{i_2} = \frac{1}{-15}$$

$$\Downarrow$$

$$i_2 = -6 \text{ cm}$$

$$M_2 = \frac{-i_2}{p_2} = \frac{3}{5}$$

FIGURE 36.15 The Action of the Second Lens in Series—2

The effect of these two lenses in series has been to take the original object and produce a virtual and inverted image appearing 6 cm to the left of Lens 2. The overall magnification of the image is

$$M_{21} = M_2 \, M_1 = -\frac{3}{10}.$$

This last result is consistent with the final image being inverted and quite diminished.

EXAMPLE [*Double-Convex–Double-Concave Lenses in Series—3*]

Consider the same lenses as in the previous examples, this time placed 15 cm apart, as illustrated in Figure 36.16.

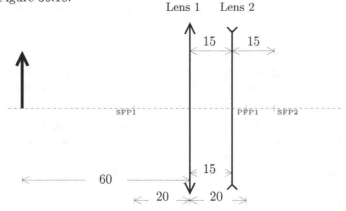

FIGURE 36.16 Biconvex–Biconcave Lenses in Series—3

Light passes through the biconvex lens as previously described. So, the object for the second lens is 15 cm to the right of the lens and is inverted and diminished in height by a factor of two. Again, Lens 2 has a virtual object. Not to worry, though: with proper care one can reliably ascertain the location and properties of the final image [as found in Figure 36.17].

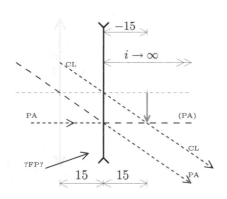

$$\frac{1}{p_2} + \frac{1}{i_2} = \frac{1}{f_2}$$

$$\frac{1}{-15} + \frac{1}{i_2} = \frac{1}{-15}$$

Therefore,

$$\frac{1}{i_2} = 0 \quad \Longrightarrow i_2 \to \infty$$

$$M_2 = \frac{-i_2}{p_2} \text{ is undefined}$$

FIGURE 36.17   The Action of the Second Lens in Series—3

It happens that this particular system of lenses has a feature, exposed in Figure 36.17 and discussed below, which prevents the formation of an image.

PA   The paraxial ray originating from the left of the second lens and directed toward the position of the virtual object [PA → [PA]] is diverted away from the optic axis, as though it had originated from the principal focal point of the biconcave lens. This ray then continues to the right [PA].

CL   The ray directed from left to right toward the virtual object that happens to pass through the centre of the second lens [CL] is not affected by its passage through the lens.

FP   It proves impossible to construct a ray from the left of the second lens which is directed at both the virtual object and the secondary focal point of the lens. [Because the [virtual] object lies at the same distance from the lens as is the focal point, a line passing through both is perpendicular to the optical axis and, therefore, cannot possibly come from the left.]

Hence, there are only two rays [rather than the expected three] and examination of the figure reveals that they lie parallel to one another. Thus, there is no point [save that at infinity] at which these rays intersect. These conclusions are corroborated by the master equations.

Simple series combinations of lenses were used to build the first prototypes [and succeeding generations] of telescopes and microscopes. It was with a plano-convex and plano-concave pair of lenses that Galileo observed irregularities of surface structure on the Moon[4] and discovered the four largest moons of Jupiter[5] (all *circa* 1609). [One of these Jovian satellites,

---

[4]This falsified the [meta-]physical Platonic presumption, then in force, that all celestial objects exhibited mathematical—spherical—perfection.

[5]This falsified the then-current maximally geocentric [Ptolemaic] model in which all celestial bodies orbited about the centre of the Earth.

Io, is considered in Chapter 1 and in the Epilogue.] An enhanced design employing pairs of double convex lenses is attributed to Kepler (*circa* 1611[6]).

Optical microscopes were developed in the late 1500s or so,[7] and were immediately put to use in scientific investigations.

ADDENDUM: The Lens Maker's Formula

Lenses are often intended for specific purposes. They are commonly used in instruments with which we extend our range of vision [*e.g.*, telescopes and microscopes], and in corrective devices with which we ameliorate defects in our vision [*e.g.*, eyeglasses]. In any case, such designs inevitably require lenses possessing particular focal lengths.

The **lens maker's formula** provides a simple expression relating the size and [symmetric] curvature of a disk-like transparent medium to its **optical power**, *a.k.a.* **refractive power**, defined as the reciprocal of the focal length. The units of optical power are diopters [dp], with dimension of inverse-length, where one diopter corresponds to $1\,\mathrm{m}^{-1}$.

Suppose that a disk-like lens has front and back radii of curvature R_f and R_b. [The convexity/concavity of the lens has bearing on the signs of these curvatures]. Further suppose that the lens is comprised of homogeneous material with index of refraction n_{lens} and is immersed in a surrounding medium with index n_{env}. In the so-called THIN LENS APPROXIMATION,

$$\text{Refractive Power} = \frac{1}{f} = \left[\frac{n_{\mathrm{lens}} - n_{\mathrm{env}}}{n_{\mathrm{env}}}\right]\left(\frac{1}{R_{1\mathrm{st}}} - \frac{1}{R_{2\mathrm{nd}}}\right),$$

where the radii of curvature are labelled according to the order in which the light passes through the surfaces.

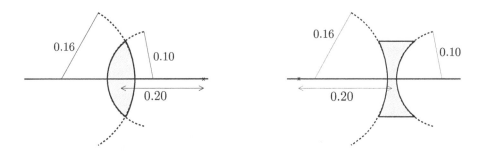

FIGURE 36.18   Optical Power of Thin Lenses

For example, in the biconvex lens shown in the left panel of Figure 36.18, $R_{1\mathrm{st}} = +0.10\,\mathrm{m}$, while $R_{2\mathrm{nd}} = -0.16\,\mathrm{m}$, taking account of the respective positions of the centres of curvature and the customary left-to-right passage of light rays. With $n_{\mathrm{lens}} \simeq \frac{17}{13}$, the refractive power of the biconvex lens, when immersed in air, $n_{\mathrm{air}} \simeq 1$, is

$$\frac{1}{f_L} = \left[\frac{17}{13} - 1\right]\left(\frac{1}{0.10} - \frac{1}{-0.16}\right) = 5\ \mathrm{m}^{-1}.$$

---

[6]Kepler's achievement came too late to aid his former patron, Tycho Brahe (1546–1601), who is considered to be the finest and the last great naked-eye observational astronomer.

[7]Some claim that the idea originates with Roger Bacon (*circa* 1267), derived from earlier medieval and Islamic sources.

Therefore, the focal length of this lens is $+0.20\,\text{m}$.

The biconcave lens possessing similar geometry and the same indices of refraction, shown in the right panel of Figure 36.18, has refractive power

$$\frac{1}{f_R} = \left[\tfrac{17}{13} - 1\right]\left(\frac{1}{-0.16} - \frac{1}{0.10}\right) = -5\,\text{m}^{-1}.$$

The focal length of this biconcave lens is $-20\,\text{cm}$.

Accounting for the thickness of the lens, d [measured along the optic axis], results in a more complicated expression:

$$\text{Optical Power} = \frac{1}{f} = \left[\frac{n_{\text{lens}} - n_{\text{env}}}{n_{\text{env}}}\right]\left(\frac{1}{R_{\text{1st}}} - \frac{1}{R_{\text{2nd}}} + \frac{(n_{\text{lens}} - n_{\text{env}})\,d}{n_{\text{lens}}\,R_{\text{1st}}\,R_{\text{2nd}}}\right).$$

Chapter 37

Temperature and Thermometry

THERMODYNAMICS   The science of thermodynamics is concerned with the production and flow of thermal energy, *a.k.a.* heat.

TEMPERATURE   The temperature associated with a thermodynamic system may be thought of as a measure of the density of heat [energy] contained in the substance comprising the system.

[We shall forego the technical definition of temperature until Chapter 47.]

Three commonly used temperature scales are:

C   Celsius [so-named in honour of Anders Celsius, *circa* 1742], the standard SI unit

K   kelvin [in honour of Lord Kelvin, *circa* 1848], the "other" standard SI unit

°F   degrees Fahrenheit [after Gabriel Daniel Fahrenheit, *circa* 1724], the Imperial unit

Here are a few words on the origin of each of the three systems of temperature units.

°F   Fahrenheit was one of the first scientists to realise that **thermometry** [reliable measurement of temperature] was possible. Fahrenheit lived in the Netherlands and calibrated his measuring devices to the local conditions. In his estimation, 0 °F was as low a temperature as one might ever encounter,[1] while 100 °F was the highest temperature that he thought he might endure.[2] The Fahrenheit scale accommodates, in a range from zero to one hundred, the temperatures most relevant to people residing in Northern Europe.

C   In contrast, the Celsius scale is imbued with a sense of *cartesian rationalism* wherein the laws of nature, assumed to be immutable, take priority over human concerns. The Celsius scale is based upon the properties of water [freezing occurs at 0 C and boiling at 100 C], and is thus somewhat independent of pesky [water-based!] life forms.

K   The **absolute temperature** scale [K] was introduced during a vainglorious era (*circa* 1840) in which it was fervently believed that [classical] physics afforded a framework for the complete understanding of all of nature. The kelvin scale is predicated upon [what was believed to be] a fundamental reconciliation of the Celsius scale with the revolutionary notion of **absolute zero** temperature.

---

[1] It is evident that Fahrenheit did not live in Northern Canada, nor did he reside in Siberia.
[2] Clearly, Fahrenheit did not swelter in the tropics, nor did he reside in Siberia.

Notwithstanding the choice of scale, it remains true that temperatures can be meaningfully assigned and compared. Those corresponding to a number of physical phenomena appear in the table below, quoted[3] in each of the three common scales.

PHENOMENON	C	K	°F
Solar Surface Temp.	~ 5500	~ 5800	~ 10000
Cookies Bake	~ 175	~ 450	~ 350
Water Boils	100	373	212
[HUMAN] Body Temperature	37	310	98.6
Room Temperature	~ 20	~ 293	~ 68
Water Freezes	0	273	32
"Minus Forty" or "Forty Below"	-40	233	-40
Absolute Zero	-273.15	0	-459.67

A Practical Comparison of the Three Standard Temperature Unit Schemes

Conversion from one temperature system to another is easily accomplished, as all three are linearly related.

	Celsius	Kelvin	Fahrenheit
Celsius	C = C	C = K - 273.15	C = (5/9) × (F - 32)
Kelvin	K = C + 273.15	K = K	K = F × 5/9 + 255.37
Fahrenheit	F = C × 9/5 + 32	F = K × 9/5 - 459.67	F = F

THERMOMETER   A thermometer is a calibrated device for measuring temperature.

Thermometers come in a variety of forms. One type is comprised of a tube partially filled with liquid, while another consists of a bi-metallic coil. Both of these types of thermometer, illustrated in Figure 37.1, rely on incommensurate thermal expansion of two or more substances for their operation. Still other physical correlates with temperature [*e.g.,* electrical resistivity[4]] can be exploited for thermometry.

---

[3]These values are accurate, but not necessarily precise. For instance, 0 C ≡ 273.15 K.
[4]See VOLUME III, Chapter 19 for a discussion of temperature-dependent resistivity.

 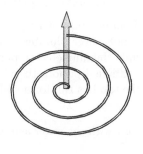

FIGURE 37.1 A Rudimentary Picture of Thermometric Devices

THERMAL EXPANSION

It is generally observed[5] that substances $\left\{\begin{array}{c}\text{EXPAND}\\\text{CONTRACT}\end{array}\right\}$ when $\left\{\begin{array}{c}\text{HEATED}\\\text{COOLED}\end{array}\right\}$.

Thermal expansion is well-modelled by the following LAWS OF LINEAR PROPORTION:

LENGTH   IF L_0 is the length at temperature T_0, THEN the length at temperature T is

$$L(T) = L_0 \left[1 + \alpha \left(T - T_0 \right) \right],$$

where α is the **Lineal Thermal Expansion Coefficient** of the particular material.

AREA   IF A_0 is the area at temperature T_0, THEN the area at temperature T is

$$A(T) = A_0 \left[1 + \widetilde{\alpha} \left(T - T_0 \right) \right],$$

where $\widetilde{\alpha}$ is the **Areal Thermal Expansion Coefficient** of the particular material.

[Truth be told, I just invented the notation "$\widetilde{\alpha}$," but will atone momentarily.]

VOLUME   IF V_0 is the volume at temperature T_0, THEN at temperature T,

$$V(T) = V_0 \left[1 + \beta \left(T - T_0 \right) \right],$$

where β is the **Volume Thermal Expansion Coefficient** of the particular material.

The SI unit for all three of the thermal expansion coefficients is inverse Celsius, $\left[\texttt{C}^{-1} \right]$.

ASIDE: For isotropic materials with lineal coefficient α, the areal and volume coefficients are approximately $\widetilde{\alpha} \approx 2\,\alpha$ and $\beta \approx 3\,\alpha$, respectively.

----

[5]The most prominent exception to this general observation is H_2O at temperatures near $0\,\texttt{C}$. Compounds exhibiting thermal shrinkage over broader temperature ranges include Zirconium Tungstate [ZrW_2O_8] and Scandium Trifluoride [ScF_3].

EXAMPLE [*Thermal Expansion of Length*]

A straight section of copper pipe with a length of 3 m [determined at 20 C] is found in one part of a household's hot-water plumbing system. The operating temperature of the hot-water tank is 50 C, while the ambient temperature inside the house is 20 C. When the water does not flow for an extended period, the pipe and the water inside it cool[6] to the ambient temperature. Once the tap has run for a while, the water in the pipe and the pipe itself are at 50 C.

[HOW heat flows along with the water and into the copper pipe is discussed in Chapter 39.]

Q: Given that the coefficient of thermal expansion for copper is $\alpha_{\mathrm{Cu}} = 1.7 \times 10^{-5}$ C^{-1}, what allowance must be made to accommodate the expected change in length of the pipe?

A: An equivalent statement of the law of linear proportion governing lineal thermal expansion is:

$$\frac{\Delta L}{L_0} = \alpha \, \Delta T \, .$$

In this case,

$$\Delta L = L_0 \, \alpha_{\mathrm{Cu}} \, \Delta T = \left(3 \text{ m} \right) \left(1.7 \times 10^{-5} \text{ C}^{-1} \right) \left((50 - 30) \text{ C} \right) = 1.02 \times 10^{-3} \text{ m} \simeq 1 \text{ mm} \, .$$

Therefore one should ensure that the installed pipe has at least one millimetre of "wiggle room" to reduce stresses on the pipe and the fittings which connect it to the rest of the system.

[Unrelieved thermally induced stress can lead to material fatigue and the development of leaks.]

EXAMPLE [*Thermal Expansion of Area*]

A straight section of copper pipe with an inner radius of 12.5 mm and an outer radius of 13.5 mm [radii determined at 20 C] is part of a hot-water household plumbing system. The range of temperatures throughout which the pipe typically operates is 20 C to 50 C.

Q: Given that the coefficient of thermal expansion for copper is $\alpha_{\mathrm{Cu}} = 1.7 \times 10^{-5}$ C^{-1}, by what factor does the cross-sectional area of the pipe change?

A: The law of linear proportion for areas is also expressed by

$$\frac{\Delta A}{A_0} = \widetilde{\alpha} \, \Delta T \, .$$

Making the reasonable approximation $\widetilde{\alpha}_{\mathrm{Cu}} = 2 \, \alpha_{\mathrm{Cu}}$,

$$\frac{\Delta A}{A_0} = \widetilde{\alpha}_{\mathrm{Cu}} \, \Delta T = \left(2 \times 1.7 \times 10^{-5} \text{ C}^{-1} \right) \left((50 - 30) \text{ C} \right) = 6.8 \times 10^{-4} \, .$$

For all intents and purposes, this has negligible effect on the flow of the water in the pipe. Thus, we were not forced to incorporate this temperature effect when discussing Fluid Dynamics in Chapters 6 and 7.

-----

[6]The phenomenology of cooling is discussed in the Addendum to Chapter 40. The three principal means by which substances may be heated or cooled are discussed in Chapters 39 and 40.

EXAMPLE [*Thermal Expansion of Volume*]

An amount of mercury [Hg] occupying two cubic centimetres at $0\,\text{c}$ is contained in a hollow glass cylinder with cross-sectional area equal to $1\,\text{mm}^2$. A line marks the height of the Hg in the tube when the system is at $0\,\text{c}$. The volume coefficients of thermal expansion for mercury and this type of glass are $\beta_{\text{Hg}} = 1.82 \times 10^{-4}\ \text{c}^{-1}$ and $\beta_{\text{glass}} = 2.7 \times 10^{-5}\ \text{c}^{-1}$, respectively.

Q: At what distance from the $0\,\text{c}$ line does the top of the Hg column lie when the temperature is $100\,\text{c}$?

A: To determine this distance, one must account for the expansion of the Hg and of the cavity in the glass in which the Hg is enclosed.

$$\Delta V_{\text{Hg}} = V_0\, \beta_{\text{Hg}}\, \Delta T = \left(2\ \text{cm}^3\right)\left(1.82 \times 10^{-4}\ \text{c}^{-1}\right)\left((100-0)\ \text{c}\right) = 3.64 \times 10^{-2}\ \text{cm}^3$$
$$= 36.4\ \text{mm}^3\,,$$
$$\Delta V_{\text{glass}} = V_0\, \beta_{\text{glass}}\, \Delta T = \left(2\ \text{cm}^3\right)\left(2.7 \times 10^{-5}\ \text{c}^{-1}\right)\left((100-0)\ \text{c}\right) = 5.4 \times 10^{-3}\ \text{cm}^3$$
$$= 5.4\ \text{mm}^3\,.$$

Thus, the [excess] volume of Hg thrust above the line is

$$\delta(\Delta V) = \Delta V_{\text{Hg}} - \Delta V_{\text{glass}} = 31.0\ \text{mm}^3\,.$$

In a column with cross-sectional area equal to one square millimetre, this fluid stands $3.1\,\text{cm}$ high. In this model, we have neglected to account for the ever-so-slight increase in the cross-sectional area of the segment of the tube into which the mercury rises.

EXAMPLE [*Thermal Expansion and Density*]

Ice floats in water, according to Archimedes' Principle, because the density of ice at $0\,\text{c}$, $917\,\text{kg/m}^3$, is less than that of liquid water.

[The specific gravity of water ice is approximately 0.917.]

The lineal coefficient of thermal expansion for ice is approximately $5.1 \times 10^{-5}\ \text{c}^{-1}$, and, as ice is an isotropic solid, $\beta_{\text{H2O}} \simeq 3\,\alpha_{\text{H2O}} = 1.53 \times 10^{-4}\ \text{c}^{-1}$.

Q: Below what temperature would ice sink in water?

A: Suppose that we start with one cubic decimetre of water ice at $0\,\text{c}$. The mass of this amount of ice is $0.917\,\text{kg}$. Neutral buoyancy occurs when the specific gravity of the sample is equal to 1. For this to occur, the volume must be reduced to 0.917 of its original amount, *i.e.*,

$$\frac{\Delta V}{V_0} = \frac{0.917 - 1.000}{1.000} = -0.083\,.$$

Also,

$$\frac{\Delta V}{V_0} = \beta_{\text{H2O}}\, \Delta T = \beta_{\text{H2O}}\,(T - 0) = \beta_{\text{H2O}}\, T\,.$$

Thus,

$$T_{\text{neutral}} = \frac{-0.083}{\beta_{\text{H2O}}} \simeq \frac{-0.083}{1.53 \times 10^{-4}\,[\text{c}^{-1}]} = -542.5\,\text{c}\,.$$

Clearly such a temperature does not exist—it lies far below absolute zero—and hence it is always the case that water ice floats in water.

ADDENDUM: A Few Words on Absolute Zero

The first thing that one should suspect about anything touted as being "absolute" is that it may be obtained via an unwarranted extrapolation. It is worth recollecting the standard disclaimer:

"Past performance is not a reliable predictor of future trends."

And yet, the existence of absolute zero seemed incontrovertible once the pressures of various gases as functions of temperature were cast in a universal form for the purpose of comparison. Figure 37.2 presents a rough picture of the experimental determination of pressure *vs.* temperature for dilute gases [*cf.* Gay-Lussac's Law mentioned in Chapter 10].

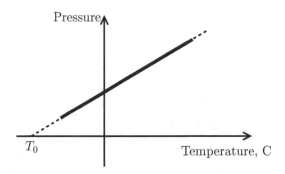

FIGURE 37.2 Pressure *vs.* Temperature for Gases

Each time that the range of experimental data was extended, by refinements in technique or technology, the new data lay along the line generated by extrapolation from the set of previously existing data. The persistence of this remarkable linear diminution of the [magnitude of the] gas pressure with decreasing temperature led physicists to entertain seriously the idea of a unique temperature, T_0, for which the pressure [of all such "universal" gases] would be exactly zero. As negative pressure is unphysical for a classical fluid, such a T_0 would provide a lower bound for temperatures. It then makes perfect sense to anchor one's temperature scale upon this value.

All of this was hotly debated until it was realised that there are two essential impediments, one classical [incontrovertably formalised by the early 1900s] and the other quantum mechanical [realised in the 1920s], to cooling a physical system to absolute zero.

CLASSICAL   The classical impediment is that absolute zero can never be attained in a finite number of refrigeration steps. This result is codified as the **Third Law of Thermodynamics**.

QUANTUM   The classical laws of nature are superseded by quantum mechanical laws as absolute zero is approached. Energy discretisation and zero-point energy effects become manifest and conspire to prevent one from cooling to the "absolute state of rest" envisaged classically.

Chapter 38

Heat

HEAT   Heat is the thermal form of energy.

- Heat is correlated with temperature.

  [Recall that temperature measures something akin to heat density.]

- The SI unit for heat is the joule. Other units are in current usage too.

  c   calorie   [1 c = 4.186 J]
  One calorie is the amount of thermal energy required to raise the temperature of one gram of water [H_2O] from 14.5 c to 15.5 c.

  C   Calorie [*a.k.a.* kilocalorie]   [1 C = 1000 c = 4186 J]
  This is the "Calorie" that one encounters in food product labelling.

  BTU   British Thermal Unit   [1 BTU = 1055 J]
  This Imperial unit is often employed in furnace and air-conditioner ratings, as well as in wholesale pricing of natural gas.

- It is an indisputable fact[1] that

  Heat is observed to flow [spontaneously] from HOT **to** COLD,
  i.e., from regions of higher temperature to regions of lower temperature.

    ASIDE: A *Gedanken* experiment involves the cup of hot coffee resting nearby on your desk. Heat from the coffee warms the cup, that part of the desk beneath the cup and the nearby air. We should be quite alarmed were the coffee to begin to boil while the cup, the desk, and the surrounding air cooled! It is not surprising that early models treated heat as a diffusing fluid substance, *cf.* Chapter 8.

SPECIFIC HEAT CAPACITY   The specific heat capacity relates changes in the temperature of a certain amount of a particular substance to changes in its thermal energy content, *i.e.,*
$$\Delta Q = M\,C\,\Delta T.$$
On the RHS, M denotes the mass of the material comprising the system, C is the **specific heat capacity** of the substance, and ΔT is the change in its temperature, in c. Since the LHS, ΔQ, is the change in the heat content of the given amount of the particular substance it must have units of J. As mass is conventionally measured in **kg**, the SI units of specific heat capacity are
$$[\,C\,] = \frac{\text{J}}{\text{kg}\cdot\text{c}}.$$
A common alternative unit is calorie [or Calorie] per gram per Celsius.

---

[1] In Chapter 47 this observation will be cast as an expression of the SECOND LAW OF THERMODYNAMICS.

ASIDE: Here are three distinct usages of the symbol "C" in quick succession: Calorie, specific heat capacity, and Celsius. Fortunately, context will always serve to disambiguate expressions involving these quantities.

Let's argue for the necessity of each of the factors on the RHS of the expression for ΔQ.

M The amount of energy required to effect a particular temperature change should be directly proportional to the amount of substance present.

C One might reasonably expect that different substances have differing degrees of thermal responsiveness.

There are two important qualifications that must be made.

o   Specific heat capacities have a [very tiny] temperature dependence, which we have chosen to ignore by declaring C to be constant.

o   The temperature response of a system to the addition of heat depends on the manner in which the heat is added. This effect is also ignored for now, pending discussion in Chapters 42 and 45.

ΔT The amount of energy needed to produce a certain temperature change should be proportional to the size and sign of the temperature increment.

HEAT CAPACITY   The heat capacity of a given amount of substance is the energy required to raise its temperature by $1\,C$. That is,

$$\text{Heat Capacity} = M\,C\,.$$

Heat capacity is a convenient shorthand quantity.

MOLAR SPECIFIC HEAT   The molar specific heat of a substance [*a.k.a.* its molar heat capacity] is the amount of heat required to raise the temperature of one `mole` of that substance by $1\,C$. Molar specific heats are commonly used in analyses of chemical reactions.

DULONG–PETIT LAW   Pierre-Louis Dulong (1785–1838) and Alexis Thérèse Petit (1791–1820) observed that the molar heat capacity of solid materials is approximately constant, with value $C_{\text{molar}} \simeq 3\,R \simeq 25\,$ `J`/(`mole·K`).

THERMAL EQUILIBRIUM   A system is in thermal equilibrium with its environment when there is zero net heat flow [on average] into or out of the system. Overall thermal equilibrium also requires that internal heat flows—those occurring among the system's constituent parts—vanish [on average]. The conditions of thermal equilibrium ensure that there is a UNIQUE temperature associated with the system and that it remains constant in time.

THERMAL ISOLATION   A thermodynamic system is said to be thermally isolated if heat energy is NOT able to flow into [out of] the system from [to] its surroundings. [In practice, systems are effectively isolated if their net heat fluxes are sufficiently small, compared to their total heat content, that their temperatures change very slowly.]

EXAMPLE [*Equilibrium Temperature of a Two-Component Mixture*]

Two samples of material are brought into thermal contact[2] in such a way that heat can flow from one to the other, but no heat leaves the composite system.

[*I.e.*, the mixture is isolated.]

The mass, specific heat capacity, and temperature of substance A are $\{M_A, C_A, T_A\}$, while the corresponding quantities for B are $\{M_B, C_B, T_B\}$, respectively.

Q: What is the final temperature of the mixture [after it has reached thermal equilibrium]?

A: As the mixture is thermally isolated, whatever heat departs from one substance must be absorbed by the other. This internal flow of heat ceases when the temperatures of the two substances are precisely equal. The temperature, T_*, may be determined by application of the thermal energy conservation constraint:

$$\Delta Q_A + \Delta Q_B \equiv 0 \qquad \text{where} \qquad \left\{ \begin{array}{l} \Delta Q_A = M_A\, C_A\, (T_* - T_A) \\ \Delta Q_B = M_B\, C_B\, (T_* - T_B) \end{array} \right\}.$$

Substituting the expressions for the changes in the thermal content of A and B into the constraint leads to

$$M_A\, C_A\, (T_* - T_A) + M_B\, C_B\, (T_* - T_B) = 0\,.$$

Isolation of T_* is straightforwardly accomplished,

$$\left(M_A\, C_A + M_B\, C_B\right)T_* = M_A\, C_A\, T_A + M_B\, C_B\, T_B\,,$$

enabling the precise determination of the final equilibrium temperature,

$$T_* = \frac{M_A\, C_A\, T_A + M_B\, C_B\, T_B}{M_A\, C_A + M_B\, C_B}\,.$$

The final temperature of the mixture is the heat-capacity-weighted average of the initial temperatures of the substances. A few special cases bolster this result.

T IF $T_A = T_B = T$, THEN

$$T_*\Big|_{T_A=T_B=T} = \frac{M_A\, C_A\, T + M_B\, C_B\, T}{M_A\, C_A + M_B\, C_B} = T\,.$$

IF the substances are initially at the same temperature, AND no heat enters or leaves the system, THEN the temperature of the mixture must be equal to the temperature common to the inputs.

C IF $C_A = C_B = C_*$, THEN

$$T_*\Big|_{C_A=C_B=C_*} = \frac{M_A\, C_*\, T_A + M_B\, C_*\, T_B}{(M_A + M_B)C_*} = \frac{M_A\, T_A + M_B\, T_B}{M_A + M_B}\,.$$

IF the two substances possess the same specific heat capacity [*e.g.*, they may be comprised of the same material], THEN the final temperature of the mixture is the mass-weighted average of the initial temperatures.

---

[2]The lawyers insist that the assumption that chemical reactions between the substances do not occur be made explicit.

M IF $M_A = M_B = M_*$, THEN, in analogy with the above case,

$$T_*\Big|_{M_A=M_B=M_*} = \frac{C_A T_A + C_B T_B}{C_A + C_B}.$$

IF the mixture is comprised of equal-mass quantities of the two substances, THEN the temperature of the mixture is the specific-heat-capacity-weighted average of the initial temperatures.

The expression relating a change in temperature to a change in heat content is valid for a single [homogeneous] substance and can be generalised to apply to mixtures.

EXAMPLE [*Specific Heat Capacity of a Two-Component Mixture*]

A mixture of two substances, A and B, is at temperature T. The mass and specific heat capacity of substance A are $\{M_A, C_A\}$, while the corresponding quantities for B are $\{M_B, C_B\}$.

Q: What is the effective specific heat capacity of the mixture?

A: Any addition of heat, ΔQ, to the mixture must be partitioned between the components so as to raise their respective temperatures by the same amount. Thus,

$$\Delta Q = \Delta Q_A + \Delta Q_B \qquad \text{AND} \qquad \left\{ \begin{array}{l} \Delta Q = M_{\text{Tot}} C_{\text{AB}} \Delta T \\ \Delta Q_A = M_A C_A \Delta T \\ \Delta Q_B = M_B C_B \Delta T \\ M_{\text{Tot}} = M_A + M_B \end{array} \right\}.$$

Substituting the expressions for the changes in the thermal content of the composite, A, and B systems into the constraint and simplifying result in the following chain of equalities.

$$M_{\text{Tot}} C_{\text{AB}} \Delta T = M_A C_A \Delta T + M_B C_B \Delta T$$
$$\implies \qquad M_{\text{Tot}} C_{\text{AB}} = M_A C_A + M_B C_B$$
$$\implies \qquad C_{\text{AB}} = \frac{M_A C_A + M_B C_B}{M_A + M_B}.$$

The specific heat capacity of the mixture is the mass-weighted average of the specific heat capacities of the constituent substances.

One often thinks of a mixture as consisting of two or more chemically distinct materials. However, one can envision mixtures comprised of coexisting **phases** of a single substance.

PHASE   The material phase or **state** of a quantity of matter is typically GAS, LIQUID, or SOLID. [Other states, *e.g.,* PLASMA, also exist, but we'll not concern ourselves with these.]

ASIDE: Magnetisation lends itself to a description in terms of phases, as well.

PHASE CHANGE   A phase change is the conversion of the physical system from one state into another.

A number of possible conversions are defined in the table below.

	→	←	
Solid	MELTING	FUSION	**Liquid**
Liquid	VAPORISATION	CONDENSATION	**Gas**
Gas	DEPOSITION	SUBLIMATION	**Solid**

One can readily imagine many instances of the fusion ↔ melting and condensation ↔ vaporisation transitions of water.

FUSION	Formation of ice on a lake, production of ice cubes
MELTING	The spring thaw, ice cubes in a beverage
CONDENSATION	Formation of raindrops, "fog" on a mirror
VAPORISATION	Humidity found near bodies of water, a steaming kettle

Sublimation and **deposition** are less familiar phenomena.

SUBLIMATION	Sublimation occurs on late winter days, when snowbanks shrink appreciably without the formation of pools of liquid water. "Dry ice," solid CO_2, is used in frozen food preparation, transportation, and storage, in part because it sublimates to gas rather than melting to liquid when placed in thermal contact with objects at room temperature. [Dry ice is also employed in theatrical productions. Sublimated, cool, and heavy CO_2 gas promotes condensation of water vapour in the surrounding air, producing a wispy ground-hugging fog.]
DEPOSITION	Deposition of thin-film coatings [some only a few atoms thick] is a vital aspect of high-tech manufacturing. The hoar frost decorating lawns, trees, and rooftops on chilly mornings is deposited water vapour.

Figure 38.1 shows the evolution of a thermally isolated sample of ice [solid H_2O], with an initial temperature well below 0 c, receiving thermal energy at a constant rate.[3] The ice warms up at a constant rate, in keeping with the notion of specific heat capacity developed earlier in this chapter. This trend persists until the temperature reaches 0 c, whereupon it temporarily stops rising.

In the plateau region at $T = 0$ c, the constituent fractions of relative phase are changing. The system passes from entirely solid, through decreasing amounts of solid together with increasing amounts of liquid, to all liquid, all the while remaining at a fixed temperature. Once the entire sample has melted, the temperature of the water rises at a constant pace governed by the specific heat capacity of liquid water. Another plateau corresponding to the phase transformation from liquid to gas occurs at 100 c.

While the phase is changing, it is evident that heat is being added with no corresponding increase in the temperature of the [two-coexisting-phase] system. The energetics of these phase change plateaus are well-modelled by writing

$$\Delta Q = M L \, .$$

---

[3]Constancy of the rate of heating means that the time parameter serves as a proxy for the total amount of energy added to the sample. Furthermore, it is implicitly assumed that the heat is added slowly enough that all parts of the sample are at [very nearly] the same temperature at each instant.

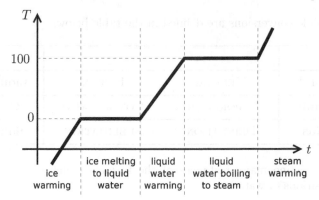

FIGURE 38.1   Temperature *vs.* Time for a Fixed Amount of Water Being Heated at a Constant Rate and Undergoing Phase Changes

Here, the heat added [the LHS] is equal to the input power multiplied by the temporal duration of the plateau. The RHS is the product of the mass of the material undergoing the phase change, M, and the substance- and transition-specific latent heat, L.

LATENT HEAT   The latent heat is the amount of heat which must be added or removed in order to effect the transformation of $1\,\mathrm{kg}$ of substance from one phase to another at the critical temperature. The SI units for latent heat are joules per kilogram [$\mathrm{J/kg}$].

CRITICAL TEMPERATURE   The critical temperature(s) for a particular material are those at which the system's phase changes. Only at the critical temperature can the two[4] phases co-exist in equilibrium.

[The critical temperature depends on properties of the system.]

For water, the latent heats of fusion and vaporisation along with the critical temperatures [under standard conditions[5]] are presented in the table below.

Phase Transition			Latent Heat [$\mathrm{kJ/kg}$]		T_C [C]
FUSION	water	$\longrightarrow$ ice	L_F	= 333	0
VAPORISATION	water	$\longrightarrow$ steam	L_V	= 2256	100

In the process illustrated in Figure 38.1, heat was added to warm and then melt the ice, to warm and then evaporate the liquid water into steam, and subsequently to further heat the steam. The reverse processes of condensation and solidification require removal of the latent heat from the system.

   ASIDE: When it snows on very calm winter days, the latent heat released by the condensation and solidification of water vapour can accumulate in the air nearby and produce a significant rise in the air temperature.

---

   [4]It is possible to have three phases, *e.g.*, ice, water, and water vapour, simultaneously present in a system at a fixed temperature.
   [5]For example, at ambient pressure equal to $1\,\mathrm{atm}$.

Chapter 39

Convective and Conductive Heat Flow

The transfer of heat from one point in space to another is of paramount significance in thermodynamics.

Q: How does heat flow from one place to another?

A: Via three distinct means: **convection**, **conduction**, and **radiation**.

CONVECTION   In convection, the heat energy borne by a moving object or flowing substance is carried along by its bulk motion.

Instances of convective heat transfer abound. Here are a few.

Volcanic Eruption   Molten rock from the depths surges to the surface and runs in red-hot rivulets.

"Forced Air" Heating Systems   Air, warmed by a furnace, is propelled by a fan through ducts and into rooms.

"Passive" Heating Systems   If you've ever lived in a house with [old-fashioned] water radiator heating, you have encountered a double convective system. The first convection is the flow of heat-laden water through the radiator pipes. [In older systems, this is accomplished by exploiting the positive and negative buoyancies of warmer (less dense) and cooler (more dense) water. The warmed water rises and is replaced by an inflow of cooler water. A steady circulation ensues. In modern systems, a pump ensures reliable flow of heated water through the system.] The second convection involves the flow of air. The hot water flowing in the pipes warms the air in the vicinity of the radiator. This warmer air then rises toward the ceiling, pulling cooler air from the lower central part of the room toward the radiator.

The Gulf Stream   The ocean current which runs from the Caribbean to Northern Europe carries sufficient heat energy to moderate the Irish–British–Scandinavian–Baltic climates.

It is not possible to develop a simple model describing convective heat transfer precisely because the heat flow is incidental to the motion of the substance bearing the heat.

CONDUCTION   In conduction, heat flows within a material without bulk movement of the substance.

Examples of conductive heat transfer include the following phenomena.

A Pot On a Stove Element   The pot handle grows hotter as heat from the stove element is conducted through the bottom and sides of the pot and into the handle.

An Exterior Door   An exterior door with southern exposure can become very warm. Some of this heat conducts through to the inner surface. In wintertime, heat from within the door flows outward to the exterior, leaving the inner surface cool to the touch.

A MODEL OF CONDUCTIVE HEAT FLOW

A homogeneous block of material with uniform cross-section is shown in Figure 39.1.

>   ASIDE: The assumptions of homogeneity of substance and uniformity of shape may be some-
>   what relaxed by consideration of a composite structure comprised of blocks which are them-
>   selves homogeneous and uniform, as we shall see later in this chapter and again in the next.

The block has two opposing faces of area A, separated by distance L, at [constant] temperatures T_C and $T_H > T_C$.

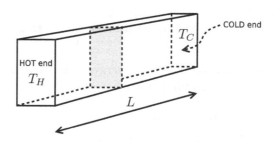

FIGURE 39.1   Conduction of Heat through a Block of Material

CLAIM: The STEADY STATE AVERAGE rate at which heat flows through this homogeneous
uniform block of substance is

$$\frac{\Delta Q}{\Delta t} = \frac{k\,A\,\Delta T}{L}.$$

The LHS is the rate at which heat conducts through the block of material **from** the HOT
end **to** the COLD end. The SI unit for the rate of heat flow is the watt [W]. On the RHS is
arrayed a collection of factors.

k The thermal conductivity of the material comprising the block is a measured
constant whose value is specific to the material. Substances with large values of
k are good conductors of heat, whereas those with small k are good insulators.

>   Contrast the metallic fins on the back of the processor chip in your computer
>   with the sweater on your back when the lecture room is cool.

A The cross-sectional area of the block proportionally influences the rate at which
heat is conducted.

>   Contrast your urge to curl up when the lecture room is cool with your incli-
>   nation to stretch out when it is too warm.

$\boldsymbol{\Delta T}$ The difference in temperature between the ends of the block directly affects
the rate of conductive heat flow.

>   It is standard energy- [and cost-]saving practice for home and business ther-
>   mostat settings to be lowered in winter and raised in summer. In each season,
>   this diminishes the temperature difference between indoors and out, thereby
>   reducing the flow of thermal energy out of or into the dwelling. Thus, less
>   energy is consumed to replace the lost heat or dispose of the excess warmth.

L   Greater separation between the HOT and COLD surfaces reduces the rate at which heat flows through the block.

> Consider your inclination to layer two or more items of clothing when cold and to remove layers when overly warm.

Now that the expression for the rate of heat conduction has been parsed and all of the dependencies [on the composition of the material and on its size] agreed upon, one must address those words STEADY STATE and AVERAGE found in the statement of the claim.

AVERAGE   The rate computed by taking the amount of heat flowing through the block within a time interval Δt, and dividing by Δt, is the **time-average rate throughout the time interval.**   [It doesn't hurt to remind ourselves of this, though.]

STEADY STATE   An implicit assumption is that the physical situation remains unchanged throughout the time interval. That the dimensions of the block remain constant is unremarkable. **The temperature difference across the block is assumed to be fixed.** *I.e.,* despite the flow of heat from the hot side to the cold side, the temperature at each end is not changing [on the time scale set by Δt]. This constancy requires there be infinite [read "large"] reservoirs of heat associated with the substances in contact with the ends of the block. The reservoir on the HOT side is at temperature T_H, while that on the COLD side is at temperature T_C.

It is also assumed that there is no heat flow through the sides of the block of material. [Most often, this is because adjacent to the particular block under consideration are other blocks extending between the same two HOT and COLD reservoirs.]

In this model of heat conduction through the block, the temperature field[1] has the properties:

x   Cross-sectional slices of the block are at uniform temperature

t   The temperatures at the ends, T_H and T_C, and throughout the block, remain constant in time

Under these assumptions,[2] heat flows at a constant rate and the system is in steady state.

THERMAL CONDUCTIVITY   The thermal conductivity, k, of a particular material is a phenomenological parameter which quantifies the relative ability of that substance to conduct heat. That is:

$$\left\{ \begin{array}{c} \text{LARGE} \\ \text{SMALL} \end{array} \right\} \text{ values of } k \text{ indicate } \left\{ \begin{array}{c} \text{GOOD} \\ \text{POOR} \end{array} \right\} \text{ heat conductivity.}$$

Select values of k are provided in the following table.

Substance	Argon	Air	Water	Glass	Ice	Lead	Copper	Diamond
k [W/(m·K)]	0.016	1/40	0.6	1.1	2	35	360	~ 1000

---

[1]The temperature field is a scalar-valued function of position in space.
[2]Constancy of ΔT is relaxed in the context of NEWTON'S LAW OF COOLING, discussed in the Addendum to Chapter 40.

THERMAL RESISTIVITY   The thermal resistivity of a particular material substance is the converse of its conductivity, while the conductivity is the reciprocal of the resistivity:

$$\rho_{\text{thermal}} = \frac{1}{k} \qquad \Longleftrightarrow \qquad k = \frac{1}{\rho_{\text{thermal}}} \, .$$

THERMAL RESISTANCE   In common usage, the thermal resistance,

$$R_{\text{thermal}} = \frac{L}{k} = \rho_{\text{thermal}} \, L \, ,$$

is often employed in expressing the rate of conductive heat transfer,

$$\frac{\Delta Q}{\Delta t} = \frac{A \, \Delta T}{R_{\text{thermal}}} \, .$$

There is nothing new in this reformulation.

> ASIDE: The insulating effect of a particular type and grade of building material is given by its so-called R–value. Thermal resistance is equivalent to R–value, except that Imperial units are used for the latter.

Q: Why bother with thermal resistances [R–values]?
A: These combine simply and naturally. That is, batts of R–20 fibreglass insulation stacked two-deep in an attic space have an effective thermal resistance of R–40.

[We'll verify this simple law of series composition in the example immediately below!]

EXAMPLE [*Thermal Conduction: Series*]

Two slabs composed of different material are arranged in **series**. It is assumed that the slabs possess the same face area and are perfectly overlapping. The slab parameters are:

$$\{k_1, \; A, \; L_1\} \qquad \text{and} \qquad \{k_2, \; A, \; L_2\} \, .$$

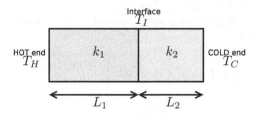

FIGURE 39.2   Series Conduction of Heat

Q: In what manner does heat flow through a series composite of materials?
A: In steady state the flow of heat through the first material must be exactly equal to that through the second material, AND the temperature at the interface between the materials must be constant at an intermediate value, $T_H > T_I > T_C$.

Q: What is the effective thermal conductivity of a series composite of materials?

A: With reference to Figure 39.2, the steady state heat flow and temperature difference constraints are

$$\frac{\Delta Q_{(12)}}{\Delta t} \equiv \frac{\Delta Q_1}{\Delta t} \equiv \frac{\Delta Q_2}{\Delta t} \qquad \text{and} \qquad \Delta T_{(12)} = \Delta T_1 + \Delta T_2.$$

These [aptly named] SAME and SUM constraints determine the net, or effective, thermal conductivity for the system. According to the model for heat conduction:

$$\frac{\Delta Q_1}{\Delta t} = \frac{k_1 \, A \, \Delta T_1}{L_1}, \qquad \frac{\Delta Q_2}{\Delta t} = \frac{k_2 \, A \, \Delta T_2}{L_2}, \qquad \text{and} \qquad \frac{\Delta Q_{(12)}}{\Delta t} = \frac{k_{(12)} \, A \, \Delta T_{(12)}}{L_{(12)}}.$$

Isolating the temperature differences in the above relations yields:

$$\Delta T_{(12)} = \frac{L_{(12)}}{k_{(12)}} \frac{1}{A} \frac{\Delta Q_{(12)}}{\Delta t}, \qquad \Delta T_1 = \frac{L_1}{k_1} \frac{1}{A} \frac{\Delta Q_1}{\Delta t}, \qquad \text{and} \qquad \Delta T_2 = \frac{L_2}{k_2} \frac{1}{A} \frac{\Delta Q_2}{\Delta t}.$$

Thus, the SUM constraint reads

$$\frac{L_{(12)}}{k_{(12)}} \frac{1}{A} \frac{\Delta Q_{(12)}}{\Delta t} = \frac{L_1}{k_1} \frac{1}{A} \frac{\Delta Q_1}{\Delta t} + \frac{L_2}{k_2} \frac{1}{A} \frac{\Delta Q_2}{\Delta t}.$$

As both materials and the aggregate have the SAME heat flux and face area,

$$\frac{L_{(12)}}{k_{(12)}} = \frac{L_1}{k_1} + \frac{L_2}{k_2}.$$

This yields an expression for the effective thermal conductivity of the series composition of two blocks of material. Rewritten in terms of thermal resistances, it reads:

$$R_{(12)} = R_1 + R_2.$$

The effective thermal resistance of an aggregate of blocks arranged strictly in series is the sum of the thermal resistances of the blocks.

Q: What is the [constant] temperature at the interface of the blocks [in steady state]?

A: The temperature differences across the blocks may be explicitly cast in terms of the [as yet unknown] intermediate temperature, T_I, *i.e.*,

$$\Delta T_1 = T_H - T_I \qquad \text{and} \qquad \Delta T_2 = T_I - T_C.$$

The intermediate temperature is determined by enforcing the putative equality of the steady state heat flows through the two materials:

$$\frac{k_1}{L_1}\left(T_H - T_I\right) = \frac{k_2}{L_2}\left(T_I - T_C\right) \qquad \Longrightarrow \qquad T_I = \frac{1}{\frac{k_1}{L_1} + \frac{k_2}{L_2}}\left(\frac{k_1}{L_1} T_H + \frac{k_2}{L_2} T_C\right).$$

Re-expressed in resistance terms,

$$\frac{(T_H - T_I)}{R_1} = \frac{(T_I - T_C)}{R_2} \qquad \Longrightarrow \qquad T_I = \left(\frac{1}{R_1} + \frac{1}{R_2}\right)^{-1}\left[\frac{T_H}{R_1} + \frac{T_C}{R_2}\right] = \frac{R_2 \, T_H + R_1 \, T_C}{R_1 + R_2}.$$

All of the quantities appearing on each of the RHSs above are known [in principle], and thus the intermediate temperature is uniquely defined and computable. The rate at which heat flows through the series composite system may be obtained by substituting this intermediate temperature into the expression for the heat flow through either of the individual slabs. The results so obtained are mutually consistent.

EXAMPLE [*Thermal Conduction: Parallel*]

Two equal-length slabs of different material are arranged in **parallel** between HOT, T_H, and COLD, T_C, heat reservoirs. The slab parameters are

$$\{k_1,\ A_1,\ L\} \quad \text{and} \quad \{k_2,\ A_2,\ L\}.$$

The slabs are taken to be thermally isolated from one another so as to prevent the crossover of heat from one block to the other *en route* between the reservoirs.

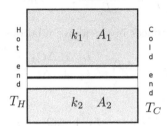

FIGURE 39.3   Parallel Conduction of Heat

Q: In what manner does heat flow through a parallel arrangement of materials?

A: In steady state the two materials spanning the same temperature difference conduct heat independently of one another.

Q: What is the effective thermal conductivity of a parallel combination of materials?

A: With reference to Figure 39.3, the rates of heat transfer through each substance and overall [the SUM] are:

$$\frac{\Delta Q_1}{\Delta t} = \frac{k_1\, A_1\, \Delta T_1}{L}, \qquad \frac{\Delta Q_2}{\Delta t} = \frac{k_2\, A_2\, \Delta T_2}{L}, \qquad \text{and} \qquad \frac{\Delta Q_{[12]}}{\Delta t} \equiv \frac{\Delta Q_1}{\Delta t} + \frac{\Delta Q_2}{\Delta t}.$$

The temperature difference is the SAME for both slabs, *i.e.*,

$$\Delta T_1 = \Delta T_2 = T_H - T_C = \Delta T_{[12]},$$

and thus also for the composite. Therefore, the total rate of heat flow is

$$\frac{\Delta Q_{[12]}}{\Delta t} = \frac{k_1\, A_1\, \Delta T_1}{L} + \frac{k_2\, A_2\, \Delta T_2}{L} = \left(k_1\, A_1 + k_2\, A_2 \right) \frac{\Delta T_{[12]}}{L}.$$

When heat flows through two or more thermally isolated channels, the total rate of heat flow is the sum of the rates through the individual channels.

The effective thermal conductivity of this parallel arrangement of materials obeys

$$\frac{\Delta Q_{[12]}}{\Delta t} = \frac{k_{[12]}\, A_{[12]}\, \Delta T_{[12]}}{L},$$

where $A_{[12]} = A_1 + A_2$. Comparison with the results just above reveals that

$$k_{[12]} = \frac{k_1\, A_1 + k_2\, A_2}{A_1 + A_2}.$$

The thermal conductivity of the strictly parallel arrangement of blocks is the area-weighted average of the thermal conductivities of the slabs.

Chapter 40

Radiative Heat Flow

In Chapter 39, we investigated two of the three modes of heat transfer: convection and [steady state] conduction. In this chapter, we'll begin to consider radiation.

RADIATION   Thermal radiation transfers heat energy via emission or absorption of electromagnetic waves.

Before we begin to discuss examples, there are aspects of this definition that invite questions. Here are three.

Q1: Aren't electromagnetic waves the same as light?

A1: Well, yes, in a manner of speaking. When people say "light" they usually mean that part of the electromagnetic spectrum detectable by unaided human eyes. In this sense, certainly light is comprised of electromagnetic waves. If one is willing to consider radio waves and gamma-rays and microwaves and X-rays to be species of light, then the distinction becomes moot.

Q2: Why are both emission and absorption mentioned?

A2: The snarky answer is that, for heat energy to flow, it must originate at one locale and end up at another.

A2: A better answer is that all objects simultaneously radiate and absorb thermal energy, and only the net flow of heat has physical significance.

Q3: Aren't electromagnetic waves involved in much more than thermal radiation?

A3: Yes. Quite a bit more. Electromagnetic radiation and absorption [the production and capture of light] are fundamental physical processes. Here, our concern is that heat energy in a local neighbourhood may be converted to electromagnetic [wave] energy, carried along with the waves to another—possibly distant—neighbourhood, and then converted back.

Instances of radiative heat transfer abound. Here are a few.

The Sun Warms the Earth   It is, to a large measure,[1] solely the flow of heat from the Sun which sustains the climate [and life, too] in the Earth's trophosphere.

> ASIDE: The interplanetary medium is too diffuse for efficient conduction, and we certainly don't want convective heating from the solar surface at $T_\odot \simeq 5800\,\mathrm{K}$.

Fireplaces   The colour of a flame or a stove element is associated with its temperature and rate of heat emission. In increasing order of temperature and power, the colour sequence is approximately as follows:

dark (no colour),   red,   red-orange,   yellow,   blue-white.

---

[1] Heat from the decay of radioactive elements deep within the Earth contributes, too.

ASIDE: Imagine basking in the warm glow produced by burning embers in a glassed-in fireplace. Here, the conductive and convective contributions to the overall heat flow are reduced by design. The warmth is quite directional in that one side of one's face may be warmed while the other remains cool.

Thermal Imaging Cameras   These "heat-sensitive" cameras have electro-optical sensors which respond to the infrared range of the electromagnetic spectrum.

Cosmic Microwave Background   The **Cosmic Microwave Background** [CMB] is nearly homogeneous and isotropic radiation permeating the universe, discovered[2] in the 1960s. It is believed to have originated about 10^5 years post-Big Bang, as the universe cooled enough to allow hydrogen atoms to form without immediately re-ionising. Since then the CMB has circulated throughout the universe, cooling as the universe expands. At present, its temperature is approximately 2.7 K. Spaceborne instruments which have studied the CMB with astounding experimental precision include COBE, [Cosmic Orbiting Background Explorer[3]], WMAP [Wilkinson Microwave Anisotropy Probe], and ESA PLANCK [European Space Agency Planck mission].[4]

A MODEL OF RADIATIVE HEAT FLOW

Consider a block of material with uniform surface properties [material, colour, roughness, *etc.*] as shown in Figure 40.1.

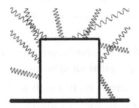

FIGURE 40.1   Thermal Radiation from the Surface of a Block at Temperature T

STEFAN–BOLTZMANN LAW   The Stefan–Boltzmann Law expresses the average rate at which heat flows by means of radiative transfer:

$$\frac{\Delta Q}{\Delta t} = e\,\sigma\,A\,T^4.$$

The LHS of the Stefan–Boltzmann Law is the average thermal radiative power. Its units are watts [joules per second]. The block both **emits** and **absorbs** radiation [concurrently]. Emission is based upon the temperature of the block, T, while absorption occurs at the temperature(s) of the radiation in which the block is bathed, T_{RAD}.

---

[2] Arno Penzias and Robert Wilson shared part of the 1978 Nobel Prize for this discovery.

[3] John Mather and George Smoot, the two principle architects of the scientific experiments carried aloft on COBE, won the 2006 Nobel Prize.

[4] WMAP and PLANCK have yielded such precise data that "experimental cosmology" has ceased to be an *oxymoron*. It is a good bet that a future Nobel Prize shall go to one or both of these consortia.

$$\text{IF}\ \begin{bmatrix} T > T_{\text{RAD}} \\ T < T_{\text{RAD}} \end{bmatrix},\ \text{THEN the block is a net}\ \begin{bmatrix} \text{emitter} \\ \text{absorber} \end{bmatrix}.$$

On the RHS there are four factors.

e The **emissivity** of the material coating the block is a substance-specific constant. All emissivities lie in the range

$$0 < e < 1.$$

$$\text{Substances with}\ \left\{ \begin{matrix} \text{LARGE} \\ \text{SMALL} \end{matrix} \right\}\ \text{emissivity are}\ \left\{ \begin{matrix} \text{GOOD} \\ \text{POOR} \end{matrix} \right\}\ \text{emitters/absorbers.}$$

> ASIDE: In springtime, snowbanks [white snow has $e \approx 0$] line the edge of a bare asphalt parking lot [black asphalt has $e \approx 1$]. In full sun, the asphalt heats up appreciably, whereas the snow temperature barely rises.[5] After the sun sets, the asphalt cools rapidly, while the snow cools slowly.

A perfect emitter/absorber, $e \equiv 1$, is an idealisation called a **blackbody**. The other extreme, $e \equiv 0$, describes a material whose surface neither absorbs nor emits thermal radiation.

σ The **Stefan–Boltzmann Constant** is a constant of nature, with SI value

$$\sigma = 5.67 \times 10^{-8}\ \frac{\text{W}}{\text{m}^2 \cdot \text{K}^4}.$$

A The rate at which heat is emitted and absorbed is directly proportional to the surface area of the block.

T^4 The absolute temperature to the fourth power appears in the expression for the rate of radiative emission. There are two important implications of this surprising dependence.

> ABSOLUTE   The net radiated heat flux does not simply depend on a difference in temperatures. Rather, one must consider emission and absorption processes separately and then take the difference to obtain the net rate.

> NON-LINEAR   This non-linearity has the effect of the limiting the temperature increase of any system which is not completely radiation isolated.

> > ASIDE: Suppose that a system starts off at temperature T_{initial} and is in thermal equilibrium with its surroundings [zero net heat flow]. Further suppose that heat is added to the system at a constant rate. The temperature of the system will increase, and as it does the rate of radiative energy loss will also increase. Further increases in temperature beget more rapidly increasing energy loss until the system reaches a steady state temperature, T_{final}. At T_{final} the rate at which heat is added is precisely equal to the rate at which it is being radiated to the surrounding environment, *i.e.*, there is no net accumulation of heat and the system's temperature remains constant. Should one wish to increase the temperature beyond T_{final}, one must either increase the rate at which heat flows into the system [subject to the law of diminishing returns, owing to the T^4 dependence in the Stefan–Boltzmann Law], or limit the ability of the system to radiate heat to its environment.

---

[5]A snarky person might quip that the snow *keeps its cool*.

EXAMPLE [*Radiative Heating and Cooling: Series*]

Solar panels on a spacecraft provide energy for its operation. The panels are customarily arrayed so that one side faces the sun directly, while the other looks out into deep space. The simplest model of a solar panel on a spacecraft orbiting close to the Sun has the following features.

*   The panel operates at a unique temperature, T_o. This assumption usually requires that the panel be thin enough [or conductive enough] to make the temperature gradient negligible.

*   The panels have equal face areas fore and aft.

*   The front and back surfaces have emissivities e_f and e_b, respectively. In the computation below, we'll take $e_f = e_b = 0.9$.

*   The Sun, with surface temperature[6] $T_\odot = 5000\,\mathrm{K}$, completely fills the entire [hemispherical] field of view of the front of the panel.

*   Deep space at a temperature[7] of $T_{\mathrm{space}} \sim 5\,\mathrm{K}$ completely fills the field of view on its side of the panel.

*   The conversion of some fraction of the incident radiative energy into electrical energy is neglected.

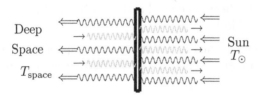

FIGURE 40.2   Series Radiation of Heat to and from a Spacecraft Panel

Were it the case that the Sun shone evenly on all sides of the panel, it would eventually heat up to the solar surface temperature, while if deep space surrounded it, it would eventually cool to that temperature. The present case is an admixture of these two extremes: the sun on one side, space on the other. One might expect the panel to reach a steady state temperature intermediate between those of the Sun and space.

    ASIDE: In the absence of a physical model, one might hazard a guess that the steady state temperature is $T_{\mathrm{guess}} = 2502.5\,\mathrm{C}$, the midpoint between $5\,\mathrm{C}$ and $5000\,\mathrm{C}$.

Q: What is the steady state temperature, T_o, of the panel?

A: The steady state operating temperature is that for which the net rate of heat absorption on the front face is exactly equal to the net rate of heat loss on the back.

[Conversion of radiant energy to electrical energy—the *raison d'être* of the panel—is neglected.]

---

    [6]We deliberately chose to underestimate the Sun's temperature in order to partially compensate for the Sun's not completely filling the hemispheric $2\,\pi$ solid angle.

    [7]This temperature is chosen to account for the relative "warmth" of the interplanetary medium *vis-à-vis* intersteller and intergalactic space.

METHOD ONE: [*Net Input and Output*]

The net power input to the panel is provided by radiation from the Sun and from deep space both striking equal-area equal-emissivity faces.

$$P_{\text{INPUT}} = e\,\sigma\,A\left(T_{\odot}^4 + T_{\text{space}}^4\right).$$

At the same time the thermal power output is emitted at the operating temperature of the panel, on both sides of the panel. Thus,

$$P_{\text{OUTPUT}} = e\,\sigma\left(2\,A\right)T_o^4.$$

The input and output powers must be equal in steady state, since otherwise T_o would be changing. Hence,

$$T_o^4 = \frac{T_{\odot}^4 + T_{\text{space}}^4}{2} \quad \Longrightarrow \quad T_o = \sqrt[4]{\frac{T_{\odot}^4 + T_{\text{space}}^4}{2}} \simeq 4204.5 \simeq 4200\,\text{K}.$$

METHOD TWO: [*Net Front and Back*]

The front face of the panel absorbs radiation from the Sun at the rate

$$P_{\odot \to \text{panel}} = e_f\,\sigma\,A_f\,T_{\odot}^4,$$

and emits radiation back toward the Sun at a different rate,

$$P_{\text{panel} \to \odot} = e_f\,\sigma\,A_f\,T_o^4.$$

Thus, the net thermal power absorbed on the front of the solar panel is

$$P_{\text{front}} = e_f\,\sigma\,A_f\left(T_{\odot}^4 - T_o^4\right).$$

The rate at which the back of the panel radiates thermal energy into space is

$$P_{\text{panel} \to \text{space}} = e_b\,\sigma\,A_b\,T_o^4,$$

while that at which it absorbs radiation from space is

$$P_{\text{space} \to \text{panel}} = e_b\,\sigma\,A_b\,T_{\text{space}}^4.$$

Thus, the net thermal power outflow from the back of the panel is

$$P_{\text{back}} = e_b\,\sigma\,A_b\left(T_o^4 - T_{\text{space}}^4\right).$$

The steady state operating temperature is that value of T_o for which the front and back powers are precisely equal in magnitude. Since we have assumed that the front and rear panel emissivities and areas are the same, we obtain

$$T_o = \sqrt[4]{\frac{T_{\odot}^4 + T_{\text{space}}^4}{2}} = 4204.5 \approx 4200\,\text{K}.$$

The temperature of the thin panel is certainly not equal to the [unweighted] average of the temperatures of the Sun and deep space.

EXAMPLE [*Radiative Cooling: Parallel*]

A SPHERICAL COW,[8] with total surface area $3\,\mathrm{m}^2$, rolls about on a grassy field at night. One half of the cow's surface area faces generally up and the other half down. The effective temperature of the clear night sky is about $-23\,\mathrm{C}$, and we shall take the temperature of the ground to be $+20\,\mathrm{C}$. The cow maintains a constant surface temperature of $35\,\mathrm{C}$.

Q1: At what rate is the cow losing energy?

Q2: How many Calories does the cow burn over the course of a night lasting 12 hours merely to maintain its body temperature against radiative heat loss?

A1: We have to guess an emissivity for the cow. Let's try $e = 0.5$. The relevant temperatures, in kelvin, are:

$$T_{\text{cow}} = 308\,\mathrm{K}\,, \qquad T_{\text{ground}} = T_{\text{grd}} = 293\,\mathrm{K}\,, \qquad \text{and} \qquad T_{\text{sky}} = 250\,\mathrm{K}\,.$$

The cow loses thermal energy to both the sky and the ground at the following net rates:

$$P_{\text{cow}\to\text{sky}} = e\,\sigma\,A_{\text{up}}\left(T_{\text{cow}}^4 - T_{\text{sky}}^4\right) = 0.5 \times 5.67 \times 10^{-8} \times 1.5 \times \left(308^4 - 250^4\right) \approx 217\,\mathrm{W}\,,$$

$$P_{\text{cow}\to\text{ground}} = e\,\sigma\,A_{\text{down}}\left(T_{\text{cow}}^4 - T_{\text{grd}}^4\right) = 0.5 \times 5.67 \times 10^{-8} \times 1.5 \times \left(308^4 - 293^4\right) \approx 69\,\mathrm{W}\,.$$

The total rate of radiative heat loss suffered by the cow is

$$P_{\text{cow, total}} = P_{\text{cow}\to\text{sky}} + P_{\text{cow}\to\text{ground}} \approx 286\,\mathrm{W}\,.$$

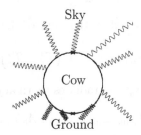

FIGURE 40.3   Parallel Radiation of Heat from a Spherical Cow

A2: There are $60 \times 60 \times 12 = 43200$ seconds in twelve hours. Thus, the cow must expend

$$E = P_{\text{cow, total}} \times (\text{time interval}) \approx 12.36\,\mathrm{MJ} \approx 2950\,\mathrm{C}$$

to maintain its body temperature overnight.

---

[8]This beastie appears in the Epilogue to VOLUME I.

ADDENDUM: Newton's Law of Cooling

Until now, we have tacitly or explicitly assumed that the flow of heat from a HOT reservoir to a COLD one was not [appreciably] changing the temperature of either.

ASIDE: This situation arises when the reservoirs are very large, the heat flux modest, and the time scales short.

The more realistic situation, in which the flow of heat is consequential, is modelled via **Newton's Law of Cooling**.

NEWTON'S LAW of COOLING When two systems at different temperatures are in thermal contact [*i.e.*, heat is being transferred from one to the other], the instantaneous time rate of change of the temperature difference is proportional to its magnitude. *I.e.*,

$$\frac{d(\Delta T)}{dt} = -\gamma \, \Delta T \,,$$

for γ, a phenomenological [substance, geometry, and heat-flow-mechanism dependent] constant. This is another manifestation of exponential behaviour in natural phenomena.

That the flow of heat is always from HOT to COLD ensures that ΔT diminishes with time. Concomitantly, so too does the rate of heat flow.

Familiar instances of this variable cooling occur when a cup of hot coffee and a glass of cold water are placed [far apart] in the same room. The respective temperatures of the beverages converge to the ambient temperature in the manner illustrated in Figure 40.4.

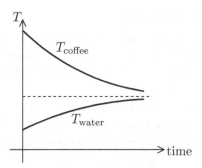

FIGURE 40.4 Cooling of Coffee and Warming of Water to Ambient Temperature

---

Newton's Law of Cooling seems exactly right for thermal conduction, in which the rate of heat flow is proportional to the temperature difference. Radiative cooling, on the other hand, depends on the difference of the fourth powers of temperatures. It might at first appear that this nonlinearity is inconsistent with Newton's Law of Cooling and will serve to reduce it to [at best] a crude approximation. Fortunately, the situation is not as bad as this, as may be shown by analysis of two different factorisations of the difference of quartic temperature factors.

METHOD ONE: [*Double Application of the Difference of Squares*]
The difference of quartics factorises twice into differences of squares:

$$T_H^4 - T_C^4 \equiv \left(T_H^2 + T_C^2\right)\left(T_H^2 - T_C^2\right) \equiv \left(T_H^2 + T_C^2\right)\left(T_H + T_C\right)\left(T_H - T_C\right)$$
$$\equiv \left(T_H^2 + T_C^2\right)\left(T_H + T_C\right)\left(\Delta T\right).$$

Remarkably, IF the temperatures T_H and T_C are slowly changing, *i.e.*, $T_H + T_C$ and $T_H^2 + T_C^2$ are both nearly constant, THEN the rate of radiative heat loss/gain is, to good approximation, linear in the temperature difference, and Newton's Law of Cooling is efficacious.

METHOD TWO: [*General Approach to the Difference of Polynomials*]
The difference of fourth degree monomials is always expressible as the product of a cubic polynomial and the linear difference:

$$T_H^4 - T_C^4 = \left(T_H^3 + T_H^2 T_C + T_H T_C^2 + T_C^3\right)\left(T_H - T_C\right),$$

IF the HOT and COLD temperatures are not so very different, $T_H \sim T \sim T_C$, THEN all of the cubic terms in the parentheses are approximately equal, and

$$T_H^4 - T_C^4 \sim 4\,T^3\,\Delta T.$$

Again, Newton's Law of Cooling is seen to accurately approximate the effects of radiative heating and cooling.

Cooling of the hot coffee by evaporation and convection is harder to incorporate into realistic models.

EXAMPLE [*Newton's Law of Cooling*]

PK pours a piping hot [approximately 85 C and too hot to sip right away] cup of coffee and places it on his desk. The thermostat in PK's office is set at 22 C. Fifteen minutes later, awakening from a physics-induced reverie, he realises that the coffee has cooled to 40 C [just slightly warmer than body temperature].

Q: What is the cooling rate, γ, for this scenario?

A: The first step in forming an estimate for γ is assuming that Newton's Law of Cooling is operative. When we denote the ambient temperature of the room by T_r, the Law becomes:

$$\frac{d(T - T_r)}{dt} = -\gamma\,(T - T_r).$$

This may be self-consistently integrated to yield

$$T(t) - T_r = (T_0 - T_r)\,e^{-\gamma t} \quad \Longrightarrow \quad \frac{T(t) - T_r}{T_0 - T_r} = e^{-\gamma t} \quad \Longrightarrow \quad \ln\left[\frac{T(t) - T_r}{T_0 - T_r}\right] = -\gamma\,t.$$

Therefore,

$$\gamma = \frac{1}{t}\ln\left[\frac{T_0 - T_r}{T(t) - T_r}\right].$$

Substituting the pertinent parameter values yields

$$\gamma = \frac{1}{15}\ln\left[63/18\right] = \frac{\ln(3.5)}{15} \simeq 0.0835\ \text{min}^{-1}.$$

Chapter 41

More Radiation

The Stefan–Boltzmann Law, introduced in Chapter 40, governs the energy flux associated with radiative heat transfers. Heat flows via the concurrent emission and absorption of electromagnetic radiation. According to the Stefan–Boltzmann Law, the one-way rate of energy flow is

$$\frac{\Delta Q}{\Delta t} = e\,\sigma\,A\,T^4\,.$$

In this formula, e is the emissivity of the surface $[0 < e < 1]$, σ is the Stefan–Boltzmann constant, A is the area of the surface, and T is the relevant temperature [expressed in kelvin]. The competing effects of emission and absorption combine to yield the net rate of radiative heat flow,

$$\left.\frac{\Delta Q}{\Delta t}\right|_{\text{net}} = e\,\sigma\,A\left(T^4_{\text{emission}} - T^4_{\text{absorption}}\right)\,.$$

The intensity of emission or absorption is the energy flux divided by the surface area:

$$I = \frac{\Delta Q/\Delta t}{A} = e\,\sigma\,T^4\,.$$

The net intensity depends solely on the local value of the emissivity and the local values of the surface temperature [for emission] and the radiation temperature [for absorption].

$$I_{\text{net}} = e\,\sigma\left(T^4_{\text{emission}} - T^4_{\text{absorption}}\right)\,.$$

As a substance's temperature increases, it is observed to progress from having no perceptible thermal colour, to glowing dull red, brighter red, orange-yellow, and finally blue-white.

SPECTRAL RADIANCE   The spectral radiance is the emitted electromagnetic wave power, in a differential range of frequencies [wavelengths] centred on a particular frequency [wavelength], per unit area of the emitter, and per unit solid angle.

[Spectral radiance is an energy intensity density within the electromagnetic spectrum.]

Graphical and algebraic forms of the spectral radiance of a blackbody at temperature T as functions of wavelength and frequency, $I(\lambda, T)$ and $I(\nu, T)$, are given later in this chapter. For now, we'll merely quote the units conventionally associated with spectral radiance:

$$[\,I(\lambda, T)\,] = \frac{\text{W/m}^2}{\text{nm} \cdot \text{sr}} \qquad \text{and} \qquad [\,I(\nu, T)\,] = \frac{\text{W/m}^2}{\text{Hz} \cdot \text{sr}}\,,$$

where a steradian, $[\,\text{sr}\,]$, is the SI unit for measurement of solid angles.

[There are $4\pi\,\text{sr}$ of solid angle surrounding each point in 3-d Euclidean space.]

Spectral radiances of a blackbody object at three rather different temperatures are plotted as functions of wavelength [on the left] and frequency [on the right] in Figure 41.1. Within each plot, two trends are evident. The first is that there is a general increase in spectral radiance at all wavelengths and frequencies as the temperature is increased.

[Although all the curves converge to zero asymptotically, they do not cross.]

The second is that the peaks [modal values] of the spectral radiances shift to smaller wavelengths and higher frequencies as the temperature increases. This effect is observed in the colour-shift from red- to white-hot, mentioned earlier in Chapter 40.

ASIDE: **Pyrometers** measure the ratios of light intensity emitted at various frequencies and wavelengths. Matching these ratios to blackbody [greybody[1]] spectral radiances enables estimation of the temperature of an incandescent light source.

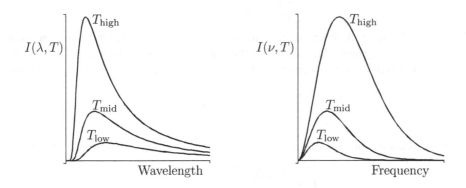

FIGURE 41.1   Blackbody Spectral Radiance *vs.* Wavelength and Frequency

In comparing the plots one must grapple with the differing parameterisations, λ *vs.* ν, and consider their mutual relation: $\lambda\nu = c$, where c represents the constant [wave] speed of light [*cf.* Chapter 21]. Two important consequences ensue.

CROSSOVER   The short-wavelength region, $\lambda \to 0$, in $I(\lambda, T)$ corresponds to the asymptotic ultra-high frequency region, $\nu \to \infty$, in $I(\nu, T)$. Concomitantly, the long-wavelength regime corresponds to the low-frequency part of the $I(\nu, T)$ spectrum.

DENSITY   The curves themselves are not simply related by inversion, *i.e.*, $\nu \leftrightarrow c/\lambda$, because the spectral radiance is an energy density. The energy intensity per unit emitting area and per unit solid angle ought to be exactly the same in corresponding wavelength and frequency intervals, so

$$I(\lambda, T)\,d\lambda = \frac{dE}{dA\,d\Omega} = I(\nu, T)\,d\nu\,, \quad \text{along with} \quad d\nu = -\frac{c}{\lambda^2}\,d\lambda\,,$$

relates the spectral radiances. Evidently, additional factors are needed to convert one into the other.

The modal frequency, $\nu_{\max}$, and the modal wavelength, $\lambda_{\max}$, at a particular temperature are not commensurate, *i.e.*,

$$\lambda_{\max}\,\nu_{\max} \neq c\,.$$

---

[1] A greybody's spectral radiance distribution is that of a blackbody scaled by a constant emissivity factor.

For solid bodies with differentiable [smooth] surfaces, the emission of radiation occurs locally from small neighbourhoods [approximately planar patches] on the surface. The emitted radiation is distributed isotropically throughout the entire 4π steradians available. The radiation which is directed inside the emitting body is presumed to be reabsorbed. The integral of the spectral radiance over the [outward] hemispheric solid angle yields the net emitted intensity per wavelength [frequency] interval. The intensity of that which is emitted outside the body experiences the expected inverse-square-law [geometric] attenuation (see Figure 41.2).

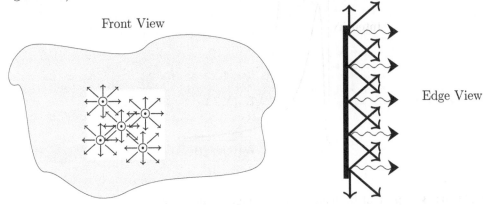

FIGURE 41.2 Thermal Radiation from a Patch on the Surface of a Body

When the solid's surface geometry is simple, the integral over the hemispheric solid angle may be trivially performed. In such cases, the integrated spectral radiance functions,

$$\mathcal{I}(\lambda, T) = 2\pi\, I(\lambda, T) \qquad \text{and} \qquad \mathcal{I}(\nu, T) = 2\pi\, I(\nu, T)\,,$$

describe the total emitted energy flux in narrow wavelength and frequency intervals, per unit area from the surface of a solid body at temperature T.

The search for an analytical expression of the spectral radiance associated with a blackbody at temperature T occupied physicists throughout the latter half of the nineteenth century.

[Experimental observation of swathes of the spectrum were in accord with Figure 41.1.]

The long-wavelength [low-frequency] behaviour was accounted for by the classical model of electromagnetic radiation[2] developed by Maxwell[3] and others in the 1860s. The spectral characteristics of thermal radiation were finally worked out by Lord Rayleigh (1842–1919) and James Jeans (1877–1946) in 1905.[4]

---

[2] This model is presented as one of the triumphs of the unification of electricity and magnetism into electromagnetism in VOLUME III.

[3] James Clerk Maxwell (1831–1879) was the very first Cavendish Professor of Physics at Cambridge. He made foundational contributions to electromagnetism (as noted) and statistical mechanics (as we shall begin to see in future chapters).

[4] In an astounding coincidence, 1905 was Einstein's *Annus Mirabilis*, in which his exposition of SPECIAL RELATIVITY, and explanation of the photoelectric effect (initiating QUANTUM MECHANICS), overturned the foundations of classical electromagnetic theory.

RAYLEIGH–JEANS LAW   The Rayleigh–Jeans Law, expressing the blackbody solid-angle-integrated spectral radiance as a function of temperature and wavelength,

$$\mathcal{I}_{\mathrm{RJ}}(\lambda, T) = \frac{8 \pi c k_B T}{\lambda^4},$$

was derived using classical electromagnetic theory. The factor c is the speed of light, and k_B is Boltzmann's constant [introduced in Chapter 10].

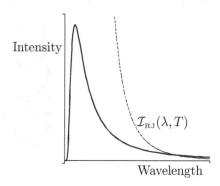

FIGURE 41.3   Spectral Radiance and Rayleigh–Jeans Law

The Rayleigh–Jeans Law provides an accurate description at long wavelengths. At shorter wavelengths, the R–J Law predicts a divergence,

$$\lim_{\lambda \to 0} \mathcal{I}_{\mathrm{RJ}}(\lambda, T) = \infty,$$

which is not observed. Even worse, the wavelength-integrated intensity blows up,

$$\int_0^\infty \mathcal{I}_{\mathrm{RJ}}(\lambda, T) \, d\lambda \to \infty,$$

for all non-zero temperatures.

> **Such predicted behaviour is impossible to reconcile with the Stefan–Boltzmann Law.**

This curious state of affairs, dubbed the **ultraviolet catastrophe**, led to disparate [and desperate] attempts to reconcile the Rayleigh–Jeans Law with reality.

Q: Might the catastrophe be mitigated by the imposition of a cutoff, say $\lambda \geq \lambda_0 > 0$?

A: Nope. There is no compelling physical argument for this. Furthermore, even if such a cutoff were imposed, the integrated powers, now finite, would still be too large to reconcile with observation.

Q: Might the location of the peak value of the spectral radiance be modelled as a function of temperature? [This would dictate the scale on which the Rayleigh-Jeans Law has completely broken down.]

A: The peak is the modal value of the radiance [in the statistical sense], and a rule determining its location was formulated by Wilhelm Wien (*circa* 1893).

WIEN DISPLACEMENT LAW The wavelength of light corresponding to the peak in the blackbody spectral radiance is inversely proportional to the absolute temperature of the source. Incorporating the modern values of k_B, the Boltzmann constant, and c, the speed of light, the Wien Displacement Law reads

$$\lambda_{\text{MAX}} = \frac{2.898\,[\text{mm} \cdot \text{K}]}{T}\,.$$

The modal blackbody frequency is determined by the expression

$$\nu_{\text{MAX}} = 5.879 \times 10^{10}\,[\text{Hz/K}] \times T\,.$$

[These wavelength and frequency maxima occur for incommensurate waves.]

The Rayleigh–Jeans debacle was formally resolved in 1900, when Max Planck exhibited formulae for spectral radiance in accord with experiment, the Stefan–Boltzmann Law, and the Wien Displacement Law. Planck's derivation relied on a mathematical trick [discretisation] for which he was unable to provide a satisfactory physical rationale, and thus his result, the **Planck Law**, was not generally accepted. Einstein realised that

Planck's derivation implies that the energy borne by light is quantised!

PLANCK LAW There exist formulae[5] for spectral radiance:

$$\mathcal{I}(\lambda, T) = \frac{2\pi h c^2}{\lambda^5} \frac{1}{e^{\frac{hc}{\lambda k_B T}} - 1} \qquad \Longleftrightarrow \qquad \mathcal{I}(\nu, T) = \frac{2\pi h \nu^3}{c^2} \frac{1}{e^{\frac{h\nu}{k_B T}} - 1}\,.$$

These expressions require the introduction of a new fundamental physical constant,

$$h = 2\pi\hbar \simeq 6.626 \times 10^{-34}\,\text{J} \cdot \text{s}\,,$$

now known as Planck's constant [*cf.* VOLUME I, Chapter 47].

EXAMPLE [*Solar Energy Modal Intensity*]

Q: At what wavelength and frequency is the Sun's energy intensity maximised?

A: The solar surface may be approximated by a blackbody at temperature 5800 K. The wavelength and frequency at which the radiation intensity is maximised are determined by the Wien Displacement Law.

$$\begin{aligned}\lambda_\odot &= \frac{2.898}{5800}\,\text{mm} \\ &= 5.0 \times 10^{-4}\,\text{mm} \\ &= 500\,\text{nm}\end{aligned} \qquad \begin{aligned}\nu_\odot &= \left(5.879 \times 10^{10}\right) \times 5800\,\text{Hz} \\ &= 3.41 \times 10^{14}\,\text{Hz}\end{aligned}$$

Two noteworthy aspects of these results are listed below.

INCOM-MENSURATE The product, $\lambda_\odot\,\nu_\odot = 1.7 \times 10^8$ m/s, is NOT equal to the speed of light. These values are not commensurate, since they are associated with different waves [as was discussed earlier in this chapter].

GREEN The modal wavelength, 500 nm, lies near the middle of the visible part of the spectrum. [It is not surprising that our eyes are most sensitive to this range of wavelengths.]

---

[5]These expressions contain the factor 2π, rather than the often-quoted 4π, owing to our decision to integrate over a hemisphere.

INTEGRATED SPECTRAL RADIANCE AND THE STEFAN–BOLTZMANN LAW

Let's compute the total, *i.e.*, integrated, intensity of radiation from a blackbody,

$$I_{\text{Tot}} = \int_0^\infty \mathcal{I}(\lambda, T)\, d\lambda \qquad \text{and} \qquad I_{\text{Tot}} = \int_0^\infty \mathcal{I}(\nu, T)\, d\nu\,.$$

Reparameterising, $x = \frac{h\nu}{kT} = \frac{hc}{\lambda kT}$, and simplifying lead in both cases to

$$I_{\text{Tot}} = \frac{2\pi k^4}{h^3 c^2}\, T^4 \int_0^\infty \frac{x^3}{e^x - 1}\, dx\,.$$

The integral appearing here is a classic and evaluates to $\pi^4/15$. Thus,

$$I_{\text{Tot}} = \frac{2\pi^5}{15} \frac{k^4}{h^3 c^2}\, T^4\,.$$

Plugging in the accepted values of the mathematical and physical constants,

$$\left\{ \pi = 3.141593\,,\ c = 2.997925 \times 10^8\,,\ h = 6.626176 \times 10^{-34}\,,\ k = 1.380662 \times 10^{-23} \right\}\,,$$

one obtains

$$I_{\text{Tot}} = 5.6703 \times 10^{-8}\, T^4\,.$$

With the understanding that this intensity is per-unit-surface-area, and that the blackbody emissivity is $e \equiv 1$, it is apparent that

we have derived the Stefan–Boltzmann Law from the Planck Law

and prescribed the value of the Stefan–Boltzmann constant in terms of $\{\pi, c, h, k\}$.

Chapter 42

Laws of Thermodynamics 0 and 1

In Kinetic Theory heat is viewed as a manifestation of the *disordered kinetic energy* of the atomic and molecular constituents of matter. This modern viewpoint, that thermodynamics emerges from the particle mechanics of vast numbers of constituent bodies, was first advanced in the mid-1800s and attained universal acceptance in the early 1900s. Previously, heat was likened to a fluid, and thermodynamics was considered a separate branch of science, albeit along a continuum between physics and chemistry. It was natural enough, then, to formulate laws expressing regularities in thermodynamic behaviour.

These laws were later shown[1] to arise as consequences of laws and principles of mechanics. And yet the thermodynamic laws retain their currency because they provide a robust framework within which to explicate everyday [and exotic] natural phenomena.

> ASIDE: A silly example serves to make the point. One is unlikely to overhear someone say, "My trick knee hurts whenever the overall collision-rate of atmospheric molecules diminishes somewhat, eh?"

It is with an eye to phenomenology that we present the **Laws of Thermodynamics**.

ZEROTH LAW of THERMODYNAMICS The Zeroth Law is an expression of the PRINCIPLE OF TRANSITIVITY applied to the conditions of thermal equilibrium. That is, IF a thermodynamic system, $\mathcal{A}$, is in thermal equilibrium with some other system, $\mathcal{B}$, AND system $\mathcal{B}$ is in thermal equilibrium with system $\mathcal{C}$, THEN systems $\mathcal{A}$ and $\mathcal{C}$ must be in thermal equilibrium also.

In notation adapted from mathematical logic, one may write the Zeroth Law as:

$$\text{IF} \quad \mathcal{A} \sim \mathcal{B}, \quad \text{AND} \quad \mathcal{B} \sim \mathcal{C}, \quad \text{THEN} \quad \mathcal{A} \sim \mathcal{C}.$$

The condition "is in thermodynamic equilibrium with" is [mathematically] an **equivalence relation**.[2]

Q: What good things doth follow from transitivity of thermal equilibrium?

A: Recall that when two systems are in thermodynamic equilibrium, there is no net heat flow from one to the other.

[Exchange of heat is permitted, provided that the net flux is zero.]

For there to be no net transfer of heat, the temperatures of the two systems must be equal. Hence, a pithy form of the Zeroth Law is

$$\text{IF} \quad T_{\mathcal{A}} = T_{\mathcal{B}}, \quad \text{AND} \quad T_{\mathcal{B}} = T_{\mathcal{C}}, \quad \text{THEN} \quad T_{\mathcal{A}} = T_{\mathcal{C}}.$$

---

[1] How mechanics undergirds thermodynamics is a subject suitable for a statistical mechanics class.
[2] Equivalence relations are reflexive, $\mathcal{A} \sim \mathcal{A}$, and symmetric, $\mathcal{A} \sim \mathcal{B} \iff \mathcal{B} \sim \mathcal{A}$, as well as transitive.

The Zeroth Law makes thermometry [measuring and comparing temperatures, *cf.* Chapter 37] possible. To measure the temperature of a system, one takes a thermometer—another physical system—and arranges for the two systems to come to thermodynamic equilibrium. The thermometer scale was calibrated by having previously brought the thermometer into equilibrium with other systems of known temperature. Employing a thermometer to assign a temperature to a system implicitly invokes the Zeroth Law.

> ASIDE: The existence of the "Zeroth" Law of Thermodynamics adds to the appreciable body of evidence favouring MURPHY'S LAW. What happened, of course, was that everyone tacitly assumed the Zeroth law, and assigned the sobriquet "First" to the thermodynamical law most directly concerned with energy. Later, it was realised that the transitivity property of thermal equilibrium is even more fundamental.

FIRST LAW of THERMODYNAMICS

The First Law generalises the mechanical energy conservation law to include transfers of thermal energy:

$$\Delta U = Q - W .$$

The LHS of the expression for the First Law is the change in the internal energy of the thermodynamic system. The RHS consists of the quantity of heat, Q, added to the system, less the amount of mechanical work, W, performed by the system.

$$\text{IF heat flows} \begin{bmatrix} \text{into} \\ \text{out of} \end{bmatrix} \text{the system, THEN} \begin{bmatrix} Q > 0 \\ Q < 0 \end{bmatrix} .$$

$$\text{IF work is done} \begin{bmatrix} \text{by the system (upon its surroundings)} \\ \text{by the environment (on the system)} \end{bmatrix}, \text{THEN} \begin{bmatrix} W > 0 \\ W < 0 \end{bmatrix} .$$

The First Law states that energy is conserved [neither created nor destroyed] and may be interconverted amongst all of its various forms, including work and heat.

THERMODYNAMIC STATE VARIABLES

Thermodynamic state variables comprise the set of well-defined physical quantities which [partially] describe or characterise [aspects of the configuration of] the physical system. State variables, with one crucial exception, always appear in conjugate **intensive/extensive** pairs.

[CAVEAT: Q and W are NOT thermodynamic state variables.]

INTENSIVE VARIABLES

Intensive variables are independent of the size of the system. That is, IF the size of the system were doubled by

1st   preparing a copy of the system possessing the same values for all of the known state variables,[3]

2nd   merging the copy with the original,

THEN the intensive variables would retain their original values.

[Temperature and pressure in a sample of gas are intensive variables.]

---

[3]The copy need not be identical in its microscopic aspects.

EXTENSIVE VARIABLES   Extensive variables scale [linearly] with the size of the system. That is, IF the size of the system were doubled [by means of the process described earlier], THEN each extensive variable in the doubled system would have twice the value that it possessed in the original system.

[Volume and internal energy in a sample of gas are extensive variables.]

THERMODYNAMIC STATE SPACE   Thermodynamic state space is a MANIFOLD [a multidimensional mathematical structure] consisting of all possible states of any system [in a given class]. Roughly speaking, every thermodynamic state variable lies along its own Cartesian axis. The state of a particular system [in equilibrium] is completely specified and represented by the point in phase space corresponding to the component values of all of its thermodynamic state variables. A transition from an initial state to a final state corresponds to a motion within state space. The manner in which the transition is effected determines the path taken. Constrained systems reside [and remain] in a particular subspace within the broader space of states.

EQUATION of STATE   An equation of state [EoS] is any physical relation, encoded in mathematical form, holding among thermodynamic variables. These may be approximations and constraints, and may involve both equalities and inequalities. All EoS depend on the particulars of the system under study, but many, exhibiting common attributes, fall into **universality[4] classes.**

**Thermodynamic variables characterise physical aspects of a system.
Equations of state encode relations among thermodynamic variables.**

Several EoS were proposed in Chapter 10 to model the behaviour of dilute gases.

Charles' Law	For fixed P, $\quad V \;=\; V_0 \left(1 + \frac{T - T_0}{273}\right)$
Ideal Gas Law	$PV \;=\; nRT$
Van der Waals Gas Law	$\left(P + a\frac{n^2}{V^2}\right)\left(V - bn\right) \;=\; nRT$

Constraints on thermodynamic transitions from an initial state, i, to a final state, f, are EoS. Common constraints for dilute gas systems are listed in the table below.

CONSTRAINT	MEANING
Quasi-static	Slow Enough
Isobaric	Constant Pressure
Isochoric	Constant Volume
Isothermic	Constant Temperature
Adiabatic	No Net Heat Flow

---

[4]Universality indicates that these EoS transcend many of the specific details of the systems to which they are applied.

These constraints are elaborated upon below and in upcoming chapters.

QUASI-STATIC   The quasi-static [*cf.* Chapter 21 of VOLUME I] constraint requires that transitions occur sufficiently slowly to enable the system to remain very close to thermal equilibrium states throughout the time interval during which the transition occurs.

   [Without this, the thermodynamic state variables are ill-defined during the transition.]

   Q: How is "sufficiently slowly" determined?

   A: The time scale under which the system changes appreciably is much longer than the time scale(s) governing the various thermalisation processes by which far-flung parts of the system equilibrate.

   Q: How is this useful?

   A: *Let me count the ways ...*[5]

   ✓   WELL-DEFINED   The thermodynamic state of the system is [approximately] well-defined at all instants.

   ✓   CONTINUITY   The system passes through a sequence of adjacent intermediate equilibrium states.

   ✓   REVERSIBILITY   Changes which are performed very slowly and in a controlled manner have much greater chance of being self-consistently undone.

The essential idea is that the **phase space trajectory**[6] of the thermodynamical system may be considered continuous.

$\Delta P = 0$ In isobaric transitions, the pressure is held constant. The system's phase space trajectory lies entirely within a pressure hyperplane.

   ASIDE: The thermal response [described by the specific heat capacity] of a quantity of material depends on how the pressure and the volume of the sample are coincidentally changing. When there is no concomitant change in pressure as heat is added or removed, one employs the **specific heat capacity at constant pressure**, C_P. Most reference tables of specific heat capacities quote C_P.

$\Delta V = 0$ In an isochoric transition, the volume of the system is held constant.

   ASIDE: This is the other common condition imposed for computation or measurement of specific heats of gases.[7] In this case, one employs the **specific heat capacity at constant volume**, C_V.

$\Delta T = 0$ The isothermic constraint, that constant temperature be maintained, is realised by calibrated addition or removal of heat.

$\Delta Q = 0$ Adiabatic processes are those for which there is little or no net flow of heat. This constraint may be met by thermal isolation or be satisfied when the transition from the initial to final states occurs on a time scale much shorter than that in which the system exchanges heat with its surroundings. It is generally possible to be adiabatic without violating the quasi-static constraint.

---

[5]With apologies to Elizabeth Barrett Browning.

[6]Every thermodynamic state variable has an axis in phase space. The thermodynamic state of the system specifies the coordinates of a unique point in phase/state space.

[7]It is difficult to keep liquids or solids at constant volume. Owing to their large bulk moduli, tremendous forces are required to overcome the effects of thermal expansion and contraction.

ADDENDUM: Two Forms of Expression for the First Law

We have chosen to write the First Law as

$$\Delta U = Q - W\,,$$

where Q is positive [negative] when heat enters [leaves] the system, and W is positive [negative] when work is done by [on] the system. This may be called the *traditionalist* approach in that the heat is viewed as an input cost and the work is seen as a beneficial output. This point of view is summarised in Figure 42.1.

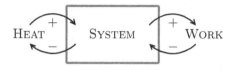

FIGURE 42.1   First Law: Distinguished Rôles of Heat and Work

While this is all consistent, it seemingly disregards the insight that heat and mechanical work are but species of energy. A more *modern* [egalitarian?] approach emphasises their commonality, *i.e.,*

$$\Delta U = Q + W\,,$$

where the quantities on the RHS are positive [negative] when energy enters [leaves] the system. This perspective on the First Law appears in Figure 42.2.

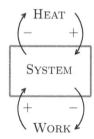

FIGURE 42.2   First Law: Heat and Work Treated on the Same Footing

Some books espouse the traditional view, others prefer the modern. We remain neutral on this issue, but compelled to make a choice, opted for the traditional presentation [augmented by careful explication so as to be intelligible to both camps].

Chapter 43

The First Law of Thermodynamics

The First Law of Thermodynamics reads:

$$\Delta U = Q - W \, .$$

The LHS is the **change in the internal energy** of the thermodynamic system under consideration. The RHS is the difference between **the net heat input to the system** and **the net mechanical work performed by the system**.

$$\text{IF heat flows } \left\{ \begin{array}{c} \text{into} \\ \text{out of} \end{array} \right\} \text{ the system, } \textsf{THEN } Q \text{ is } \left\{ \begin{array}{c} \text{positive} \\ \text{negative} \end{array} \right\}.$$

$$\textsf{IF net work is done } \left\{ \begin{array}{c} \text{by} \\ \text{on} \end{array} \right\} \text{ the system [interacting with its environment], } \textsf{THEN } \left\{ \begin{array}{c} W > 0 \\ W < 0 \end{array} \right\}.$$

The widespread applicability of thermodynamics makes it worthy of study in the abstract. On the other hand, it is essential to work with a model system, because it is only in concrete cases that one can readily see and begin to appreciate the subject.

Our model system shall consist of a fixed amount of ideal gas confined in a fancy box. The adjective "fancy" is meant to suggest that precise control may be exercised over

V the volume available to the system

Q exchanges of heat between the system and its environment[1]

W the mechanical interactions between the system and its surroundings

The state of any system may be described in terms of a collection of its thermodynamic state variables. The set of values assumed by the state variables specifies a **configuration** of the system. Look to the chart below for a whimsical example of demographic state variables and configurations.

STATE VARIABLE [TRAIT]	CONFIGURATIONS
Sex	{ Male | Female }
Eye Colour	{ black | brown | hazel | green | blue | grey }
Hair Colour	{ black | brown | red | blond | grey | magenta | cyan }
Height [cm]	{ $(0, 140]$ | $(140, 160]$ | $(160, 180]$ | $(180, 200]$ | 200+ }
Weight [kg]	{ $(0, 50]$ | $(50, 70]$ | $(70, 90]$ | $(90, 110]$ | 110+ }
$\cdots$	$\cdots$

---

[1] In particular, the box is thermally insulated against all heat flows not directly under our control.

ASIDE: The set of state variables need not provide an exhaustive description of every aspect of every constituent of the system.

Suitable thermodynamic state variables for the gas in the box might include:

[INTENSIVE]   pressure, P, and temperature, T

[EXTENSIVE]   volume, V, quantity of gas, n (or N), and internal energy, U

It was remarked [in Chapter 42] that neither Q nor W is a thermodynamic state variable. This seems perplexing until one realises that the First Law pertains to changes in the internal energy of a system accompanying changes in its state.

A person not properly apprehending the nature of heat might reason *falsely* thus:

"The situation with heat is reminiscent of potential energy, in that only its increments are physically meaningful. *Bona fide* potential energy functions were constructed by assembling potential energy differences [computed with respect to a commonly chosen reference position]. Perhaps we might do the same for heat."

This bootstrap procedure fails for heat. First, a potential energy function exists only when the work done by the associated conservative force is independent of the path taken from the initial to the final position. For non-conservative forces the work done is path-dependent and no such potential energy function can exist. Second, the heat gradient does not behave in the manner of a conservative force. [Its integral depends on the path taken from the initial to the final state.] This makes it impossible, in principle, for there to be a unique amount of heat associated with each and every state, and therefore Q is not a thermodynamic state variable.

Next, we'll re-define work in a more general way and realise with incontrovertible certainty that compatibility with the First Law requires that W, also, is not a thermodynamic state variable.

WORK   The work done by a system [while it undergoes a transition from an initial to a final thermodynamic state] may be [approximately] expressed in terms of a conjugate pair of intensive [x] and extensive [X] variables,[2] via

$$W \approx x\,\Delta X\,.$$

Three salient aspects of this formula are reviewed below.

W1   For transitions in which x remains constant while X changes, the expression yields the exact value of the work done [*i.e.*, $\approx \longrightarrow =$]. However, these are exceptional cases. In general, one expects both of the conjugate variables to change in the course of the system's transition.

Q: What does one do when both of the conjugate variables vary?

A1: Partition the transition from the initial state to the final state into a collection of smaller steps marching through intermediate states[3] such that x is held approximately constant throughout each. [See Figure 43.1.]

---

[2]In general, more than one conjugate pair may be required. Here we consider but one such set.
[3]These are well-defined [and guaranteed to exist] by the quasi-static presupposition.

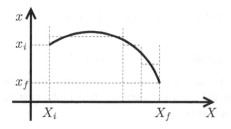 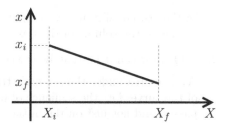

FIGURE 43.1 Computing Work When Intensive and Extensive Variables Both Vary

The march through the [piecewise x-constant] intermediate states produces a Riemann Sum. Refining the partition leads one to conclude that the work done by the system is equal to the area under the curve describing the system's path in state space *en route* from the initial to the final state. *I.e.*,

$$W_{if}[x] = \int_{X_i}^{X_f} x \, dX \, ,$$

which generalises the notion of mechanical work presented in VOLUME I.

W2   In the context of the gas-in-a-box system, $W \approx P \, \Delta V$.

Q: Does this make sense?

A: Why yes, it does. Here's the argument.

Think of the box as a parallelepiped with six rigid walls. Five of these remain immobile, while the sixth moves slowly [in the z-direction, remaining parallel to itself], as depicted in Figure 43.2.

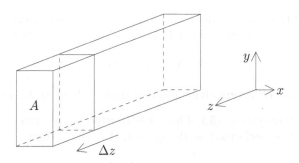

FIGURE 43.2 The Expanding Gas Does Mechanical Work on the Moving Wall

The pressure within the gas is manifest as the force per unit area exerted upon the walls. The net force exerted by the gas on the moving wall is

$$F_{\text{NET}} = P \, A \, .$$

IF this force remains constant while the wall moves a distance of Δz, THEN the net work done by the gas is

$$W_{if}[\text{NET}] = F_{\text{NET}} \, \Delta z = (P \, A) \, \Delta z = P \, (A \, \Delta z) = P \, \Delta V \, .$$

In the last equality it was recognised that the combination $A\,\Delta z$ is precisely the change in the volume of the box.

W3   **Q:** How can it be that W is NOT a thermodynamic state variable?

A: The area under the system's trajectory in state space depends on the shape of the curve, *i.e.,* the sequence of intermediate states[4] through which the system passes, and not just on the initial and final states.

The internal energy of the system is a thermodynamic variable, and hence it is entirely dependent on the state of the system and not on the means by which the state was attained. Thus, changes in the internal energy, $\Delta U = U_f - U_i$, are well-defined for all pairs of [allowed] initial, i, and final, f, states. Under any transition, the LHS of the First Law is unambiguous, and so too must be the RHS.

While the two terms on the RHS, Q and W, are each ill-defined,
all ambiguity is dispelled from their difference!

An example helps to make this clear.

EXAMPLE [*Internal Energy, Work, Heat, and All That*]

Two systems characterised by the same set of thermodynamic state variables[5] both underwent a transition from their common initial state to, ultimately, the same final state. The initial and final internal energies happened to be 50 J and 40 J, respectively. The change in the internal energy of each system was

$$\Delta U = U_f - U_i = -10\,\text{J}\,.$$

- In one of the systems, the transition occurred with concomitant absorption of 10 J of heat and performance of 20 J of mechanical work. For this system,

$$Q - W = 10 - 20 = -10\,\text{J}\,.$$

  The First Law of Thermodynamics is verified to hold in this instance.

○ For the other system, 15 J of heat was emitted by the system, $Q = -15\,\text{J}$, while 5 J of work was performed on the system, $W = -5\,\text{J}$. Here,

$$Q - W = -15 + 5 = -10\,\text{J}\,,$$

  and the First Law is seen to be operative in this case also.

The initial and final states were the same, while the thermodynamic paths differed. The essential idea is that Q and W are not uniquely specified[6] for any particular transition. Hence, NEITHER Q NOR W, alone, can be a state variable.

----

[4] The quasi-static constraint ensures that these intermediate states are, in fact, possible equilibrium states for the physical system.

[5] In colloquial terms, we say that such systems are **identical**.

[6] The difference, $\Delta U = Q - W$, is determined; the sum, $Q + W$, is not.

THERMODYNAMICS OF THE GAS-IN-A-BOX SYSTEM

Dilute gas systems confined to our fancy box admit description by the ideal gas law EoS. This law, as it is usually[7] expressed, reads $PV = nRT$, but here it is recast as:

$$\frac{PV}{T} = nR = N k_B.$$

The LHS is a combination of readily measured state variables, while the RHS is a proxy for the amount of gas/substance in the box. If the box is sealed so that gas may neither escape nor enter [and the gas is chemically inert[8]], the RHS is constant with its value computable by means of the expression on the LHS.

ASIDE: Back in the day, when the molecular nature of matter was much disputed, the recast version of the ideal gas law provided a reasonably effective quantification[9] of the size of the sample. Recalling Chapter 10, N represents the number of constituent gas molecules in the system; $k_B = 1.38 \times 10^{-23}$ J/(molecule·K) is the Boltzmann constant (Ludwig Boltzmann: 1844–1906); n is the number of moles of gas in the system;[10] and $R = 8.31$ J/(mole·K) is the Universal Gas Constant (the Universe: Big Bang–present day).

With this elaborate lead-in, we are now ready to investigate implications of the First Law on various constrained thermodynamic transitions involving ideal gas systems.

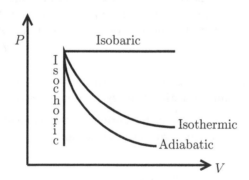

FIGURE 43.3   Transitions from a Fixed Initial State to Various Final States

• **QUASI-STATIC**   The quasi-static condition ensures that the transition $i \to f$ occurs slowly enough that the system is always in [close to] some intermediate equilibrium state, and hence the thermodynamic state variables are well-defined throughout.

ASIDE: Contrast the decompressions of gas in a box by *deliberate vs. sudden* motions of a moving wall. In the former case, the system's pressure exhibits virtually no spatial dependence while it and the volume slowly change in time. In the latter case, the local pressures vary considerably during the brief time interval that the wall is in motion, rendering ill-defined the very notion of a single [global] pressure.

---

[7]The ideal gas law EoS was discussed in Chapter 10.
[8]Introduction of **chemical potentials** [intensive variables] along with the molar amounts of reactants and products [extensive variables] enables the thermodynamic treatment of chemical processes, too.
[9]In the modern vernacular, the RHS representing the quantity of gas is named *Ennar* or *'Nkebi*.
[10]There are Avogadro's Number, $N_A = 6.022 \times 10^{23}$, molecules in a mole.

- **ISOBARIC**   An isobaric process occurs at constant pressure, and hence the work done by the gas can be determined exactly via $W = P \Delta V$.

  V+ IF the volume increases, THEN the system does a net amount of work on its environment. The energy to perform such work must come from conversion of internal energy and/or input heat.

  V− IF the volume decreases, THEN net work is done on the system. The energy associated with this work may be stored as internal energy and/or flow out of the system as heat.

- **ISOCHORIC**   An isochoric process occurs at constant volume. In these transitions, no work is done on or by the system, *i.e.*, $W = P \Delta V \equiv 0$.

  Q+ Adding heat to a gas at constant volume increases its internal energy and pressure.

  Q− Removing heat from a gas at constant volume decreases its internal energy and pressure.

- **ISOTHERMIC**   An isothermic process occurs at constant temperature. In these transitions the internal energy remains constant, *i.e.*, $\Delta U \equiv 0$. Hence, $Q = W$, according to the First Law. Heat may flow into or out of the system, as long it is balanced by a counterflow of work.

  Q+ Addition of heat at constant temperature leads to the performance of an equal amount of mechanical work by the system on its environment.

  W+ Mechanical work done on an isothermic system engenders a flow of heat from the system in equal measure to the work input.

- **ADIABATIC**   In the course of an adiabatic transition, heat does not flow into or out of the system [on the time scale set by the process], and thus $Q = 0$ and $\Delta U = -W$.

  W+ IF positive work is done by the system in the absence of heat flow, THEN the system's store of internal energy must have been drawn down, AND the temperature of the gas is reduced.

  W− IF work is done adiabatically on the system, THEN the system's internal energy must increase AND its temperature rise.

Some processes are approximately adiabatic because the time scale for the process is much shorter than that needed for heat to flow.

> ASIDE: The next time that you employ a hand pump to inflate a bicycle tire, note that the hose connecting to the valve stem becomes warm in the process. Rapidly moving the pump handle down adiabatically compresses the air in the pump barrel and forces it through the hose and into the tire. In response to this compression, the air temperature increases. A portion of this heat then conducts into the connector hose.

Chapter 44

First Law Encore

The First Law of Thermodynamics states that

$$\Delta U = Q - W.$$

The Ideal Gas Law [EoS] may be rearranged to read

$$\frac{PV}{T} = nR = Nk_B.$$

The work done by the system on its environment may be approximated using

$$W \approx P\,\Delta V.$$

The quasi-static trajectory of the system in state space, leading from the initial to the final state, may have to be partitioned in order to compute the work [approximately or exactly]. However, this is a mere technicality.

We are unable to directly compute Q. Moreover, Q is not a thermodynamic state variable, and its value for any transition between states depends crucially on the path taken. To circumvent this obstacle, we rely on an internal energy EoS for the ideal gas:[1]

$$U = \frac{\eta}{2}\,nRT = \frac{\eta}{2}\,Nk_B T.$$

Therefore, for a fixed quantity[2] of monatomic gas[3] the change in the internal energy is

$$\Delta U = \frac{3}{2}\,nR\,\Delta T = \frac{3}{2}\,Nk_B\,\Delta T.$$

Five pertinent comments follow directly below.

ΔU   Internal energy is a thermodynamic state variable, and thus the ΔU associated with transitions between specified initial and final states is uniquely determined.

EoS-1   The expression for ΔU was obtained independently of the First Law.

EoS-2   The ideal gas law EoS, also in effect, enables self-consistent removal of the explicit temperature dependence from the internal energy EoS:

$$\left\{ \begin{array}{l} PV = nRT \\ U = \frac{\eta}{2}\,nRT \end{array} \right\} \quad \Longrightarrow \quad U = \frac{\eta}{2}\,PV \quad \Longrightarrow \quad \Delta U = \frac{\eta}{2}\,\Delta\big[PV\big].$$

---

[1]This equation of state amounts to an expression of the EQUIPARTITION PRINCIPLE. This principle will be encountered again in the next chapter and discussed at greater length in Chapters 49 and 50.

[2]*I.e.*, $nR = Nk_B =$ constant. This restriction can be relaxed, allowing gas to exit and enter the system, and even to undergo chemical reaction!

[3]For a monatomic gas, the number of accessible microscopic degrees of freedom, η, is 3.

$\mathbf{\Delta T = 0}$ In Chapters 42 and 43 it was mentioned that, under isothermic transitions, the internal energy of the system is unchanged.

[The flows of heat and work precisely cancel.]

The expression for the change in internal energy for the fixed-amount sample of ideal gas is consistent with this observation.

$\mathbf{Q, W}$ The heat flow and the work done are both path-dependent, and thus Q and W are not thermodynamic state variables.

The First Law, along with the expression for the change in the internal energy and the specification of mechanical work, fixes the quantity of heat, Q, required in a particular thermodynamic transition involving a fixed amount of monatomic gas. This determination yields:

$$Q = \frac{3}{2} n R \Delta T + W = \frac{3}{2} N k_B \Delta T + W \,,$$

written both in molar and molecular terms.

These tools,

$$W \approx P \Delta V \,, \qquad Q = \frac{3}{2} n R \Delta T + W \,, \quad \text{and} \quad PV = nRT \,,$$

shall be employed to analyse thermodynamic transitions in the suite of examples below.

EXAMPLE [*Isobaric Expansion*]

The initial and final states

$$\left\{ P_i = 40\,\text{Pa} \quad V_i = 2\,\text{m}^3 \quad T_i = 400\,\text{K} \right\} \quad \text{and} \quad \left\{ P_f = 40\,\text{Pa} \quad V_f = 3\,\text{m}^3 \quad T_f = 600\,\text{K} \right\}$$

and all intermediate states are illustrated in Figure 44.1.

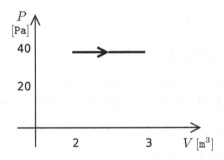

FIGURE 44.1 Isobaric Expansion

The temperature and the volume both increased [in a correlated manner] so as to maintain constant pressure throughout the transition. Verifying that the amount of gas is fixed provides a consistency check on the initial and final states:

$$n R = \frac{PV}{T} \qquad \Longrightarrow \qquad \left\{ \begin{array}{l} \dfrac{P_i V_i}{T_i} = \dfrac{40 \times 2}{400} = \dfrac{1}{5} \\[2ex] \dfrac{P_f V_f}{T_f} = \dfrac{40 \times 3}{600} = \dfrac{1}{5} \end{array} \right\} \,.$$

In the transition from $i \to f$, the work done by the system is

$$W_1 = P \Delta V = 40 \left(3 - 2\right) = 40 \, \text{J} \,,$$

while the heat added is

$$Q_1 = \frac{3}{2} n R \Delta T + W = \frac{3}{2} \times \frac{1}{5} \times \left(600 - 400\right) + 40 = 60 + 40 = 100 \, \text{J} \,.$$

Thus, for the system of gas in the fancy box to make the transition from i to f, following this particular path in state space, it is necessary to add $100 \, \text{J}$ of heat energy and for the system to do $40 \, \text{J}$ of [potentially useful] work on its environment.

> ASIDE: In Chapter 42 it was remarked that the quasi-static condition was one of the factors[4] affecting **reversibility**. Under the reverse transition, $f \to i$, $40 \, \text{J}$ of external work must be performed on the gas and $100 \, \text{J}$ of heat flow out of the system.

EXAMPLE [*Isochoric Cooling*]

The initial and final states

$$\left\{ P_i = 40 \, \text{Pa} \quad V_i = 3 \, \text{m}^3 \quad T_i = 600 \, \text{K} \right\} \quad \text{and} \quad \left\{ P_f = 20 \, \text{Pa} \quad V_f = 3 \, \text{m}^3 \quad T_f = 300 \, \text{K} \right\}$$

along with all intermediate states are illustrated in Figure 44.2.

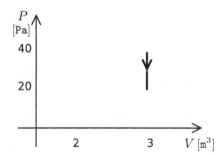

FIGURE 44.2 Isochoric Cooling

The temperature and the pressure both diminish while the volume remains constant. The amount of gas in the box remains fixed during the transition:

$$n R = \frac{P V}{T} \qquad \Longrightarrow \qquad \left\{ \begin{array}{l} \dfrac{P_i V_i}{T_i} = \dfrac{40 \times 3}{600} = \dfrac{1}{5} \\[2mm] \dfrac{P_f V_f}{T_f} = \dfrac{20 \times 3}{300} = \dfrac{1}{5} \end{array} \right\} \,.$$

In passing from $i \to f$, no work is done,[5]

$$W_2 = P \Delta V = P \left(3 - 3\right) = 0 \, \text{J} \,,$$

---

[4]Other factors are that no unobserved heat flows occur and that no dissipative forces act.

[5]The expression for work appears ill-defined, since the pressure, P, is changing throughout the isochoric transition. Not to worry, though: as ΔV is zero by assumption, so too is the work.

while heat flows out of the system,

$$Q_2 = \frac{3}{2} n R \Delta T + W = \frac{3}{2} \times \frac{1}{5} \times (300 - 600) + 0 = -90 \, \text{J}.$$

Thus, for the gas in the fancy box to make this isochoric transition from i to f, 90 J of heat energy must be expelled and zero work done by or on the system.

ASIDE: For $f \rightarrow i$, no work is done and 90 J of heat flow into the system.

EXAMPLE [*Isobaric Compression*]

The initial and final states

$$\left\{ P_i = 20 \, \text{Pa} \quad V_i = 3 \, \text{m}^3 \quad T_i = 300 \, \text{K} \right\} \quad \text{and} \quad \left\{ P_f = 20 \, \text{Pa} \quad V_f = 2 \, \text{m}^3 \quad T_f = 200 \, \text{K} \right\}$$

and all intermediate states are shown in Figure 44.3.

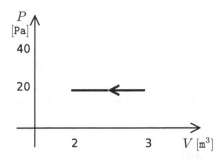

FIGURE 44.3 Isobaric Compression

The temperature and the volume both decrease in such a manner that the pressure remains constant. The fixed amount of substance provides a constraint on the values of the thermodynamic variables along the trajectory and at the endpoints:

$$n R = \frac{P V}{T} \quad \Longrightarrow \quad \left\{ \begin{array}{l} \dfrac{P_i V_i}{T_i} = \dfrac{20 \times 3}{300} = \dfrac{1}{5} \\[2mm] \dfrac{P_f V_f}{T_f} = \dfrac{20 \times 2}{200} = \dfrac{1}{5} \end{array} \right\}.$$

In the isobaric transition $i \rightarrow f$ the work done is

$$W_3 = P \Delta V = 20 \, (2 - 3) = -20 \, \text{J},$$

while the heat input is

$$Q_3 = \frac{3}{2} n R \Delta T + W = \frac{3}{2} \times \frac{1}{5} \times (200 - 300) - 20 = -30 - 20 = -50 \, \text{J}.$$

Thus, to effect this transition from i to f [along this particular path in state space], 20 J of work must be performed on the system and 50 J of heat energy removed.

ASIDE: Under the reverse transition, 50 J of heat flow in and 20 J of mechanical work are performed by the system on its environment.

EXAMPLE [*Isochoric Heating*]

The initial and final states

$$\left\{ P_i = 20\,\text{Pa} \quad V_i = 2\,\text{m}^3 \quad T_i = 200\,\text{K} \right\} \quad \text{and} \quad \left\{ P_f = 40\,\text{Pa} \quad V_f = 2\,\text{m}^3 \quad T_f = 400\,\text{K} \right\}$$

and all intermediate states are displayed in Figure 44.4.

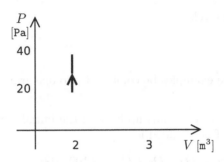

FIGURE 44.4 Isochoric Heating

The temperature and the pressure both increase, while the volume and amount of gas remain constant:

$$nR = \frac{PV}{T} \quad \Longrightarrow \quad \left\{ \begin{array}{l} \dfrac{P_i V_i}{T_i} = \dfrac{20 \times 2}{200} = \dfrac{1}{5} \\[2mm] \dfrac{P_f V_f}{T_f} = \dfrac{40 \times 2}{400} = \dfrac{1}{5} \end{array} \right\}.$$

In the transition $i \to f$ no work is done [owing to the constancy of the volume, and irrespective of the changing pressure],

$$W_4 = P\,\Delta V = P\,(2 - 2) = 0\,\text{J},$$

while the heat added to the system is

$$Q_4 = \frac{3}{2}\,nR\,\Delta T + W = \frac{3}{2} \times \frac{1}{5} \times (400 - 200) + 0 = 60\,\text{J}.$$

Thus, for this isochoric transition, 60 J of heat energy are absorbed and zero work is done by or on the system.

ASIDE: Under reversal, no work is done and 60 J of heat flow out of the system.

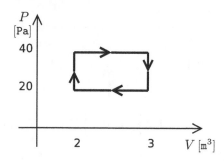

FIGURE 44.5   A Box Cycle

EXAMPLE [*Box Cycle*]

Q: Might all four of these examples be combined into one grand **cycle**?
A: You betcha!

Cyclicity entails that the system ends up back in the initial state from whence it started. The total [net] quantity of heat added is

$$Q_{\text{total}} = Q_1 + Q_2 + Q_3 + Q_4 = +100 - 90 - 50 + 60 = 20 \text{ J},$$

and the net work output over the course of the cycle is

$$W_{\text{total}} = W_1 + W_2 + W_3 + W_4 = +40 + 0 - 20 + 0 = 20 \text{ J}.$$

That the net amount of heat added is equal to the net work performed is a necessary consequence of energy conservation [as the system returns to its original state at the end of the cycle].

> ASIDE: One might be experiencing *déjà vu* reading this last observation. The same line of reasoning was employed in the context of driven DHOs [studied in Chapter 19], where it was realised that the TIME AVERAGED[6] applied power [INPUT] and drag power [OUTPUT] must cancel, *i.e.*,
> $$\langle P_A \rangle + \langle P_D \rangle \equiv 0.$$
> In similar fashion, a thermodynamic system cannot both cycle [*i.e.*, follow an oscillatory trajectory in phase space] and accrue energy.

BOX CYCLE   A box cycle is a closed sequence of isobars and isochores.

$\Delta_{\text{Box}}U$ There must be zero net change in the internal energy, since the system returns to its initial state.

$\Delta_{\text{Box}}W$ The net amount of work performed is equal in magnitude to the area enclosed within the system trajectories in phase space. Positive work [performed by the system on its environment] results when the cycle is traversed in a clockwise sense. The work is negative [absorbed by the system] for the anti-clockwise cycle.

$\Delta_{\text{Box}}Q$ The quantity of heat needed exactly matches the work done, $Q = W$. The heat is input when the system does net work, and is output when work is absorbed by the system.

---

[6]The time average over a number of complete cycles is equivalent to that throughout a single cycle.

Chapter 45

Isotherms and Adiabats

In this chapter, we continue our investigation of constrained thermodynamic transitions and consider in detail isothermic and adiabatic processes.

ISOTHERMIC PROCESSES

Isothermic processes occur at constant temperature. The terms appearing in the First Law, specialised to isothermic ideal gas systems, are enumerated below.

ΔU Owing to the constancy of the temperature, the internal energy of the system remains unchanged, *i.e.*, $\Delta U = 0$.

Q Thus, according to the First Law, the flows of work and heat must cancel:

$$0 = \Delta U = Q - W \qquad \Longrightarrow \qquad Q = W \, .$$

W The work done by an ideal gas expanding isothermically from an initial volume, V_i, to a final volume, V_f, was computed in Chapter 10:

$$W_{if}[P] = n\,R\,T\,\ln\left[\frac{V_f}{V_i}\right] \, .$$

The ideal gas law and the internal energy EoS remain operative and hence the tools that we shall use for our analyses of isothermic transitions of an ideal gas system are

$$W = n\,R\,T\,\ln\left[\frac{V_f}{V_i}\right] \, , \qquad Q = W \, , \qquad \text{and} \qquad P\,V = n\,R\,T \, .$$

ADIABATIC PROCESSES

Adiabatic processes occur with no heat flow. The application of this constraint to an ideal gas system has several implications.

FIRST LAW   In the absence of heat flow, changes in the internal energy of the system are identical in magnitude and opposite in sign to the work done by the system:

$$\Delta U = -W \, .$$

Any work done by the gas is accompanied by a concomitant reduction in the system's internal energy; work input to the system is transformed into internal energy.

TRAJECTORY   The phase space trajectory of an ideal gas undergoing an adiabatic transition [quasi-statically] satisfies

$$PV^\gamma = \text{CONSTANT},$$

where γ is a parameter dependent on properties of the gas, and the constant factor is system- and context-dependent. For monatomic gasses [_e.g._, He, Ne, _etc._], $\gamma \simeq 5/3$. For air at room temperature, $\gamma \simeq 7/5$.

> ASIDE: The exponential factor governing the shape of the adiabatic phase trajectory, γ, is the ratio of the molar specific heats of the gas at constant pressure, $\widetilde{C}_P$, and at constant volume, $\widetilde{C}_V$,
>
> $$\gamma \equiv \frac{\widetilde{C}_P}{\widetilde{C}_V} = \frac{C_P}{C_V}.$$
>
> It is possible to show that molar specific heats of ideal gasses obey $\widetilde{C}_P = \widetilde{C}_V + R$, where R is the universal gas constant.
>
> The EQUIPARTITION PRINCIPLE [Chapter 49] states that the internal energy of each molecule in the system receives a contribution of $\frac{1}{2} k_B T$ from each accessible degree of freedom. Denoting the number of degrees of freedom by η, one obtains $U = \eta \times \frac{1}{2} N k_B T = \eta \times \frac{1}{2} n R T$ for the total internal energy. Thus, the total internal energy per mole is $\widetilde{U} = \frac{\eta}{2} R T$.
>
> The molar specific heats, $\widetilde{C}_P$ and $\widetilde{C}_V$, relate increases in the temperature of one mole of gas to increases in its internal energy. At constant volume, no work is done on or by the gas, and the heat flow contributes directly to the internal energy: $\widetilde{C}_V = \frac{\eta}{2} R$. Combining these observations for the molar specific heats into the expression for γ reveals that
>
> $$\gamma = \frac{1 + \frac{\eta}{2}}{\frac{\eta}{2}} = \frac{2 + \eta}{\eta}.$$
>
> Hence, the value assumed by the adiabatic exponent, γ, is determined by the available number of degrees of freedom.
>
> **1**   A monatomic gas can only translate in three coordinate directions, and hence possesses three degrees of freedom. Thus $\eta = 3$, and $\gamma = 5/3$.
>
> **2**   At all temperatures, a diatomic molecule is able to rotate about two axes,[1] as well as translate, and so $\eta = 5$ and $\gamma = 7/5$. At high temperatures, the diatomic molecule may also vibrate, whence[2] $\eta = 7$ and hence $\gamma = 9/7$.
>
> **3+**   Molecules with three or more constituent atoms which are not purely lineal have three rotational modes, hence $\eta = 6$ and $\gamma = 4/3$. When these systems also vibrate it becomes a sticky wicket to accurately count the accessible degrees of freedom.

WORK   The work done by the system in the course of an adiabatic transition from an initial state, $(P_i, V_i; T_i)$, to a final state, $(P_f, V_f; T_f)$, is

$$W_{if}[P] = \int_{V_i}^{V_f} P \, dV = \int_{V_i}^{V_f} \frac{\text{CONSTANT}}{V^\gamma} \, dV = -\frac{\text{CONSTANT}}{\gamma - 1} \left[\frac{1}{V_f^{\gamma-1}} - \frac{1}{V_i^{\gamma-1}} \right].$$

---

[1] Why only two axes are possible is best explained from the perspective of quantum mechanics.
[2] Subtleties enter into the counting of the degrees of freedom associated with vibration. It suffices for us to acknowledge [on the basis of our study of SHO energetics] that oscillator energies are [on average] equally split between kinetic and potential.

The value of the CONSTANT is particular to the trajectory and may be specified in terms of its boundary [initial or final] conditions. Specialising to the case in which $\gamma = 5/3$, the work done by the gas during the adiabatic transition from the initial to final states is

$$W = -\frac{3}{2} P_i V_i^{5/3} \left[V_f^{-2/3} - V_i^{-2/3} \right] .$$

The tools for analysis of adiabatic transitions of a monatomic ideal gas system are:

$$W = -\frac{3}{2} P_i V_i^{5/3} \left[V_f^{-2/3} - V_i^{-2/3} \right] , \qquad Q = 0 , \qquad \text{and} \qquad PV = nRT .$$

EXAMPLE [*Isothermal Expansion*]

The initial and final states

$$\left\{ P_i = 243\,\text{Pa} \quad V_i = 8\,\text{m}^3 \quad T_i = 900\,\text{K} \right\} \text{ and } \left\{ P_f = 243/2\,\text{Pa} \quad V_f = 16\,\text{m}^3 \quad T_f = 900\,\text{K} \right\},$$

along with all intermediate states, are illustrated in Figure 45.1. [Perhaps this process was effected with the system in thermal contact with a reservoir at $T_H = 900\,\text{C}$.]

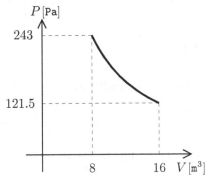

FIGURE 45.1 Isothermal Expansion

The volume and pressure changes are correlated so as to maintain constancy of the temperature and amount of matter in the system. *I.e.,*

$$\frac{P_i V_i}{T_i} = \frac{243 \times 8}{900} = 2.16 \qquad \text{and} \qquad \frac{P_f V_f}{T_f} = \frac{243/2 \times 16}{900} = 2.16 .$$

The quantity of work done is

$$W_1 = n\,RT \ln\left[\frac{V_f}{V_i}\right] = 8 \cdot 243 \ln(2) = 1944 \ln(2) = 2^3\,3^5 \ln(2) \text{ J} .$$

An amount of heat equal to the work done by the expanding gas,

$$Q_1 = W = 2^3\,3^5 \ln(2) \text{ J} ,$$

flows into the system so as to hold its temperature constant.

EXAMPLE [*Adiabatic Expansion*]

The initial and final states

$$\left\{ P_i = 243/2 \text{ Pa} \quad V_i = 16\,\text{m}^3 \quad T_i = 900\,\text{K} \right\} \text{ and } \left\{ P_f = 16\,\text{Pa} \quad V_f = 54\,\text{m}^3 \quad T_f = 400\,\text{K} \right\},$$

along with all intermediate states, are illustrated in Figure 45.2. The work done by the expanding gas is accomplished by drawing upon the system's store of internal energy. The temperature of the gas is reduced from $T_{\text{H}} = 900\,\text{C}$ to $T_{\text{C}} = 400\,\text{C}$, and the pressure is diminished.

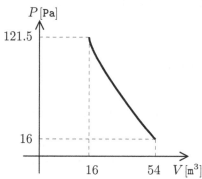

FIGURE 45.2   Adiabatic Expansion

The volume and pressure changed adiabatically, with $\gamma = 5/3$. The amount of gas is fixed:

$$\frac{P_i V_i}{T_i} = \frac{243/2 \times 16}{900} = 2.16 \qquad \text{and} \qquad \frac{P_f V_f}{T_f} = \frac{16 \times 54}{400} = 2.16.$$

The quantity of work done by the gas while it expands is

$$W_2 = -\frac{3}{2}\,(16)\,(54)^{\frac{5}{3}}\left[(54)^{-\frac{2}{3}} - (16)^{-\frac{2}{3}} \right] = -\frac{3}{2}(16 \times 27)\left[1 - \left(\tfrac{3}{2}\right)^2 \right] = +2^2\,3^4\,5 = +1620\,\text{J},$$

while no amount of heat flows into the system, $Q_2 \equiv 0$, by assumption of adiabaticity.

EXAMPLE [*Isothermal Compression*]

The initial and final states

$$\left\{ P_i = 16\,\text{Pa} \quad V_i = 54\,\text{m}^3 \quad T_i = 400\,\text{K} \right\} \text{ and } \left\{ P_f = 32\,\text{Pa} \quad V_f = 27\,\text{m}^3 \quad T_f = 400\,\text{K} \right\},$$

along with all intermediate states, are shown in Figure 45.3. Isothermal compression of the gas requires that the external work input to the system be precisely offset by a flow of heat out of the system.

The quantity of gas in the system, $n\,R$, remains unchanged in the course of the transition, i.e.,

$$\frac{P_i V_i}{T_i} = \frac{16 \times 54}{400} = 2.16 \qquad \text{and} \qquad \frac{P_f V_f}{T_f} = \frac{32 \times 27}{400} = 2.16,$$

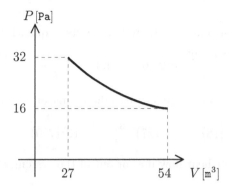

FIGURE 45.3 Isothermal Compression

while the temperature, T, also is constant. Work in the amount of

$$W_3 = n\,RT \ln\left[\frac{V_f}{V_i}\right] = 32 \cdot 27 \ln(1/2) = -864\,\ln(2)\,, = -2^5\,3^3\,\ln(2)\ \text{J}$$

is done by[3] the gas. To preserve the value of the internal energy of the gas, an equal amount of heat,

$$Q_3 = W = -864\,\ln(2) = -2^5\,3^3\,\ln(2)\ \text{J}\,,$$

is added to[4] the system.

EXAMPLE [*Adiabatic Compression*]

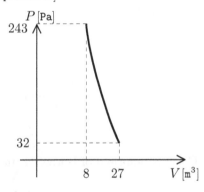

FIGURE 45.4 Adiabatic Compression

The initial and final states

$$\left\{P_i = 32\,\text{Pa} \quad V_i = 27\,\text{m}^3 \quad T_i = 400\,\text{K}\right\} \quad \text{and} \quad \left\{P_f = 243\,\text{Pa} \quad V_f = 8\,\text{m}^3 \quad T_f = 900\,\text{K}\right\},$$

along with all intermediate states, are displayed in Figure 45.4. The adiabatic compression, with $\gamma = 5/3$, is accompanied by the uptake of work, $W < 0$, in the absence of net heat flow, $Q = 0$. Unsurprisingly, the internal energy, temperature, and pressure all increase.

---

[3]The negative value obtained here indicates that net positive work was done on the gas system.
[4]This negative value indicates that heat was removed from the gas.

The state data are consistent with there being a fixed quantity of gas in the system, *i.e.*,

$$\frac{P_i V_i}{T_i} = \frac{32 \times 27}{400} = 2.16 \qquad \text{and} \qquad \frac{P_f V_f}{T_f} = \frac{243 \times 8}{900} = 2.16.$$

The work done by the gas is negative [since the volume of the gas is reduced], *i.e.*,

$$W_4 = -\frac{3}{2}(243)(8)^{5/3}\left[(8)^{-2/3} - (27)^{2/3}\right] = -(16)(729)\left[\tfrac{1}{4} - \tfrac{1}{9}\right] = -2^2\,3^4\,5 = -1620 \text{ J},$$

and again there is no net heat flow into or out of the system, $Q_4 \equiv 0$.

EXAMPLE [*Carnot Cycle*]

Q: Might these four examples be combined into a [closed] cycle?
A: Certainly!
The +'s in Figure 45.5 mark the start/end points of the four legs enumerated above.

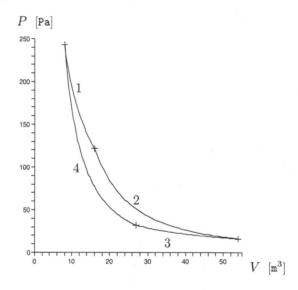

FIGURE 45.5 A Carnot Cycle (Comprised of Isotherms and Adiabats)

Having passed through the sequence of isothermic and adiabatic transitions comprising a Carnot cycle, the system ended up back in its original state. The total amount of heat added was

$$Q_{\text{total}} = Q_1 + Q_2 + Q_3 + Q_4 = 1944 \ln(2) + 0 - 864 \ln(2) + 0 = 1080 \ln(2) \text{ J}.$$

The total amount of work done by the gas in the course of the cycle was

$$W_{\text{total}} = W_1 + W_2 + W_3 + W_4 = 1944 \ln(2) + 1620 - 864 \ln(2) - 1620 = 1080 \ln(2) \text{ J}.$$

The equality of the net amount of heat added and the net work done is entirely consistent with conservation of energy.

Remarkable features of Carnot cycles will be described in the next couple of chapters.

Chapter 46

Thermodynamic Cycles and Heat Engines

Thermodynamic cycles have a number of essential features.

$\Delta_{\text{CYCLE}}U$ There must be no net change in the internal energy, since the system returns to its initial state.

$\Delta_{\text{CYCLE}}W$ The net work performed is equal in magnitude to the area enclosed within the cycle. The work is positive or negative depending on whether the traversal is clockwise or anti-clockwise.

$\Delta_{\text{CYCLE}}Q$ The heat needed exactly matches the work done, $i.e.$, $Q = W$. The net heat is positive when the system does positive net work and is negative when work is absorbed by the system.

BOX CYCLE   A box cycle is a closed phase-space trajectory comprised of isobars and isochores.

CARNOT CYCLE   A Carnot[1] cycle is a closed sequence of isotherms and adiabats.

OTTO CYCLE   An Otto cycle consists of adiabats and isochores.

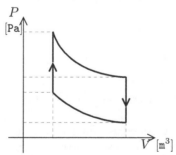

FIGURE 46.1   Otto Cycle (Consisting of Isochores and Adiabats)

The Otto cycle crudely describes the workings of an automobile[2] engine.

> ASIDE: We can map the workings of an internal combustion engine onto the Otto cycle, starting in the lower right corner of Figure 46.1 and proceeding clockwise. First, the [intake and] compression stroke, with fuel vapour and air mixture in the cylinder, occurs rapidly and adiabatically. Next, the fuel and air mixture ignites[3] and the

---

[1] Sadi Carnot (1796–1832) was a pioneer in the study of thermodynamics.
[2] Perhaps this should instead be called an *Otto-mobile*, eh?
[3] The technical term for the burning of the fuel is **deflagration**.

temperature and pressure rise rapidly in the small nearly-constant volume. Third, adiabatic expansion of the hot gas provides the engine's power-stroke. And last, the pressure drops isochorically as the exhaust gases flow out of the cylinder and are replaced with a fresh batch of fuel-air mixture.

Let's recall the box cycle considered at the very end of Chapter 44 and the Carnot cycle investigated in Chapter 45.

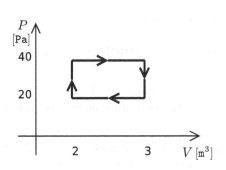
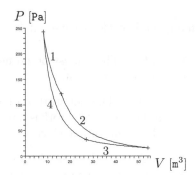

FIGURE 46.2 Box and Carnot Cycles

Closure of the cyclic paths in state space ensures that each system ends up in exactly the same state in which it began. Separate analyses of each of the four segments of the box cycle enable determination of the total quantities of heat added and work output:

$$Q_{\text{Box, total}} = Q_1 + Q_2 + Q_3 + Q_4 = +100 - 90 - 50 + 60 = +20\,\text{J}\,,$$
$$W_{\text{Box, total}} = W_1 + W_2 + W_3 + W_4 = +40 + 0 - 20 + 0 = 20\,\text{J}\,.$$

The Carnot cycle analysis is similar:

$$Q_{\text{Carnot, total}} = Q_1 + Q_2 + Q_3 + Q_4 = +1944\,\ln(2) + 0 - 864\,\ln(2) + 0 = +1080\,\ln(2)\,\text{J}\,,$$
$$W_{\text{Carnot, total}} = W_1 + W_2 + W_3 + W_4$$
$$= +1944\,\ln(2) + 1620 - 864\,\ln(2) - 1620 = +1080\,\ln(2)\,\text{J}\,.$$

That the net amount of heat added is equal to the net work performed is a necessary consequence of energy conservation. [No (internal) energy accrues to the system, because it cycles back to its original thermodynamic state.]

Thinking practically for a moment,[4] one may recast the cycle inputs in the following terms.

INPUT COST   The quantity of heat which flowed into the system from the environment was:

$$Q_{\text{Box, input}} = Q_1 + Q_4 = +100 + 60 = +160\,\text{J}$$
$$Q_{\text{Carnot, input}} = Q_1 = +1944\,\ln(2) \simeq +1347.5\,\text{J}$$

Presumably, there was a cost associated with this input heat that had to be paid.

---

[4]The snarky among us might contrast this with *practically thinking*.

USEFUL
OUTPUT   The net work performed by the system on its environment was:

$$W_{\text{Box, total}} = W_1 + W_2 + W_3 + W_4 = +40 + 0 - 20 + 0 = 20 \text{ J}$$
$$W_{\text{Carnot, total}} = W_1 + W_2 + W_3 + W_4$$
$$= +1944 \ln(2) + 1620 - 864 \ln(2) - 1620 = 1080 \ln(2) \simeq 748.6 \text{ J}$$

One supposes that this work was somehow beneficial.

WASTE
OUTPUT   The quantity of heat ejected from the system into the environment was:

$$Q_{\text{Box, output}} = -Q_2 - Q_3 = +90 + 50 = +140 \text{ J}$$
$$Q_{\text{Carnot, output}} = -Q_3 = 864 \ln(2) \simeq +598.9 \text{ J}$$

This excess heat was dispelled by the system as waste [*a.k.a.* effluent] energy.

HEAT ENGINE   A thermodynamic system operated in a cyclic manner with the net effect of converting some amount of heat into work is a heat engine. Heat engines perform interconversions of heat and work. Our assumptions, that all processes occur quasistatically and without friction or drag, ensure that the heat engines considered here are reversible.[5]

EFFICIENCY   The efficiency of a reversible heat engine is the ratio of its net work output to its heat input:

$$e = \frac{W}{Q_{\text{input}}} = \frac{Q_{\text{input}} - Q_{\text{output}}}{Q_{\text{input}}} = 1 - \frac{Q_{\text{output}}}{Q_{\text{input}}}.$$

Efficiencies always fall in the range $0 \leq e < 1$.

$e = 0$ Zero efficiency means that no useful work is performed.

$e \to 1$ The efficiency may, in principle, approach 1 by diminution of the ratio of waste to input heats. However, contrary to what one might hope for, the waste heat can never be completely[6] eliminated.

Furthermore, **no real heat engine** [involving irreversible processes or dissipative forces] **can ever be as efficient as a reversible one.**

EXAMPLE [*Efficiency of a Box Cycle Engine*]

In the box cycle example of Chapter 44, the heat input, output, and net work were

$$Q_{\text{input}} = 160 \text{ J}, \qquad Q_{\text{output}} = 140 \text{ J}, \qquad W_{\text{total}} = 20 \text{ J}.$$

Thus the efficiency of this box cycle process is

$$e = \frac{W}{Q_{\text{input}}} = \frac{20}{160} = \frac{1}{8} \quad \text{or, equivalently,} \quad e = 1 - \frac{Q_{\text{output}}}{Q_{\text{input}}} = 1 - \frac{140}{160} = \frac{1}{8}.$$

---

[5] Reversible heat engines are commonly called Carnot engines. Beware, however, as some authors apply this moniker exclusively to systems undergoing Carnot cycles.

[6] This is yet another statement of the SECOND LAW OF THERMODYNAMICS.

EXAMPLE [*Efficiency of a Carnot Cycle Engine*]

In the Carnot cycle example of Chapter 45, the heat input, output, and net work were

$$Q_{\text{input}} = 1944 \ln(2) \text{ J}, \qquad Q_{\text{output}} = 864 \ln(2) \text{ J}, \qquad W_{\text{total}} = 1080 \ln(2) \text{ J}.$$

Thus the efficiency of this Carnot cycle process is

$$e = \frac{W}{Q_{\text{input}}} = \frac{1080}{1944} = \frac{5}{9} \quad \text{or, equivalently,} \quad e = 1 - \frac{Q_{\text{output}}}{Q_{\text{input}}} = 1 - \frac{864}{1944} = \frac{5}{9}.$$

For the Carnot cycle [only], it is possible to show that

$$\frac{Q_C}{Q_H} = \frac{T_C}{T_H},$$

and thus the efficiency may be expressed in terms of the temperatures of the HOT and COLD reservoirs:

$$e_{\text{Carnot}} = 1 - \frac{T_C}{T_H}.$$

ASIDE: In the example just above, the HOT and COLD reservoirs are at temperatures 900 C and 400 C respectively, and

$$e = 1 - \frac{400}{900} = \frac{5}{9},$$

in complete agreement with the aforementioned result.

To enhance the efficiency of a Carnot engine, one must decrease the ratio T_C/T_H. In practice, it is most often the case that T_H and T_C are separately prescribed.

[*I.e.*, $T_H \sim$ the combustion temperature of gasoline, while $T_C \sim$ the ambient air temperature.]

All heat engines operate as illustrated in Figure 46.3. The notched arrow connotes the flow of heat into the device [from the HOT reservoir]. Some of the input energy is converted to useful [output] work and the rest is expelled [into the COLD reservoir].

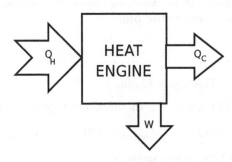

FIGURE 46.3 The Working of a Heat Engine

HEAT IN [FROM HOT] $\longrightarrow$ HEAT OUT [TO COLD] + WORK OUT

As was [repeatedly] mentioned in Chapter 44, each cycle segment was presumed to be reversible. Under reversal, both heat and work flows change sign. Ultimately, should the entire box cycle be traversed in reverse direction, the net effect would be the conversion of work into heat. Surprisingly, the net flow of heat in this case is from the COLD reservoir to the HOT reservoir, as illustrated in Figure 46.4. This "uphill" flow can only occur in conjunction with net external work's being done on the system, *i.e.*, $W < 0$. The input heat and net work are dumped into the HOT side as effluent.

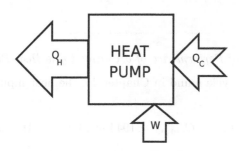

FIGURE 46.4   The Working of a Heat Pump

HEAT OUT [TO HOT] ⟵ HEAT IN [FROM COLD] + WORK IN

A heat engine operated in reverse mode[7] acts as a heat pump.

The most commonly encountered instances of heat pump technology are in refrigeration and air-conditioning devices.

COEFFICIENT of PERFORMANCE   A heat pump's coefficient of performance is the ratio of the heat drawn from the COLD reservoir to the net amount of work needed to accomplish this feat.

$$\text{CofP} = \frac{\text{Heat pumped}}{\text{Work needed}} = \frac{Q_C}{-W}.$$

Again, this is cast in terms of *What you get* versus *What you pay for.*

[Commercially available heat pumps can have performance coefficients on the order of ~ 5.]

For a Carnot heat pump [*i.e.*, a Carnot engine run in reverse mode], the performance coefficient depends only on the temperatures of the HOT and COLD reservoirs between which the heat pump is operating:

$$\text{CofP}_{\text{Carnot}} = \frac{T_C}{T_H - T_C} = \frac{T_C}{\Delta T}.$$

---

[7]Heat engines operating in reverse are not obligated to emit *"beep, beep, beep, ..."* noises.

EXAMPLE [*Coefficient of Performance of a Box Cycle Heat Pump*]

For the **reversed** box cycle example of Chapter 44, the heat input, output, and net work are

$$Q_{\text{input}} = 140 \text{ J} , \qquad Q_{\text{output}} = 160 \text{ J} , \qquad W_{\text{total}} = -20 \text{ J} .$$

The coefficient of performance associated with this heat pump [box cycle process] is

$$\text{CofP}_{\text{Box}} = \frac{Q_{\text{input}}}{W} = \frac{140}{20} = 7 .$$

EXAMPLE [*Coefficient of Performance of a Carnot Cycle Heat Pump*]

In the **reversed** Carnot cycle found in Chapter 45, the heat input, output, and net work are

$$Q_{\text{input}} = 864 \ln(2) \text{ J} , \qquad Q_{\text{output}} = 1944 \ln(2) \text{ J} , \qquad W_{\text{total}} = -1080 \ln(2) \text{ J} .$$

The coefficient of performance of this Carnot heat pump is

$$\text{CofP}_{\text{Carnot}} = \frac{Q_{\text{input}}}{W} = \frac{864}{1080} = \frac{4}{5} ,$$

or, equivalently,

$$\text{CofP}_{\text{Carnot}} = \frac{T_C}{\Delta T} = \frac{400}{900 - 400} = \frac{4}{5} .$$

We end this chapter with the following observation:

**While work and heat are both forms of energy,
a given amount of work may be transformed completely into heat,
whereas a given amount of heat cannot be transformed entirely into work.**

This is yet another statement of the Second Law of Thermodynamics.

Chapter 47

The Second Law of Thermodynamics

Before we begin, let's read an excerpt [the first stanza, emphasis added] from a poem by Yeats.

THE SECOND COMING

Turning and turning in the widening gyre
The falcon cannot hear the falconer;
Things fall apart; the centre cannot hold;
Mere anarchy is loosed upon the world,
The blood-dimmed tide is loosed, and everywhere
The ceremony of innocence is drowned;
The best lack all convictions, while the worst
Are full of passionate intensity.

.....

William Butler Yeats   (1865–1939)
Nobel Prize for Literature (1923)

Yeats' poesy calls to mind **entropy** and the SECOND LAW OF THERMODYNAMICS.

[Yeats was born very shortly after the Second Law was first clearly enunciated.]

SECOND LAW of THERMODYNAMICS   There are many expressions of the Second Law. While seeming rather different, the following are all equivalent.

| 2 | Heat flows spontaneously from HOT to COLD.

| 2 | It is impossible for a heat engine, working in a cycle, to convert all of its input heat into mechanical work with no waste heat.

| 2 | No irreversible heat engine operating between HOT and COLD reservoirs, at T_H and T_C, respectively, can have greater efficiency than a reversible engine operating between the same reservoirs.

| 2 | The Carnot cycle engine operating between reservoirs at T_H and T_C has maximal efficiency. *I.e.*, $e \leq e_{\text{Carnot}} < 1$.

| 2 | It is impossible for a refrigerator to extract heat from a COLD reservoir and deposit it into a HOT reservoir without the consumption of external work.

| 2 | The coefficient of performance of a heat pump is necessarily $< \infty$.

The preferred formulation of the Second Law of Thermodynamics is in terms of **entropy**.

| 2 | The total entropy of a system and its environment remains unchanged when the system undergoes a reversible process and increases during an irreversible process.

ENTROPY   Entropy, S, is a thermodynamic state variable.[1] Changes in entropy occur in conjunction with flows of heat.

- The temperature and the change in entropy combine to [approximately] yield the quantity of heat added to a system,

$$Q \approx T\,\Delta S.$$

  This is analogous to the determination of work via $W \approx P\,\Delta V$.

- Temperature and entropy form a pair of conjugate [intensive/extensive] thermodynamic state variables.

- The units of entropy are joules per kelvin, J/K.

  [The Boltzmann constant, k_B, has units of entropy.]

TEMPERATURE (technical definition)   The expression relating heat and entropy can be recast as

$$\frac{1}{T} \approx \frac{\Delta S}{Q}.$$

The reciprocal of the temperature [*a.k.a.* the inverse temperature] is defined by taking the limit in which the quantity of heat added is infinitesimal, *i.e.*,

$$\frac{1}{T} = \lim_{Q \to 0} \frac{\Delta S}{Q}.$$

Hence, **the inverse temperature is the instantaneous rate at which the system's entropy changes under the addition of heat.**

Consider the descriptions of HIGH and LOW temperatures.

HIGH-T   Adding a small quantity of heat to a system already at high temperature yields little concomitant increase in its entropy. Removing heat from such a system barely decreases its entropy.

LOW-T   Adding a small quantity of heat to a system which is at a very low temperature produces a large increase in entropy. Removing heat from such a system appreciably decreases its entropy.

For isothermic [T = constant] transitions the approximate formula relating the change in entropy to the quantity of heat transferred becomes exact:

$$Q = T\,\Delta S, \quad \text{and thus} \quad \Delta S\Big|_{\text{isothermic}} = \frac{Q}{T}.$$

Maintaining a fixed temperature may require flow of heat [and possibly work, too] into or out of the system, thereby changing its entropy.

---

[1]Some authors employ the term **thermodynamic potential** for entropy and a few other state variables.

EXAMPLE [*Entropy Change and Isothermic Expansion of an Ideal Gas*]

An ideal gas system experiences an isothermic transition from an initial state specified by[2] $(V_i, P_i; T)$ to a final state, $(V_f, P_f; T)$. The quantity of work associated with this particular transition [as was computed in Chapter 10] is

$$W = nRT \ln\left[\frac{V_f}{V_i}\right].$$

For the internal energy to remain fixed, as it must for isothermic transitions, an offsetting flow of heat, $Q = W$, must occur. Thus, the change in entropy is

$$\Delta S = \frac{Q}{T} = nR \ln\left[\frac{V_f}{V_i}\right].$$

IF ideal gas $\left\{\begin{array}{c}\text{expandS}\\\text{Contracts}\end{array}\right\}$ isothermally, THEN its entropy $\left\{\begin{array}{c}\text{INCREASES}\\\text{DECREASES}\end{array}\right\}$.

As not all pairs of phase space points can be joined by isothermal paths, and for those that are so related there exists an infinitude of other paths between the same states, one is tempted to ask the following question.

Q: Might this constrained computation of entropy change have, at best, marginal utility?

A: Such a dim view is not warranted, because **phase transitions**, discussed in Chapter 38, occur at a constant temperature.

EXAMPLE [*Entropy Change Associated with Phase Transitions*]

Consider a system melting from a solid to a liquid state. As revealed in the exposition on latent heats in Chapter 38, the system remains at its critical temperature throughout the time interval during which the two phases coexist. There is a latent heat associated with the phase transition, *viz.*,

$$Q = ML_F,$$

where M is the mass of the material and L_F is the latent heat of fusion particular to the substance. For melting, as assumed here, heat must be added to the system. Hence the entropy of the system is increased.

ASIDE: If instead the substance were fusing [solidifying], then heat would be liberated, and the entropy of the system would decrease. This decrease in entropy is consistent with the increased constitutive order of a solid *vis-à-vis* the amorphous disorder of a liquid.

---

[2] *A priori*, the state space has as many dimensions as there are thermodynamic state variables. For any particular physical system, not all of these variables are independent, nor may they assume any possible value, and thus the accessible states lie within a subspace. For a fixed amount of ideal gas, equations of state and constraints permit reconstruction of the state from its projection onto the (V, P) plane. The temperature, a dependent variable, is explicitly noted to emphasise that its value remains fixed.

We next prove the equivalence of two statements of the Second Law:

Heat flows spontaneously from hot to cold.
The total entropy of the universe increases under an irreversible process.

Suppose that a tiny quantity of heat, δQ, passes from a hot reservoir at T_H to a cold reservoir at T_C. Provided that δQ is sufficiently small, there should be no appreciable change in the temperature of either reservoir, and hence the approximate expressions for the changes in entropy become exact:

$$\Delta S_H = \frac{-\delta Q}{T_H} \qquad \text{and} \qquad \Delta S_C = \frac{+\delta Q}{T_C}.$$

Heat flowing $\begin{Bmatrix} \text{OUT OF} \\ \text{INTO} \end{Bmatrix}$ the $\begin{Bmatrix} \text{HOT} \\ \text{COLD} \end{Bmatrix}$ reservoir generates $\begin{Bmatrix} \text{negative} \\ \text{positive} \end{Bmatrix}$ entropy.

Q: There seems to be both production and destruction of entropy. Granted that this heat flow is irreversible, must the entropy always increase?

A: Yes. The net change in the entropy of the two reservoirs is:

$$\Delta S_{\text{NET}} = \Delta S_H + \Delta S_C = \frac{-\delta Q}{T_H} + \frac{+\delta Q}{T_C} = \delta Q \left[\frac{1}{T_C} - \frac{1}{T_H} \right].$$

Thus $\Delta S_{\text{NET}} > 0$, IFF $T_H > T_C$. The entropy change is always positive because the hot reservoir is at higher temperature than the cold reservoir.

Entropy increase is often portrayed in terms of increasing disorder or randomness.

$S \uparrow$ A dropped glass may shatter into a large number of shards. Prior to falling, the glass possessed a great deal of geometric structure and [dare one say this!] an *ontological* existence as a drinking vessel. The detritus remains glass in the sense of "amorphous solid silicate material," but it is no longer a glass. Furthermore, the glass can shatter in a multitude of ways which are [practically] indistinguishable.

$$\text{ordered glass} \rightarrow \text{disordered shards} \quad \Longrightarrow \quad S_{\text{Universe}} \uparrow$$

Does one expect to ever see a shattered glass reconstitute itself?

$S \uparrow$ Cream poured gently into a cup of coffee forms filamentous tendrils which elongate and swirl until all of the liquid becomes a uniform *cappuccino* colour. The cream, originally concentrated, is dispersed amidst the coffee.

$$\text{concentrated cream} \rightarrow \text{diffuse cream} \quad \Longrightarrow \quad S_{\text{Universe}} \uparrow$$

Would one expect to ever see the cream re-conglomerate?

$S \uparrow$ In order to clean something, it is inevitable that something else be made dirty.

In the story *The Cat in the Hat Comes Back* [by Dr. Seuss], while the children [Sally and me] were shoveling snow, the CAT re-entered their home and proceeded to bathe, leaving a *big long pink cat ring* in the bathtub. The efforts to remedy the ring produced a sequence of stain-transfers among a variety of objects [with ever greater dispersal of colour], culminating in a scene of snowbanks far-and-wide tinted pink.

stain on item → stain on another item ⟹ S_{Universe} ↑

Does a cleaning cloth forever remain unsoiled (unless it remains unused)?

ADDENDUM: A Plethora of Potentials

The First Law of Thermodynamics, introduced in Chapters 42 and 43, reads

$$\Delta U = Q - W.$$

With refinements [$W \approx P\,dV$, from Chapters 43 and 44, and $Q \approx T\,dS$, from earlier in this chapter, along with a term to accommodate changes in the number of particles present in the thermodynamics system], one may write this as

$$\Delta U = T\,\Delta S - P\,\Delta V + \mu\,\Delta N.$$

The **chemical potential**, μ, is the rate at which the internal energy changes when particles are added to the system,[3] *i.e.,*

$$\mu = \frac{\Delta U}{\Delta N}.$$

To make this more rigorous, we take the continuum limit [possible when all of the thermodynamic quantities are enormously large on the scale set by the granularity of the physical system] and reformulate these conditions in terms of partial derivatives. Hence, the chemical potential is formally defined by

$$\mu = \left.\frac{\partial U}{\partial N}\right|_{S,V}.$$

It has been explicitly noted that partial differentiation with respect to the number of particles is performed with the entropy and volume held constant. In this telling, the differential change in the internal energy is

$$dU = T\,dS - P\,dV + \mu\,dN.$$

Implicit in this expression is the notion that the internal energy on the LHS, and each of the coefficients appearing on the RHS, are functions of the entropy, volume, and number of particles associated with the thermodynamic system:

$$U(S,V,N), \qquad T(S,V,N), \qquad P(S,V,N), \qquad \mu(S,V,N).$$

This is all very fine, provided that the entropy, the volume, and the number of particles are apt descriptors of the physical system and are under experimental control. However, instances abound in which the system is best described in terms of other [conjugate] variables. Three of these follow just below.

---

[3]For simplicity, we assume that there is but one species of particle in the system. Should there be a multiplicity of particle types, then each is enumerated separately and possesses a distinct chemical potential.

H   ENTHALPY: $H = U + PV$

[When the volume is fixed and the pressure variable, enlist enthalpy.]

A differential shift in the enthalpy receives contributions from

$$dH = dU + P\,dV + V\,dP.$$

Considering the First Law expansion of dU, quoted above, this becomes

$$dH = T\,dS + V\,dP + \mu\,dN.$$

Accordingly, the enthalpy, temperature, volume, and chemical potential are functions of the entropy, pressure, and number of particles:

$$H(S, P, N), \qquad T(S, P, N), \qquad V(S, P, N), \qquad \mu(S, P, N).$$

A   HELMHOLTZ[4] FREE ENERGY: $A = U - TS$

[When the temperature is at hand, heed Helmholtz.]

The differential variation in A is

$$dA = dU - T\,dS - S\,dT \qquad \Longrightarrow \qquad dA = -S\,dT - P\,dV + \mu\,dN.$$

Accordingly, the Helmholtz free energy, entropy, pressure, and chemical potential are functions of the temperature, volume, and number of particles:

$$A(T, V, N), \qquad S(T, V, N), \qquad P(T, V, N), \qquad \mu(T, V, N).$$

G   GIBBS[5] FREE ENERGY: $G = U - TS + PV$

[When temperature and pressure gain currency, grab Gibbs.]

The differential variation in G is

$$dG = dU - T\,dS - S\,dT + P\,dV + V\,dP \qquad \Longrightarrow \qquad dG = -S\,dT + V\,dP + \mu\,dN.$$

The Gibbs free energy, entropy, volume, and chemical potential are functions of the temperature, pressure, and number of particles:

$$G(T, P, N), \qquad S(T, P, N), \qquad P(T, P, N), \qquad \mu(T, P, N).$$

---

[4]Hermann von Helmholtz (1821–1894) began his scientific career in medical physiology. His abiding interest in foundational aspects of metabolism and perception led him to significant discoveries in diverse areas of thermodynamics and electromagnetism. Perhaps his greatest contribution to science was his advocacy of the energetic point of view for analysis of physical phenomena.

[5]Josiah W. Gibbs (1839–1903) was the recipient of the very first Engineering Ph.D. awarded by an American university (Yale, 1863). After postdoctoral studies in Europe, he returned to a career in mathematical physics at Yale. Gibbs, along with Maxwell and Boltzmann, helped to show, incontrovertibly, that classical thermodynamics is derivable from the fundamental precepts of Newtonian mechanics, augmented with tools of statistical analysis.

A great number of comments are begging to be made. We'll settle for three.

1. The shift from parametric dependence on a particular state variable, to dependence upon its thermodynamic conjugate, is accomplished by means of a **Legendre transformation**. The four thermodynamic potentials considered here are particular Legendre transforms of one another.

2. The Ideal Gas EoS directly determines coefficients in the free energies, whereas an implicit functional dependence remains when the internal energy and enthalpy are considered.

$$\text{Helmholtz:} \qquad P(V,T,N) = k_B \frac{NT}{V}$$

$$\text{Gibbs:} \qquad V(P,T,N) = k_B \frac{NT}{P}$$

$$\text{Internal Energy:} \quad P(S,V,N) = k_B \frac{N}{V} T(S,V,N)$$

$$\text{Enthalpy:} \qquad V(S,P,N) = k_B \frac{N}{P} T(S,P,N)$$

> ASIDE: In our analyses of the box and other thermodynamic cycles we insisted on reversibility, $\Delta S = 0$, and implicitly adopted a Helmholtzian, temperature-dependent, point of view.

3. The value obtained for the bulk modulus is dependent on the particular manner in which it is defined.

The **isothermal** bulk modulus is derived from the Gibbs Free energy:

$$V = \frac{\partial G}{\partial P}\bigg|_{T,N} \qquad \Longrightarrow \qquad -\frac{1}{V}\frac{\partial V}{\partial P} = \frac{1}{B_{\text{isothermal}}} = -\frac{1}{V}\frac{\partial^2 G}{\partial P^2}\bigg|_{T,N} .$$

A sufficient condition for constancy of entropy is the absence of heat flow,

$$\Delta S \approx \frac{Q}{T}, \qquad \text{so} \quad \Delta S = 0 \iff Q = 0.$$

Therefore, the **adiabatic** bulk modulus stems from the internal energy:

$$-P = \frac{\partial U}{\partial V}\bigg|_{S,N} \qquad \Longrightarrow \qquad -V\frac{\partial P}{\partial V} = B_{\text{adiabatic}} = V\frac{\partial^2 U}{\partial V^2}\bigg|_{S,N} .$$

Mathematically, the internal energy and the Gibbs free energy are different functions of differing sets of variables. Hence, the isothermal and adiabatic bulk moduli for the same physical system are expected to differ.

Chapter 48

Entropy Musings and the Third Law

Now is an opportune time to recall that <u>all</u> of the thermodynamic processes discussed thus far in this VOLUME have been reversible.

Q: What about **irreversible processes**? [Most thermodynamic transitions are irreversible owing to the existence of some form of dissipation, or incomplete thermal isolation, or departure from equilibrium.]

A: The answer comes in two stages.

SYSTEM ENTROPY   Thermodynamic states are characterised by their configurations [*i.e.*, the set of specific values assumed by the thermodynamic variables].

> NEITHER the configurations NOR the states themselves possess [strong[1]] dependence on the thermodynamic history[2] of the system.

Hence, any changes in the values of the thermodynamic state variables can only depend on the initial and final states themselves and not on the precise means by which the transition was effected.

Thus, one can determine the change in entropy of a system undergoing an irreversible transition by positing a reversible path from the initial to the final states, and computing the entropy change along the fictitious trajectory using

$$Q_{\text{reversible}} \approx T \, \Delta S_{\text{system}}.$$

The various considerations signalled by the usage of "$\approx$" are still in force.

> ASIDE: Even though Q is path-dependent, so too is T, in that the intermediate thermodynamic states are characterised by particular and unique values of temperature. This occurs in such a way as to ensure that the changes in entropy are uniquely determined, as befits a *bona fide* thermodynamic state variable. For comparision, recall that work depends on the trajectory-dependent value of the pressure while the change in volume, $\Delta V = V_f - V_i$, is fixed by the initial and final states.

TOTAL ENTROPY   While the system entropy changes by the amount determined just above, the environment endures an equal and opposite change.

[Total entropy, *i.e.*, system + environment, is unchanged under a reversible transition.]

However, the dissipative aspect of irreversible processes inevitably introduces an additional energy or heat flow beyond that associated with reversible processes linking the initial and the final states.

---

[1]The weasel word "strong" is to allow for **hysteresis**, in which prior states of the system do exert some influence on its present configuration.

[2]Or the future of the system, for that matter.

[Recall—again—that Q is NOT a state variable.]

As a consequence of this additional heat flow, the change in entropy of the environment no longer exactly compensates for that of the system.

−   IF the entropy of the system decreases, THEN the entropy of the environment must increase by a greater amount.

+   IF the entropy of the system increases, THEN the corresponding decrease in the entropy of the environment is by a lesser amount. [The entropy of the environment may even increase, too.]

The net effect in both cases is an increase in the entropy of the universe.

THERMODYNAMIC ARROW OF TIME

That [according to the Second Law] the total entropy of the universe is an ever non-decreasing state variable seems to endow the universe with a "thermodynamic arrow of time." Some eminent thinkers have gone so far as to make the speculative claim that:

The Second Law of Thermodynamics provides the sole means of distinguishing between the PAST and the FUTURE.

Newton's Laws [in the absence of dissipative forces] are invariant under time reversal. In other words, the solutions[3] of a given set of dynamical equations[4] are transformed into solutions of the same equations, with different boundary/initial conditions, under the action of $t \to -t$.

Invoking the Second Law to explain the inexorable forward flow of time is not a panacea. Several concerns remain unassuaged. For instance, one wonders why the universe began in a sufficiently low entropy initial state so as to allow the formation and persistence of highly ordered subsystems including galaxies, stars, planets, ..., and us!

STATISTICAL THOUGHTS AND KINETIC THEORY

The ideal gas system, in thermal equilibrium, has its configuration specified in terms of the values assumed by the thermodynamic state variables:

$$\{P, V, n, T, S, U, \dots\}.$$

The gas in the box consists of a vast number, N, of molecules, each of which, in principle, possesses a [classically] well-defined[5] position and momentum. This system is exhaustively described by the collection,

$$\left\{\{\vec{r}_\alpha, \vec{p}_\alpha, x_\alpha\}, \text{ for } \alpha \in 1 \dots N\right\},$$

of individual particle positions, momenta, and whatever labels x_α are necessary.

The detailed description of the individual particles comprising the system is called a **microstate**, and that in terms of thermodynamic state variables is a **macrostate**.

---

[3]The solutions are the time- (and particle property-)dependent trajectories.

[4]These are expressions of $\frac{d\vec{p}}{dt} = \vec{F}_{\text{NET ext'l}}$.

[5]Let us agree to treat the molecules as particles *à la* VOLUME I.

Boltzmann made several inferences concerning microstates and macrostates.

* Should two of the gas molecules in the box be exchanged, the microstate of the system is distinctly different, and yet the thermodynamic state variables describing the macrostate are [reasonably] expected to remain unchanged.

  One concludes that there are many more microstates than macrostates.

* There is a **many-to-one mapping** from the microstates of a physical system to its macrostates. The number of such microstates consistent with a given macrostate [labelled by i] is its **degeneracy** and is denoted by Ω_i.

  Consistency requires that the sum of all of the macrostate degeneracies precisely equals the total number of microstates,

$$\Omega_{\text{Total}} \equiv \sum_{\text{macrostates, } i} \Omega_i \,.$$

CAVEAT: Technical issues may make these degeneracies difficult to define in classical physics [*i.e.*, particle positions and momenta take their values in a continuum]. However, the formal sum makes sense provided that each microstate is associated with only one macrostate.

* Boltzmann posited that all microstates are equally likely to be realised, and hence, **the probability of finding the system in a particular macrostate is directly proportional to its degeneracy**. Normalising the total probability to 1, the probability of finding the system in the ith macrostate is

$$\mathcal{P}_i = \frac{\Omega_i}{\Omega_{\text{Total}}} \,.$$

* Boltzmann's greatest triumph was to show that entropy admits a statistical formulation. In Boltzmann's reckoning,

$$S_i = k_B \ln\left(\Omega_i\right),$$

where k_B denotes Boltzmann's constant.[6]

> ASIDE: A comment on normalisation is in order. It proves elegant to express the entropy of the ith macrostate in terms of its probability, *i.e.*,
>
> $$S_i = k_B \ln\left(\mathcal{P}_i\right) = k_B \ln\left(\frac{\Omega_i}{\Omega_{\text{Total}}}\right) = k_B \ln\left(\Omega_i\right) - k_B \ln\left(\Omega_{\text{Total}}\right).$$
>
> The final term on the RHS is constant, and thus it cancels in the computation of changes of entropy, making it possible to compute the entropy directly from the degeneracy, as was stated.

Before further pursuing the ideal gas example, let's consider Boltzmann's reasoning in the context of two sorts of games and amusements.

---

[6]This expression shows most clearly the necessity that k_B bear units of entropy.

When playing a board game with a pair of dice,[7] one finds that rolls of 2 and 12 occur less frequently than all others, while 7 occurs most frequently. These disparate facts can be interpreted from the point of view of microstates and configurations as illustrated in the table below. With two dice there are a total of 36 microstates [distinguishable die outcomes[8]] and 11 macrostates [player moves[9]].

Player Move	Die Combinations	Number	Frequency
2	$\{(1,1)\}$	1	1/36
3	$\{(2,1),(1,2)\}$	2	1/18
4	$\{(3,1),(2,2),(1,3)\}$	3	1/12
5	$\{(4,1),(3,2),(2,3),(1,4)\}$	4	1/9
6	$\{(5,1),(4,2),(3,3),(2,4),(1,5)\}$	5	5/36
7	$\{(6,1),(5,2),(4,3),(3,4),(2,5),(1,6)\}$	6	1/6
8	$\{(6,2),(5,3),(4,4),(3,5),(2,6)\}$	5	5/36
9	$\{(6,3),(5,4),(4,5),(3,6)\}$	4	1/9
10	$\{(6,4),(5,5),(4,6)\}$	3	1/12
11	$\{(6,5),(5,6)\}$	2	1/18
12	$\{(6,6)\}$	1	1/36
Configuration	Microstates	Degeneracy	Probability

Careful inspection of the entries in the table affirms the *fairness* of the dice in that all microstates appear once and are equally likely. The configurations—the player's moves—are not equally likely. The Frequency [Probability] column was obtained by taking the Number (of die combinations) [Degeneracy (of the macrostate)] divided by the total number of microstates. The table is consonant with our experiences playing board games:

☐ 12's and 2's are equally scarce, 11's are as rare as 3's, 10's occur as often as 4's, 9's and 5's turn up with the same frequency, 8's and 6's are equally likely, 7 is the most probable [modal] roll.

☐ rolling a 7 is six times more likely than rolling a 2, *etc.*

An aspect of the game *Yahtzee*™ requires summing five dice. As each die has six sides, there are $6^5 = 7776$ possible microstates. The set of macrostates, sums of 5 dice, consists of the numbers from $5 \rightarrow 30$, with cardinality 26. These facts, and more, are illustrated in the following table and figure.

---

[7]A standard die is cubical with faces numbered from 1 to 6 possessing equal likelihood for selection.
[8]Two independent and distinguishable dice yield 6 × 6 potential combinations when thrown.
[9]The cardinality of the set of sums of two dice, $\{2, 3, ..., 12\}$, is 11.

CONFIGURATION	DEGENERACY	PROBABILITY		CONFIGURATION	DEGENERACY	PROBABILITY
5	1	$1/7776$		30	1	$1/7776$
6	5	$5/7776$		29	5	$5/7776$
7	15	$5/2592$		28	15	$5/2592$
8	35	$35/7776$		27	35	$35/7776$
9	70	$35/3888$		26	70	$35/3888$
10	126	$7/432$		25	126	$7/432$
11	205	$205/7776$		24	205	$205/7776$
12	305	$305/7776$		23	305	$305/7776$
13	420	$35/648$		22	420	$35/648$
14	540	$5/72$		21	540	$5/72$
15	651	$217/2592$		20	651	$217/2592$
16	735	$245/2592$		19	735	$245/2592$
17	780	$65/648$		18	780	$65/648$

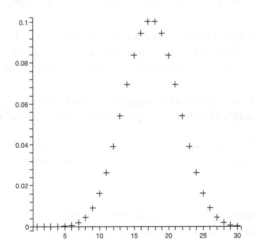

FIGURE 48.1 Relative Frequencies of Five Dice Sums

From the table and Figure 48.1 we observe that the probability distribution is nicely symmetric, and the modal probabilities [for sums of 17 or 18] are slightly greater than 0.1.

The connection between these games of chance and the gas in the box system is less tenuous than one might initially think. Consider the following thought [*Gedankan*] experiment.

A box of gas is partitioned into two equal halves: L and R.

1. IF there is one molecule of gas in the box, THEN there is probability of one-half that it resides in L.

2   IF there are two molecules of gas in the box, AND each molecule's behaviour is independent of the presence of the other, THEN there is a one-in-four chance that both reside in L.

3   IF there are three molecules in the box, THEN the probability that all three are in L is one-eighth.

$\vdots$

$\mathcal{N}$ IF there are $\mathcal{N}$ [independently acting] molecules in the box, THEN there is only a one in $2^{\mathcal{N}}$ chance of finding all of the molecules in L.

There are two object lessons to be gleaned from this example.

•   When the number of molecules is large, there is an infinitesimally small likelihood of finding all of the gas spontaneously concentrated in one half of the box.

•   IF the gas were initially confined in one half of the box, AND then this confining constraint relaxed, allowing the gas to occupy the whole of the interior volume of the box, THEN the gas would [irreversibly] disperse itself through the larger volume.

[The expansion of the gas increases its entropy.]

One can fruitfully think of the Second Law of Thermodynamics as giving rise to an **entropic force** driving systems in phase space in a manner akin to the way a mechanical system responds to a conservative force by attempting to lower its potential energy!

The Second Law of Thermodynamics drives systems toward those configurations with the largest number of associated microstates.

ASIDE: The shattered glass in Chapter 47 provides an illustration of this notion of entropic driving. The glass shatters *because* there are many more ways in which the shards can disperse than ways in which the glass remains intact!

Q: The Second Law is simultaneously significant, sweeping, and sobering. How can the Third Law of Thermodynamics follow an act like the Second?

A: Well, here it is.

THIRD LAW of THERMODYNAMICS   A statement of the Third Law is that it is impossible to reduce the temperature of a thermodynamic system to absolute zero in a finite number of thermodynamic steps.

The Third Law is a ZENOish statement. One can design clever ways to extract heat from a system [thereby lowering its temperature], but one can never quite manage to remove ALL of the heat from a physical system.

Chapter 49

The Canonical Ensemble

The Canonical Ensemble, in the present context, is NOT the name of a group of musicians. Instead, an **ensemble** is a vast assemblage of equivalent thermodynamic systems. Information of a statistical nature about the system is obtained by surveying the ensemble. The distinction **canonical** refers to the particular set of constraints imposed on the members of the ensemble. These are elaborated upon below.

CANONICAL ENSEMBLE   The canonical ensemble consists of a set of identical thermodynamic systems, each of which is maintained in thermal equilibrium with an external heat reservoir held at a common temperature. Aside from being able to exchange heat [via weak coupling] with their reservoirs, the systems are otherwise isolated. An example is an impermeable box, containing a fixed amount of gas, whose sides are maintained at a fixed temperature.

MICROCANONICAL ENSEMBLE   In the microcanonical ensemble, the system has fixed values of extensive thermodynamic state variables [*e.g., N, V, etc.*] and an immutable total quantity of heat. The mixtures of substances of differing temperature [subject to the constraint that no heat flowed into or out of the system] studied in Chapter 38 are instances of microcanonical ensembles.

GRAND CANONICAL ENSEMBLE   A grand canonical ensemble is a collection of systems and their reservoirs allowing exchanges of heat and constituent particles.

Let's investigate some of the properties of a canonical ensemble of a particular thermodynamic system. Recall that the distinction between system [sys] and environment/reservoir [res] quantities may always be made. Furthermore, any extensive thermodynamic quantity X may be partitioned into its system and reservoir parts,

$$X_{\text{Total}} = X_{\text{sys}} + X_{\text{res}}.$$

This is so, in particular, for the total energy, *i.e.*,

$$E_{\text{Total}} = E_{\text{sys}} + E_{\text{res}}.$$

Let us suppose that the system is in a microstate with a unique and precisely determined energy, E_s [whatever this happens to be]. The energy content of the reservoir is taken to be appreciably greater than that of the system. While the system is in a unique state, the macrostate corresponding to the condition of the reservoir is expected to have a [large] degeneracy:

$$\Omega(E_{\text{res}}) = \Omega(E_{\text{Total}} - E_s).$$

First taking the logarithm and then expanding to first-order in a Taylor series about zero system energy [it's actually a Maclaurin series] yields

$$\ln\left[\Omega(E_{\text{res}})\right] = \ln\left[\Omega(E_{\text{Total}} - E_s)\right]$$

$$\simeq \ln\left[\Omega(E_{\text{Total}})\right] - \left(\frac{\partial\ln\left[\Omega(E)\right]}{\partial E}\bigg|_{E=E_{\text{Total}}}\right) E_s + \text{higher-order terms}.$$

Two features emerge from this analysis.

- According to Boltzmann [Chapter 48], the logarithm of the number of accessible microstates is proportional to the entropy of the reservoir.[1] *I.e.*,

$$\ln\left[\Omega(E)\right]\bigg|_{E=E_{\text{Total}}} = \frac{1}{k_B}\,S_{\text{res}}\,.$$

- The reciprocal temperature [Chapter 47] is defined by the rate of change of entropy as heat is added or removed. Heat is but a form of energy.

Thus, the first-order term in the expansion admits the re-expression:

$$-\left(\frac{\partial\ln\left[\Omega(E)\right]}{\partial E}\bigg|_{E=E_{\text{Total}}}\right) E_s = -\left(\frac{1}{k_B}\frac{\partial S}{\partial E}\bigg|_{E=E_{\text{Total}}}\right) E_s = -\left(\frac{1}{k_B}\frac{1}{T_{\text{res}}}\right) E_s = -\frac{E_s}{k_B\,T}\,.$$

The last equality relies on the system and reservoir being in thermal equilibrium. In the remainder of this chapter and the next, we shall use the reciprocal energy-scaled temperature, **beta**,

$$\beta \equiv \frac{1}{k_B\,T}\,.$$

Returning to the Maclaurin expansion of the degeneracy of states in the reservoir, and recognising that this degeneracy can be ascribed to the state of the system, we write

$$\ln\left[\Omega(E_{\text{res}})\right] = \ln\left[\Omega(E_s)\right] \simeq \ln\left[\Omega(E_{\text{Total}})\right] - \beta\,E_s + \text{higher-order terms}.$$

Neglecting the higher-order terms[2] and rearranging results in

$$\ln\left[\frac{\Omega(E_s)}{\Omega(E_{\text{Total}})}\right] = -\beta\,E_s \qquad \Longleftrightarrow \qquad \frac{\Omega(E_s)}{\Omega(E_{\text{Total}})} = e^{-\beta\,E_s}\,.$$

Insofar as the system is concerned, $\Omega(E_{\text{Total}})$ is an enormously large-valued constant, setting the scale for the degeneracy corresponding to the system state with energy E_s. According to Boltzmann and Maxwell, the probability that the system and reservoir are in their combined macrostate is proportional to the degeneracy, and thus the probability that the system is in the[3] state with E_s is

$$\mathcal{P}_s \propto e^{-\beta\,E_s}\,.$$

It is essential that true probabilities be normalised, *i.e.*, that the sum of probabilities be equal to 1. A direct way to ensure this normalisation is by writing

$$\mathcal{P}_s = \frac{e^{-\beta\,E_s}}{\sum_n e^{-\beta\,E_n}}$$

---

[1] Here, the entropy is evaluated when the reservoir holds all of the energy. This approximation is valid for cases in which the canonical ensemble is meaningful.

[2] This may be justified but we shall not do so here.

[3] With a bit of care, this analysis extends to degenerate system states, too.

for the state probabilities. The sum in the denominator is over all possible states n of the system [assumed to be non-degenerate] with corresponding energy E_n. This expression for the probabilities applies to any small thermodynamic system with discrete energy states in contact with a large thermal reservoir at temperature T.

PARTITION FUNCTION   The sum presented just above is the partition function of the system:

$$Z = \sum_n e^{-\beta E_n} \, .$$

ASIDE: It turns out that the partition function is related to the Helmholtz Free Energy mentioned in Chapter 47 by $A(T, V, N) = -k_B T \ln(Z)$.

EXAMPLE [*Canonical Ensemble for a Two-State System*]

The simplest non-trivial system that we might consider has but two possible states, $\{0, 1\}$, with energies $E_0 > 0$ and $E_1 > E_0$. Let $E_1 = E_0 + \epsilon$ for some $\epsilon > 0$. Note that ϵ, the level splitting or energy gap, is an intrinsic property of the system. Let us place such a system in thermal equilibrium with a reservoir at inverse temperature β.

First, we shall compute the partition function for this system [and reservoir]:

$$Z = \sum_{n=0}^{1} e^{-\beta E_n} = e^{-\beta E_0} + e^{-\beta E_1} = e^{-\beta E_0} \left(1 + e^{-\beta \epsilon}\right) \, .$$

The form of the final equality facilitates the next several steps in the analysis.

Second, the probability of the system's being in each of its allowed states is straightforwardly determined.

0   The probability that the system is in the state 0 is $\mathcal{P}_0 = \dfrac{e^{-\beta E_0}}{Z} = \dfrac{1}{1 + e^{-\beta \epsilon}}.$

1   Similarly, $\mathcal{P}_1 = \dfrac{e^{-\beta E_1}}{Z} = \dfrac{e^{-\beta E_0 - \beta \epsilon}}{Z} = \dfrac{e^{-\beta \epsilon}}{1 + e^{-\beta \epsilon}} = \dfrac{1}{1 + e^{\beta \epsilon}}.$

[It may be shown that these probabilities are properly normalised.]

The control parameter [determining the respective probabilities] is $\beta \epsilon$, the size of the energy splitting measured in proportion to the ambient temperature. Let's consider the two limits: high temperatures—for which the system's energy gap is relatively insignificant—and low temperatures for which the opposite is true.

0   As $\beta \epsilon \to 0$ [*i.e.*, in the high-temperature limit], both $\mathcal{P}_0$ and $\mathcal{P}_1$ approach $\frac{1}{2}$, the former from above and the latter from below. When the energy gap is insignificant, the system is equally likely to be found in either of its two states.

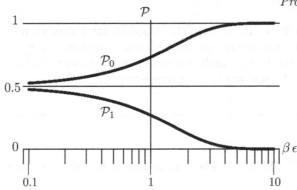

FIGURE 49.1   State Probabilities for a Two-Level System in Contact with a Thermal Reservoir

∞   As $\beta\epsilon \to \infty$ [*i.e.*, for low temperatures], $\mathcal{P}_0 \to 1^-$ and $\mathcal{P}_1 \to 0^+$. When the energy gap is large, it is difficult for the system to borrow enough energy from the reservoir to transition into the higher-level state. And when it does, it is unlikely to remain there for long. Hence, it approaches certainty that the system will be found in state 0.

The behaviour of the probabilities between these two limits is shown in Figure 49.1.

Third, let's compute the expected [ensemble average] value of the energy of the system. Formally, this is the probability-weighted average energy:

$$\langle E \rangle = \mathcal{P}_0\,E_0 + \mathcal{P}_1\,E_1 = \frac{1}{1+e^{-\beta\epsilon}}\,E_0 + \frac{1}{1+e^{\beta\epsilon}}\,(E_0 + \epsilon) = E_0 + \frac{\epsilon}{1+e^{\beta\epsilon}}.$$

At a glance, one can see that this result is reasonable. Its lower bound is E_0, while its upper bound is less than E_1.

> ASIDE: Any particular system will EITHER be in state 0 with energy E_0, OR be in state 1 with energy E_1. In an ensemble of identical systems, some fraction of the systems will be in one state and the remainder in the other. That none of the systems in the ensemble have energy equal to the average value is as much a worry to us as was the realisation [made in VOLUME I, Chapter 30] that the centre of mass of two point-like particles may lie in the empty space between them (along the line joining their centres).

Let's consider the behaviour of the average energy in the usual limits.

0   As $\beta\epsilon \to 0$ [the high-temperature limit], $\langle E \rangle \to E_0 + \frac{\epsilon}{2}$ [the mid-point energy]. This is as one might expect, since the system is equally likely to be in either of the two accessible states.

∞   As $\beta\epsilon \to \infty$ [for low temperatures], $\langle E \rangle \to E_0^+$. This conforms with our expectation that, as the relative size of the gap grows large, the overwhelming majority of the systems in the ensemble will be found in the lower-energy state.

Another formal claim bears mentioning. The ensemble average energy of the system may be computed directly from the partition function, via

$$\langle E \rangle = -\frac{\partial \ln(Z)}{\partial \beta}.$$

For the two-level system considered just above, we obtain exactly the same result as was obtained by direct computation,

$$-\frac{\partial \ln(Z)}{\partial \beta} = \frac{\partial (\beta E_0)}{\partial \beta} - \frac{e^{-\beta \epsilon}}{1 + e^{-\beta \epsilon}} (-\epsilon) = E_0 + \frac{\epsilon}{1 + e^{\beta \epsilon}}.$$

This illustrates that the partition function is more than merely the normalisation factor for probabilities.

EXAMPLE [*Canonical Ensemble for an Infinite Ladder System*]

Suppose that our physical system may abide in any one of an infinite number of equally-spaced discrete energy states: $E_n = E_0 + n\,\epsilon$, for $n \in \{0, 1, 2, \ldots\}$.

ASIDE: This comports with the quantum mechanical description of the simple harmonic oscillator!

Suppose that such a physical system is in thermal equilibrium with a heat reservoir at a common temperature described by β.

First, the partition function for this system is

$$Z = \sum_{n=0}^{\infty} e^{-\beta E_n} = e^{-\beta E_0} \left(\sum_{n=0}^{\infty} e^{-n\beta \epsilon} \right).$$

Second, the probability of the system's being in the state numbered s is

$$\mathcal{P}_s = \frac{e^{-\beta E_s}}{Z} = \frac{e^{-\beta E_0 - s\beta \epsilon}}{Z} = \frac{e^{-s\beta \epsilon}}{\sum_{n=0}^{\infty} e^{-n\beta \epsilon}}.$$

These probabilities are normalised [despite there being an infinite number of them]. The behaviours of the probabilities in the high-temperature limit are not so straightforwardly ascertained as in the two-level system because the ladder ensures that there are always [an infinitude of] states whose energy is appreciably larger than the scale set by the temperature. The best that one can say is that, at high temperatures, a set of low-energy states become essentially equivalent. The low-temperature limit, in which $\beta \epsilon \to \infty$, applies to the infinite ladder in the same manner as in the two-state system. Here, all of the probabilities for $s \geq 1$ states vanish [*i.e.*, approach 0^+], while that for its lowest-energy state becomes a near certainty: $\mathcal{P}_0 \to 1^-$.

Third, the expected value of the energy of the system in this case is

$$\langle E \rangle = \sum_{s=0}^{\infty} \mathcal{P}_s E_s = \left(\sum_{s=0}^{\infty} \mathcal{P}_s \right) E_0 + \left(\sum_{s=0}^{\infty} \mathcal{P}_s s \right) \epsilon = E_0 + \sum_{s=0}^{\infty} \frac{e^{-s\beta \epsilon}}{\sum_{n=0}^{\infty} e^{-n\beta \epsilon}} s$$

$$= E_0 + \frac{\sum_{s=0}^{\infty} s\,\epsilon\, e^{-s\beta \epsilon}}{\sum_{n=0}^{\infty} e^{-n\beta \epsilon}}.$$

The sums in these expressions may be approximated by integrals.[4] To wit,

$$\text{denominator} = \sum_{n=0}^{\infty} e^{-n\beta\epsilon} \simeq \int_0^\infty e^{-n\beta\epsilon}\, dn = \frac{-1}{\beta\epsilon}\, e^{-n\beta\epsilon}\Big|_0^\infty = \frac{1}{\beta\epsilon}$$

and

$$\text{numerator} = \epsilon \left(\sum_{s=0}^{\infty} s\, e^{-s\beta\epsilon}\right) \simeq \epsilon \int_0^\infty s\, e^{-s\beta\epsilon}\, ds = \ldots == \epsilon\left(\frac{1}{\beta\epsilon}\right)^2 = \frac{1}{\beta^2\epsilon}.$$

With these results, the expectation value of the system energy is

$$\langle E \rangle = E_0 + \frac{\frac{1}{\beta^2\epsilon}}{\frac{1}{\beta\epsilon}} = E_0 + \frac{1}{\beta} = E_0 + k_B\, T.$$

Remarkably, the expected energy of the system in thermal equilibrium with the reservoir is $E_0 + k_B\, T$, irrespective of the level-spacing on the ladder!

One might worry that this exceptional result arose as an artifact of the method used, or by the approximation of the sums by integrals. Let's try obtaining the average energy directly from the partition function:

$$\langle E \rangle = -\frac{\partial \ln[Z]}{\partial \beta} = -\frac{\partial}{\partial \beta} \ln\left[e^{-\beta E_0}\left(\sum_{n=0}^{\infty} e^{-n\beta\epsilon}\right)\right] = -\frac{\partial \ln\left[e^{-\beta E_0}\right]}{\partial \beta} - \frac{\partial}{\partial \beta} \ln\left(\sum_{n=0}^{\infty} e^{-n\beta\epsilon}\right).$$

The first term in the final expression for $\langle E \rangle$ reduces to E_0. The sum in the second term is equal to the denominator factor appearing in the expression for the ensemble-average energy. The integral approximation to this sum, $\frac{1}{\beta\epsilon}$, was computed just above. Hence,

$$\langle E \rangle = E_0 - \frac{\partial}{\partial \beta} \ln\left(\frac{1}{\beta\epsilon}\right) = E_0 + \frac{\partial}{\partial \beta} \ln\left(\beta\epsilon\right) = E_0 + \frac{1}{\beta\epsilon}\,\epsilon = E_0 + \frac{1}{\beta} = E_0 + k_B\, T,$$

the same result as was obtained earlier by more direct computation.

There is a deeper significance to these results. Recall that the underlying system possessing the infinite ladder of states is a simple harmonic oscillator and that the energetic analysis of Chapters 15 and 16 revealed that the averaged SHO potential and kinetic energies are equal. Therefore, one might posit that the SHO in contact with the thermal reservoir has

$$\langle K \rangle = \frac{1}{2} k_B\, T \qquad \text{and} \qquad \langle U \rangle = \frac{1}{2} k_B\, T.$$

This last conclusion is entirely consistent with the PRINCIPLE OF EQUIPARTITION acting in this instance.

EQUIPARTITION PRINCIPLE (simplified)   Subject to certain technical caveats, the Equipartition Principle states that a system with η independent energetic degrees of freedom, placed in contact with a thermal reservoir at temperature T, will possess, on average, $\eta \times \frac{1}{2} k_B\, T$ thermal energy.

So much for discrete systems. In the next chapter, we shall analyse a system with a continuous set of energy levels.

---

[4]This works best when the granularity of the summand is small, *i.e.*, in the high-temperature limit.

ADDENDUM: The Infinite Ladder System *sans* Integral Approximations

The sums that we encountered in evaluating the partition function and the expected energy for the harmonic oscillator system are not intractable. Here we shall work them out.

[In so doing, we shall also verify the validity of the integral approximations.]

Let's recast the sum in the denominator term:

$$\text{denominator} = \sum_{n=0}^{\infty} e^{-n\beta\epsilon} = \sum_{n=0}^{\infty} \left(e^{-\beta\epsilon}\right)^n .$$

Formally, this is a **geometric series** in $e^{-\beta\epsilon}$.

EVALUATION OF GEOMETRIC SERIES

Consider a base factor a, with $0 < a < 1$, and construct the [finite] series

$$A_n = \sum_{i=0}^{n} a^i = 1 + a + a^2 + a^3 + \ldots + a^n .$$

The product $A_n\,(1-a)$, taken termwise, simplifies to a binomial expression:

$$
\begin{aligned}
A_n\,(1-a) &= (1-a) + a\,(1-a) + a^2\,(1-a) + \ldots + a^n\,(1-a) \\
&= 1 - a + a - a^2 + a^2 - \ldots + a^{n-1} - a^n + a^n - a^{n+1} \\
&= 1 - a^{n+1} .
\end{aligned}
$$

Hence, it is incontrovertible that $A_n = \dfrac{1 - a^{n+1}}{1 - a}$.

ASIDE: In Chapter 40, when we justified the application of Newton's Law of Cooling to cases involving radiative heat transfer, a variant of this result was used to re-express the difference of quartics, $T_{\text{H}}^4 - T_{\text{C}}^4$, in terms of $\Delta T = T_{\text{H}} - T_{\text{C}}$ and a polynomial of degree three in $\{T_{\text{H}}, T_{\text{C}}\}$.

Since it has been assumed that $a < 1$, there is no obstacle to taking the $n \to \infty$ limit of the expression for the finite sum. Doing this yields

$$A_{\infty} = \lim_{n\to\infty} A_n = \lim_{n\to\infty} \frac{1 - a^{n+1}}{1 - a} = \frac{1}{1 - a} .$$

Since the base factor in the ladder sum is less than 1 for all $\beta\epsilon$ [positive finite beta times positive finite ladder spacing], the denominator term converges to

$$\text{denominator} = \frac{1}{1 - e^{-\beta\epsilon}}.$$

There is still the numerator to consider. It is not in the form of a geometric series. Instead, it may be written as

$$\frac{\text{numerator}}{\epsilon} = \sum_{s=0}^{\infty} s\, e^{-s\beta\epsilon}.$$

Evaluating this sum commences with recognition that

$$s\, e^{-s\beta\epsilon} \equiv -\frac{\partial\, e^{-s\beta\epsilon}}{\partial(\beta\epsilon)}.$$

With this observation, the sum in the numerator may be re-expressed as

$$\frac{\text{numerator}}{\epsilon} = \sum_{s=0}^{\infty} -\frac{\partial}{\partial(\beta\epsilon)} e^{-s\beta\epsilon} = -\frac{\partial}{\partial(\beta\epsilon)}\left(\sum_{s=0}^{\infty} e^{-s\beta\epsilon}\right).$$

The clever switcheroo of sum and derivative is justified because the exponential is smooth [the derivative always exists], and the sum converges, and each term is finite. The quantity in parentheses is the denominator computed a moment ago. Thus,

$$\frac{\text{numerator}}{\epsilon} = -\frac{\partial}{\partial(\beta\epsilon)}\left(\frac{1}{1-e^{-\beta\epsilon}}\right) = \left(\frac{1}{1-e^{-\beta\epsilon}}\right)^2 \left(-e^{-\beta\epsilon}\right)(-1)$$

$$= \frac{1}{1-e^{-\beta\epsilon}}\frac{e^{-\beta\epsilon}}{1-e^{-\beta\epsilon}} = \frac{1}{1-e^{-\beta\epsilon}}\frac{1}{e^{\beta\epsilon}-1}.$$

Combining these results yields the expected value of the energy,

$$\langle E \rangle = E_0 + \frac{\epsilon}{e^{\beta\epsilon}-1}.$$

Neglecting the energy of the bottommost state, E_0, and focusing exclusively on the thermal part, we may aspire to reconcile this with our previous—approximate—result. Consider the series expansion of the exponential:

$$e^x = \frac{x^0}{0!} + \frac{x^1}{1!} + \ldots + \frac{x^n}{n!} + \ldots = 1 + x + \text{higher-order terms}.$$

Therefore, in the limit of high temperatures [where we can expect to get agreement with the approximate results derived earlier],

$$e^{\beta\epsilon} - 1 \simeq \beta\epsilon\left(1 + O(\beta\epsilon)\right),$$

and hence

$$\frac{\epsilon}{e^{\beta\epsilon}-1} \simeq \frac{1}{\beta(1+O(\beta\epsilon))} \simeq \frac{1}{\beta}\left[1 + O(\beta\epsilon)\right] \simeq \frac{1}{\beta} = k_B T.$$

This justifies the analysis in the high-temperature limit. It is almost a shame that the more exact results deviate from the approximation, because the latter exhibited the equipartition principle most perfectly. The deviation comes about because the temperature cannot be made arbitrarily small without our encountering the quantum mechanical aspects of the physical system.

Chapter 50

Maxwell–Boltzmann Distribution Derived, Dulong and Petit Revisited

Let's apply the canonical ensemble formalism to a dilute monoatomic gas consisting of $N+1$ identical particles, held in a fancy insulated box with volume V at temperature T.

Q: Wait! What is the system? What is the reservoir? If the box is insulated then isn't this a MICROCANONICAL ENSEMBLE?

A: These are all excellent questions. The system is a single particle [atom or molecule] of gas and the reservoir consists of the remaining N particles. The single atom interacts [weakly] through collisions with the other atoms and is thus able to gain and lose [thermal] energy. [We implicitly assume that thermal energy can be gained in any amount—eschewing quantisation— and that the distributions governing particle position and velocity are continuous.] As has been our practice, we assume that there are no long-range interactions among the gas atoms and that the effects of gravity are negligible. [Therefore the physical situation is both homogeneous and isotropic provided that N, V, and T are sufficiently large.]

The state of the system is fully described by the [3-d] position, $\vec{r}$, and [3-d] velocity,[1] $\vec{v}$, of the single gas particle. The kinetic energy of the system is $K = \frac{1}{2} m v^2$. Potential energy terms, should they exist, are irrelevant to this system [by assumption].

The probability that the particle is to be found within a region of space bounded by $\vec{r}$ and $\vec{r} + d\vec{r}$ and possessing velocity in the range $[\vec{v}, \vec{v} + d\vec{v}]$ is denoted by $\mathcal{P}(\vec{r}, \vec{v})$. The canonical ensemble analysis of the previous chapter informs us that

$$\mathcal{P}(\vec{r}, \vec{v}) \propto e^{-\frac{\beta}{2} m v^2} \, d^3r \, d^3v \, .$$

There are two things to observe. The first is that position and velocity are [in classical dynamics] continuous kinematical properties of the particle.

> ASIDE: Formally, there is exactly ZERO probability that the [point] particle is at any [infinitely precisely specified] point. The same is true, analogously, for velocity or momentum.

The canonical ensemble's relative probability factor is re-interpreted as a **probability density**, weighting a [tiny] volume of position–velocity phase space. The factors $d^3r = dx\, dy\, dz$ and $d^3v = dv_x\, dv_y\, dv_z$ represent the [infinitesimal] ranges for position and velocity. Second, this probability density has no explicit dependence on position. The lone particle is equally likely to be anywhere in the fancy box. Equivalently, the assumptions made at the outset ensure that the gas is everywhere homogeneous.

Thus, we dispose of [integrate out] the position dependence [accepting merely that the system is somewhere inside the box] in order to focus exclusively on the velocity distribution.[2] *I.e.,*

$$\mathcal{P}(\vec{v}) \propto e^{-\frac{\beta m}{2} v^2} \, d^3v \, .$$

---

[1] In more advanced presentations of [statistical] mechanics the state is described in terms of its position and momentum. Here we will employ the velocity, in order to realise more directly the goals of this chapter.

[2] At the risk of sowing confusion, we use the same symbol, $\mathcal{P}$, for the canonical ensemble probability [density], the velocity distribution, and the Maxwellian speed distribution.

Noteworthy aspects of the analysis thus far are that (1) we continue to work in Cartesian coordinates, and (2) we will eventually have to normalise the probabilities.

Working in Cartesian coordinates ensures separation of the component terms, *i.e.*,

$$v^2 = \vec{v} \cdot \vec{v} = v_x^2 + v_y^2 + v_z^2,$$

leading to factorisation of the [velocity-space] probability density,

$$e^{-\frac{\beta m}{2} v^2} d^3 v = \left[e^{-\frac{\beta m}{2} v_x^2} dv_x \right] \left[e^{-\frac{\beta m}{2} v_y^2} dv_y \right] \left[e^{-\frac{\beta m}{2} v_z^2} dv_z \right].$$

This expression of the probability factor is the product of three separate Gaussian distributions: one for each component of the velocity. The common shape of these distributions is illustrated in Figure 50.1. There, the velocity-component axis is scaled in units of the **standard deviation**, $\sigma = \frac{1}{\beta m}$, a relative measure of the width of the Gaussian distribution. The velocity distribution exhibits forward–backward symmetry [consistent with the assumed isotropy of the system]. The MODAL value of each velocity component is 0.

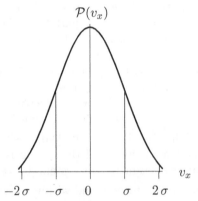

FIGURE 50.1   The Gaussian Distribution of Component Velocities

To ascertain the normalising factor for the component probabilities we need only sum over [integrate] the Gaussian factors.

INTEGRATING WITH A GAUSSIAN DISTRIBUTION

At first blush, the integral $\int_{-\infty}^{\infty} e^{-\frac{x^2}{2\sigma}} dx$ appears impervious to closed-form solution. However, being equal to the area under the curve in Figure 50.1, it has a precise numerical value which we shall denote by C^{-1}. An exceedingly clever trick begins by taking two such integrals at once,

$$C^{-2} = \left[\int_{-\infty}^{\infty} e^{-\frac{x^2}{2\sigma}} dx \right] \left(\int_{-\infty}^{\infty} e^{-\frac{y^2}{2\sigma}} dy \right) = \int_{-\infty}^{\infty} \int_{-\infty}^{\infty} e^{-\frac{x^2+y^2}{2\sigma}} dx \, dy.$$

Even though the x and y coordinates in this expression are independent,[3] one can transform the integrand to circular polar coordinates, (r, θ) [*cf.* Chapter 8 in VOLUME I]. *I.e.*,

$$r^2 = x^2 + y^2 \qquad \text{and} \qquad \tan(\theta) = y/x.$$

---

[3]Or perhaps one ought to say: "Because the x and y coordinates are independent..."

Transforming the integral requires conversion of the $dx\,dy$ factor [a.k.a. the measure] and consistent replacement of the limits. Here,

$$dx\,dy \longrightarrow r\,dr\,d\theta \qquad \text{and} \qquad \left\{\begin{matrix} x \in (-\infty, \infty) \\ y \in (-\infty, \infty) \end{matrix}\right\} \longrightarrow \left\{\begin{matrix} r \in [0, \infty) \\ \theta \in [0, 2\,\pi) \end{matrix}\right\}.$$

The polar-transformed double-integral factorises:

$$C^{-2} = \left(\int_0^{2\pi} d\theta \right) \left(\int_0^\infty r\,e^{-\frac{r^2}{2\sigma}}\,dr \right).$$

The integration over θ yields the expected factor of $2\,\pi$. The radial integral may be computed by first transforming to the scaled variable $u = \frac{r}{\sqrt{2\sigma}}$, and then to $w = u^2$. Doing so yields

$$C^{-2} = (2\,\pi)\,\sigma \int_0^\infty 2\,u\,e^{-u^2}\,du = (2\,\pi)\,\sigma \int_0^\infty e^{-w}\,dw = \left(\frac{2\,\pi}{\beta\,m} \right) \left[-e^{-w} \Big|_0^\infty \right]$$

$$= \left(\frac{2\,\pi}{\beta\,m} \right) \big(0 - (-1) \big) = \frac{2\,\pi}{\beta\,m}.$$

Thus, the normalisation factor appearing in the Gaussian distribution for each velocity component is

$$C^{-1} = \sqrt{\frac{2\,\pi}{\beta\,m}} \qquad \Longleftrightarrow \qquad C = \sqrt{\frac{\beta\,m}{2\,\pi}}.$$

With separate [equal] normalisation factors for each component, the probability that the single-particle system is in its particular velocity state is

$$\mathcal{P}(\vec{v})\,d^3v = \left(\frac{\beta\,m}{2\,\pi} \right)^{3/2} e^{-\frac{\beta\,m}{2}\,v^2}\,d^3v.$$

The speed of the particle is the magnitude of its velocity. Owing to isotropy, the direction of its velocity is moot. The integration over the 3-d Cartesian velocity components may be re-expressed in terms of speed [radial] and directional [angular] parts. *I.e.*, $d^3v = v^2\,dv\,d\Omega$, where the unadorned v represents speed and $d\Omega$ the differential solid angle in 3-d velocity space. Integration over the solid angle[4] yields $4\,\pi$ **sr**, and hence

$$\mathcal{P}(v)\,dv = 4\,\pi \left(\frac{\beta\,m}{2\,\pi} \right)^{3/2} v^2\,e^{-\frac{\beta\,m}{2}\,v^2}\,dv.$$

The LHS represents the probability that the speed of the particle in a particle–gas[5] system chosen at random from the ensemble of such systems will lie in the range $[v, v + dv]$.

There is a figure–ground *Gestalt* at work here. We chose one particle from within the box of gas to be the system and inferred ensemble-average properties. These results apply directly to the other particles comprising the gas! Thus, $\mathcal{P}(v)$ is the fraction of gas particles [identical, possessing mass m] with speed in the interval $[v, v + dv]$ in a system held in a fancy box at temperature T. This quantitative description of the distribution of speeds of gas particles [contained in a fancy box held at fixed temperature] is known as the **Maxwell–Boltzmann distribution**. Representative $\mathcal{P}$ curves for two rather different temperatures are sketched in Figure 50.2.

---

[4] As mentioned in Chapter 41, the unit of solid angle is the steradian: **sr**.
[5] The gas is comprised of the $\mathcal{N}$ other particles.

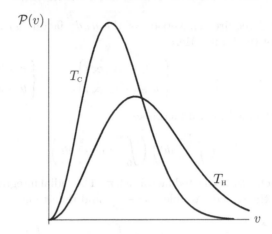

FIGURE 50.2   Maxwellian Distributions of Speed for Monoatomic Gases in Boxes

T_{H} For hotter gasses, *e.g.*, T_{H} in Figure 50.2, the distribution is broader and flatter. The higher temperature means that there is overall a higher thermal kinetic energy, and hence an enhanced higher-speed tail in the distribution of speeds.

T_{C} For cooler gasses, the speeds are more strongly peaked at smaller values.

Irrespective of temperature, these distributions have several common properties.

0   The M–B distribution goes smoothly to zero for $v \to 0$ [for temperatures $T > 0$]. The interpretation of heat as disordered kinetic energy, and the Second Law, imply that—even at very cold temperatures—there is never an appreciable fraction of the gas at rest.

∞   The M–B distribution goes smoothly to zero for $v \to \infty$. The finite amount of heat in any system and its dispersal among the particles comprising the system ensure that no particle has arbitrarily large speed.

MAX   ROLLE'S THEOREM [a specialisation of the MEAN VALUE THEOREM] informs us that the two behaviours quoted just above are sufficient to guarantee [for a continuous function] that $\mathcal{P}(v)$ has at least one maximum. Thus, there is a MODAL value for the M–B distributed speeds. Its value is computed to be:

$$v_{\text{MODE}} = \left(\frac{2}{\beta \, m} \right)^{1/2}.$$

The modal speed increases with temperature, *i.e.*, $T \propto \beta^{-1}$, and is reduced for heavier gas particles. Both of these dependencies are reasonable.

1   The area under every M–B curve is equal to 1.

Knowledge of the distribution of something [here, the speeds of gas particles; in VOLUME I, the mass density of solid objects] makes it possible—and irresistibly tempting—to compute its moments.

0   The zeroth moment [the area under the curve] is 1 for proper normalisation, *i.e.*,

$$1 \equiv \int_{0}^{\infty} \mathcal{P}(v) \, dv = \int_{0}^{\infty} 4\pi \left(\frac{\beta \, m}{2\,\pi} \right)^{3/2} v^2 \, e^{-\frac{1}{2} \beta \, m \, v^2} \, dv.$$

1   The first moment is equal to the average speed of the gas particles, *i.e.*,

$$\langle v \rangle = \int_0^\infty v \, \mathcal{P}(v) \, dv = \int_0^\infty 4\pi \left(\frac{\beta m}{2\pi} \right)^{3/2} v^3 \, e^{-\frac{1}{2}\beta m v^2} \, dv = \frac{2}{\sqrt{\pi}} \, v_{\text{MODE}} \, .$$

2   The second moment is related to the RMS[6] velocity of the particles in the gas:

$$\langle v^2 \rangle = \int_0^\infty v^2 \, \mathcal{P}(v) \, dv = \int_0^\infty 4\pi \left(\frac{\beta m}{2\pi} \right)^{3/2} v^4 \, e^{-\frac{1}{2}\beta m v^2} \, dv = \frac{3}{2} \, v_{\text{MODE}}^2 \, .$$

Therefore, $v_{\text{RMS}} = \sqrt{3/2} \, v_{\text{MODE}}$.

The modal, average, and RMS speeds for a M–B gas at a particular temperature are illustrated in Figure 50.3. Here, the speed is scaled in units of the modal speed of the gas particles.

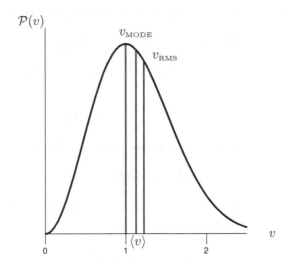

FIGURE 50.3   Maxwellian Distribution Showing Modal, Average, and RMS Speeds

It wasn't only for grins that we computed the moments of the M–B speed distribution. A quantity of supreme interest is the average energy of the particles in the gas. As our assumptions essentially ruled out any potential energy contributions, the mechanical energy of the gas is purely kinetic. The average energy per gas particle is

$$\langle E \rangle = \langle K \rangle = \langle \tfrac{1}{2} m v^2 \rangle = \tfrac{1}{2} m \, \langle v^2 \rangle = \tfrac{1}{2} m \left(\tfrac{3}{\beta m} \right) = \tfrac{3}{2} \beta^{-1} = \frac{3}{2} \, k_B T \, .$$

This is exactly the value that one obtains via appeal to the EQUIPARTITION PRINCIPLE!

---

[6]See VOLUME II, Chapter 42 *et seq.*, for a discussion of RMS quantities in the context of AC-Circuits.

RECOVERING DULONG–PETIT FROM EQUIPARTITION

Now that we have been introduced to the EQUIPARTITION PRINCIPLE and have seen that it is verifiable [in certain contexts] we might seek other instances of its utility.

The atoms in a metallic solid occupy fixed lattice sites. This arrangement is predicated upon each lattice site's being a local minimum for a [complicated, 3-d] potential energy function. Two simplifications occur when the temperature of the solid is not too low[7] [and not too high[8] either]. The first is that it is always possible to factorise an atom's [local] 3-d potential energy well into a product of three 1-d wells (one for each spatial direction). These three wells need not be equivalent. The second is that, in the near vicinity of the minimum, [almost[9]] any potential energy function can be ably approximated by a simple harmonic oscillator [quadratic] potential. Therefore, we might envision the single atom in a metallic lattice as possessing a triple ladder of energy states, *viz.*,

$$E_{n_x n_y n_z} = E_0 + n_x\,\epsilon_x + n_y\,\epsilon_y + n_z\,\epsilon_z\,.$$

ASIDE: The language *the single atom in a metallic lattice* indicates the CANONICAL division into a one-atom system and N-atom reservoir.

In thermal equilibrium, each SHO is expected to possess [on average] $2 \times \frac{1}{2} k_B T$ thermal energy. Therefore,
$$\langle E \rangle = E_0 + 3\,k_B\,T = E_0 + E_{\text{thermal}}\,.$$

Since each lattice atom behaves this way, a mole of such atoms is expected to have a total thermal energy content of

$$Q_{\text{molar}} = N_A\,E_{\text{thermal}} = 3\,N_A\,k_B\,T\,.$$

Recalling the definition of the universal gas constant, $R = N_A\,k_B$, this becomes

$$Q_{\text{molar}} = 3\,R\,T\,.$$

Q: How does the thermal energy content per mole change with temperature?
A: It changes in proportion given by the MOLAR SPECIFIC HEAT, of course! *I.e.,*

$$\Delta Q_{\text{molar}} = C_{\text{molar}}\,\Delta T\,.$$

Therefore,
$$C_{\text{molar}} = 3\,R\,.$$

This is precisely the [empirical] LAW OF DULONG AND PETIT mentioned in Chapter 38.

And that, dear reader, is the long and the short [DULONG AND PETIT] of our brief discursion into deriving thermodynamical relations from statistical mechanics principles.

---

[7] Were the temperature to be very low, quantum mechanical effects would intrude, thereby invalidating our classical analysis.

[8] As the Wicked Witch of the West said (in an unrelated context), "I'm melting!"

[9] This CAVEAT must be stated, because higher-order even-power potentials can be only roughly approximated using the SHO, quadratic model. *Cf.* Chapter 16 in VOLUME III for elaboration of this point.

Epilogue

Io, one of Jupiter's moons, exemplifies many of the phenomena we have surveyed throughout this VOLUME.

- ·○   Io's discovery was made possible by the development of optical instruments exploiting the refractive properties of glass. [Later, more precise observations were made with reflecting telescopes.]

- ○   Io's motion is quasi-oscillatory: it revolves [quickly, in a very tight orbit] around Jupiter.

- ○   Io rotates with the same period as it revolves. It has become tidally locked by immensely strong forces. [Fundamental physics is at work here.]

- ⊙   Io is flexed and kneaded [elastically] by the tides.

- ⊙   Dissipation occurs. Io's resistance to flexing and kneading can be attributed to viscosity. We recall that viscosity can act in the manner of a linear damper.

- ○   Thus the Jupiter–Io system is a driven DHO. It consumes input mechanical work and dissipates it into heat. The First and Second Laws of Thermodynamics apply to the interconversion of energy and its degradation into heat.

- ○   The deposited thermal energy has accumulated, warming Io. It is now a ball of molten rock. Melting is a phase change.

- ○·   The heat deposited deep in the interior is convected to the surface, producing spectacular displays of VULCANISM. Heat also conducts toward the surface, and the surface radiates heat into space.

There is no escape without a physics tale.

Whilst walking briskly across campus one morning, Bertal met a bleary-eyed classmate emerging from the library. The fellow stood blinking in the bright sunlight for a moment, clutching his bookbag in trembling sleep-deprived hands.

Bertal, in sympathetic salutation, cheerily said,

"Hey buddy, what's new?"

This elicited the agitated response,

"Nu? ... Nu?
It's cee over lambda.
It's omega over two pi.
It's one over the period.
It's kinematic viscosity.
It's ...
Oh. Sorry. Hi Bertal. I think I need some rest."

Part II

Materials Problems

E

Elasticity Problems

Useful Data

Whenever the value of the acceleration due to gravity at the Earth's surface is needed, take it to be $g = 10\,\mathrm{m/s^2}$.

Standard atmospheric pressure is $101.3\,\mathrm{kPa}$. Use this value unless otherwise specified.

Densities $[\mathrm{kg/m^3}]$ and Elastic Moduli $[\mathrm{GPa}]$ for Select Elements								
Element	**Density**		**Young's**		**Shear**		**Bulk**	
ALUMINIUM	ρ_{Al} =	2,700	Y_{Al} =	70	$M_{S,\mathrm{Al}}$ =	26	B_{Al} =	76
COPPER	ρ_{Cu} =	8,940	Y_{Cu} =	120	$M_{S,\mathrm{Cu}}$ =	48	B_{Cu} =	140
GOLD	ρ_{Au} =	19,300	Y_{Au} =	79	$M_{S,\mathrm{Au}}$ =	27	B_{Au} =	180
IRIDIUM	ρ_{Ir} =	22,560	Y_{Ir} =	528	$M_{S,\mathrm{Ir}}$ =	210	B_{Ir} =	320
MOLYBDENUM	ρ_{Mo} =	10,280	Y_{Mo} =	329	$M_{S,\mathrm{Mo}}$ =	126	B_{Mo} =	230
NICKEL	ρ_{Ni} =	8,910	Y_{Ni} =	200	$M_{S,\mathrm{Ni}}$ =	76	B_{Ni} =	180
PLATINUM	ρ_{Pt} =	21,450	Y_{Pt} =	168	$M_{S,\mathrm{Pt}}$ =	61	B_{Pt} =	230
TUNGSTEN	ρ_{W} =	19,250	Y_{W} =	411	$M_{S,\mathrm{W}}$ =	161	B_{W} =	310
URANIUM	ρ_{U} =	19,100	Y_{U} =	208	$M_{S,\mathrm{U}}$ =	111	B_{U} =	100

Densities and Elastic Moduli for Select Materials				
Material	**Density**		**Young's Modulus**	**Bulk Modulus**
Marshmallow	$\rho_{\mathrm{m\text{-}m}}$ =	$0.36\ \mathrm{g/cm^3}$	$Y_{\mathrm{m\text{-}m}}$ = 29 kPa	$B_{\mathrm{m\text{-}m}}$ = 80 kPa
Nylon Rope	ρ_{nylon} ≃	$1.15\ \mathrm{g/cm^3}$	Y_{nylon} ≃ 4 GPa	
Silicon Carbide	ρ_{CSi} ≃	$3,210\ \mathrm{kg/m^3}$	Y_{CSi} ≃ 450 GPa	B_{CSi} ≃ 220 GPa
Steel	ρ_{steel} ≃	$8,000\ \mathrm{kg/m^3}$	Y_{steel} ≃ 200 GPa	B_{steel} ≃ 160 GPa

E.1 Concrete is a conglomerate formed from cement, sand, and gravel. It is much stronger under tensile compression than tensile extension. Why do you think concrete exhibits this asymmetry?

E.2 An earring and an electrical sensor used to detect small magnetic fluctuations are both made from pure (24 kt.) gold. Compute the change in length occurring in each of the following scenarios.

(a) The earring post is cylindrical, with length 1.0 cm and diameter 0.8 mm. One end of the post is fixed and a tensile force of 50 N [the weight of a five kilogram bag of sugar] is applied to the other.

(b) The sensor is disk-shaped, with diameter 1.0 cm and thickness 0.3 mm. One face of the disk is firmly affixed and the other is subjected to a tensile force of 50 N.

E.3 An earring and an electrical sensor used to detect small magnetic fluctuations are both made from pure (24 kt.) gold. Compute the angle of shearing in each of the following scenarios.

(a) The earring post is cylindrical, with length 1.0 cm and diameter 0.8 mm. One end of the post is fixed and a shear force of 50 N is applied transversely on the other.

(b) The sensor is disk-shaped, with diameter 1.0 cm and thickness 0.3 mm. One face of the disk is firmly affixed and the other is subjected to a transverse force of 50 N.

E.4 PK's new desk is fabricated from 1 cubic metre of uranium. Including the 900 kg of pens, pencils, and paper tucked into the drawers, the desk has a total mass of 2.0×10^4 kg, and thus a weight of approximately 2.0×10^5 N. The desk, being too heavy to haul up the building stairs, is lifted up to PK's third-floor office by means of a crane. The steel cable on the crane's boom is 3.1 cm in diameter and has an initial length of 50 m.

(a) Determine the stress on the cable when the crane supports the entire weight of the desk (and the desk is at rest).

(b) Determine the (i) strain in and (ii) length of the cable under the conditions described in (a).

(c) How do your results for (a–b) change if the desk is accelerating upward at 1 metre per second-squared?

(d) How do your results for (a–b) change if the desk is accelerating downward at $2 \, \text{m/s}^2$?

E.5 PK's former desk was made from just under two cubic metres of uranium. It had a total mass of 3.8×10^4 kg, and thus a weight of approximately 3.8×10^5 N. A crane with boom length 40 m and cable diameter 5 cm was employed to remove the desk from the window of his third-floor office.

(a) Determine the stress on the cable when the crane supports the entire weight of the desk (and the desk is at rest).

(b) Determine the (i) strain in and (ii) change in length of the cable under the conditions described in (a).

(c) Work out the results for (a–b) were the desk to be accelerating upward at $2 \, \text{m/s}^2$.

E.6 A cylindrical aluminum peg has diameter 0.5 cm and protrudes 10 cm directly outward from a [rigid] wall. By how much does the peg shear in each of the following cases?

(a) An automobile with mass $M_a = 1200$ kg is suspended from the end of the peg.

(b) A bookbag of mass $M_b = 20$ kg is slung over the end of the peg.

E.7 A cylindrical rod has mass M. When unstressed, its radius and length are R_0 and L_0. Under a tensile force, F_A, the length of the rod increases to $L_0 + \Delta L$. Assuming that the density remains constant, estimate the change in the radius of the rod.

E.8   A rod of square cross-section with sides l_0 and length L_0 is acted upon by a tensile compressive force which causes its length to be reduced by a small amount, ΔL. Assuming that the rod geometry remains uniform with square cross-section, estimate the change in the length of the square sides.

E.9
Icicles which form on the eaves of houses in winter-time, and stalactites found in subterranean caverns, tend to fracture nearer to their bases than to their tips. This can be understood in the context of tensile stress and concomitant strains.

An idealised model of an icicle presumes its homogeneity (constant density ρ_0) and ascribes to it a regular conical geometry (circular cross-section in planes perpendicular to its axis of symmetry, and open-half-angle θ).

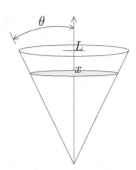

(a) The ice lying below the imaginary surface at distance x from the tip is held in place by the tensile force exerted through the surface, which cancels its weight. (i) Compute the volume of ice lying within distance x from the tip.   (ii) Determine the weight of this ice (in a locally constant gravitational field).

(b) The force counter to the weight determined in (a) is distributed across the cross-sectional face of the icicle (at distance x from the tip). Compute the tensile stress and verify that it is proportional to x.

Since the tensile stress increases with distance from the tip, so too does the likelihood of passing through the elastic limit and into the regime in which the ice fractures.

E.10   A very heavy-duty drill bit consists of a 3 cm thick silicon carbide cap rigidly affixed to a 12 cm solid steel base.

(a) Compute the effective Young's modulus of the drill bit assembly.

(b) Suppose that the drill bit operates under a pressure of approximately 2000 atm [1 atm $\simeq 10^5$ Pa]. Estimate the amount by which the bit is shortened while it is in use.

(c) What should the respective lengths of steel and silicon carbide be in a 15 cm combination that mimics the compressive response of molybdenum?

E.11   Two rods possess the same length, L, and cross-sectional area, A. The Young's moduli of the rods are Y and $3Y$ respectively. Determine the effective Young's modulus of the two rods in   (a) series and   (b) parallel.

E.12   Two rods possess the same length, L, and cross-sectional area, A. The Young's moduli of the rods are Y and nY, where $n > 1$. Determine the effective Young's modulus of the rods in   (a) series and   (b) parallel.

E.13   Two rods possess the same length, L, and cross-sectional area, A. The Young's moduli of the rods are Y and Y/n, where $n > 1$. Determine the effective Young's modulus of the rods in   (a) series and   (b) parallel.

E.14   Two blocks of material with vastly different Young's moduli, $Y_2 \gg Y_1$, different lengths, and the same face area are combined in series. Describe (mathematically and in words) the set of possible values for $Y_{(12)}$.

E.15   Two blocks of material with vastly different Young's moduli, $Y_2 \gg Y_1$, the same length, and differing face areas are combined in parallel. Describe (mathematically and in words) the set of possible values for $Y_{[12]}$.

E.16   A homogeneous and uniform aluminium bar has width $W = 1$ cm, height $H = 2$ cm, and length $L = 80$ cm. Suppose that the bar is cut in half and the two pieces are placed side-by-side to form a composite bar with width and height both equal to 2 cm and length $\tilde{L} = 40$ cm. Ascertain the Young's modulus of the new bar.

E.17 Three rectangular parallelepipeds, $\{A, B, C\}$, each with total length L and cross-sectional area $\mathcal{A}$, are comprised of nickel at one end and copper at the other. For A, Ni comprises two-thirds of the length of the bar and Cu the remaining third, while for C, these proportions are reversed. Half of bar B is nickel and the other half is copper. [Assume that stress is applied in the direction corresponding to the long axis of the composite material.]

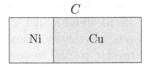

(a) Without recourse to computation, order these bars in terms of their effective Young's moduli, from least to greatest.

(b) Compute the Young's modulus of   (i) Bar A,   (ii) Bar B, and   (iii) Bar C.

(c) Do your computed results comport with your expectations in part (a)?

(d) Suppose that $L = 2.0$ m and $\mathcal{A} = 0.25$ mm^2. Determine the amount by which each bar is extended by a tensile force of 1000 N.

E.18 Three rectangular parallelepipeds, $\{A, B, C\}$, each with total length L and cross-sectional area $\mathcal{A}$, are comprised of nickel on one side and copper on the other. For A, the Ni extends two-thirds of the way up the face of the bar and Cu covers the remaining third, while for C these proportions are reversed. Half of bar B is nickel and the other half is copper. [Assume that stress is applied in the direction corresponding to the long axis of the composite material.]

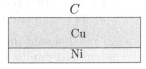

(a) Without recourse to computation, order these bars in terms of their effective Young's moduli, from least to greatest.

(b) Compute the Young's modulus of   (i) Bar A,   (ii) Bar B, and   (iii) Bar C.

(c) Do your computed results match your expectations in part (a)?

(d) Suppose that $L = 2.0$ m and $\mathcal{A} = 0.25$ mm^2. Determine the amount by which each bar is extended by a tensile force of 1000 N.

E.19
A composite bar, with length L and cross-sectional area $\mathcal{A}$, is constructed from two rectangular wedges of pure nickel and copper, as indicated in the adjacent figure.
Derive an expression for the effective Young's modulus of this composite bar.

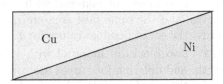

[HINTS: The tensile force acting on the bar is constant throughout. Set $x = 0$ at the Cu end of the bar and $x = L$ at the Ni end. The cross-sectional area fractions of Cu and Ni vary in a simple x-dependent manner. Determine the amount dl of stretch that the element of bar (with length dx) located at x experiences under the stress (felt equally throughout the bar). Integrate these differential stretches over the extent of the bar, *i.e.*, from $x = 0$ to $x = L$, to get the total amount of stretch, and hence infer the total strain produced by the tensile force acting on the composite bar. The ratio of the stress and total strain is the effective Young's modulus of the composite bar.]

E.20 Compute the effective Young's modulus of a bar comprised of three smaller bars in series. The three bars have different lengths and Young's moduli, while possessing the same cross-sectional area:

$\{Y_1, L_1, A\}$
$\{Y_2, L_2, A\}$
$\{Y_3, L_3, A\}$

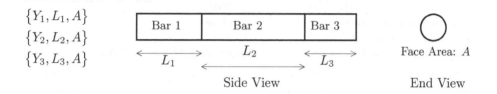

E.21
Compute the effective Young's modulus for a bar comprised of three (annular) bars in parallel. These three bars have different cross-sectional areas and Young's moduli, while possessing the same length.

$\{Y_1, L, A_1\}$
$\{Y_2, L, A_2\}$
$\{Y_3, L, A_3\}$

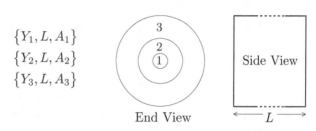

E.22 Two laminated wood–steel composite 10 cm cubic blocks are made. In one the ratio of steel to wood is 1 : 3, while in the other it is 3 : 1. Steel and wood have the elastic moduli quoted in the table below.

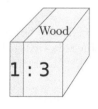

Material	Young's	Shear
Steel	$Y_{\text{steel}} = 200\,\text{GPa}$	$M_{S,\text{steel}} = 80\,\text{GPa}$
Wood	$Y_{\text{wood}} = 10\,\text{GPa}$	$M_{S,\text{wood}} = 10\,\text{GPa}$

(a) (i) Consider the two distinct orientations of the **1 : 3** block and compute the effective Young's modulus for each. (ii) Consider the two distinct orientations of the **3 : 1** block and compute the effective Young's modulus for each.

(b) (i) Consider the two distinct orientations of the **1 : 3** block and compute the effective shear modulus for each. (ii) Consider the two distinct orientations of the **3 : 1** block and compute the effective shear modulus for each.

E.23 Two blocks with top face area A and height h are fabricated. One is made of pure tungsten (W), and the other of aluminum (Al).

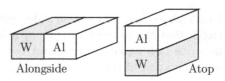

Determine the effective shear modulus when the blocks are glued   (a) alongside and
(b) atop   one another.

E.24
An ink/pencil combination eraser consists of two
rectangular blocks of rubber glued together with
a strong adhesive. The harder block, with length
$L_i = 1.5$ cm, width $W_i = W = 1$ cm, and height
$H_i = H = 0.7$ cm, erases ink. The softer block, with
length $L_p = 3.5$ cm and the same width and height as
the ink block, is for pencil.

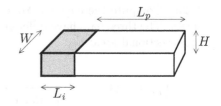

The shear moduli of the blocks are: $M_{S,i} = 500$ kPa and $M_{S,p} = 300$ kPa.

(a) Suppose that one of the $H \times W$ ends is held firmly, while the other end is rubbed
vigorously against a piece of paper.   (i) Does the shear response of the entire eraser
depend upon which end is in use?   (ii) Compute the effective shear modulus for the
eraser in this instance.

(b) Suppose that the $W \times (L_i + L_p)$ faces are held and rubbed. Does the effective shear
modulus in this instance differ from that determined in (a)?

E.25   In PK's old dorm room, pizza boxes formed a stack 1.2 m high and 50 cm both wide
and deep. The effective shear modulus of the stack was 20 N/m². Suppose that PK accidently
brushed against the pile, exerting a horizontal force of   (a) 2 N and   (b) 5 N   directly
on the topmost box. Compute the angle at which the stack tilts and ascertain whether the
pile collapses.

E.26   [See also E.25.] Rigorous investigations have revealed that the shear modulus of pizza
boxes depends on how the lid is oriented. Under shear acting in the direction in which the
box is hinged, the shear modulus is approximately 15 N/m². In the transverse direction, the
box is somewhat stiffer, with a shear modulus of 30 N/m². Given that each box is 5 cm thick,
and that the 1.2 m high pile has an effective shear modulus of 20 N/m², determine the number
of boxes in the stack lying in each particular orientation.

E.27   A homogeneous and uniform cube of iridium has edge length $L = 10$ cm. Suppose
that the cube is halved along its height and the two pieces are put alongside each other
(forming a rectangular parallelepiped with edges in the ratio 1 : 2 : 4). Ascertain the
effective shear modulus of the iridium parallelepiped.

E.28   Large marshmallows are cylindrically shaped, with face area 3.0 cm² and height 2.5 cm.
At sea level the average density of these marshmallows is $\rho_m = 0.36$ g/cm³ [as quoted at the
beginning of this section].

(a) PK wants to make a platform supported by a layer of marshmallows [standing on
their circular ends] which will sag no more than 1 mm when he stands upon it. PK and
the rigid base surface he intends to use have total weight 870 N. Determine the number
of marshmallows needed for this design and [assuming that each marshmallow occupies a
2 cm by 2 cm square] the minimum area of the [rectangular] platform.

(b) PK takes one of the marshmallows to a tall mountain where the atmospheric pressure
is reduced from 101.3 kPa to 91.3 kPa. Determine the marshmallow's new   (i) volume,
(ii) dimensions [assuming proportional expansion], and   (iii) density.

E.29 The TENSILE STRENGTH of a material is the minimum stress at which the material, subjected to an elongating force, fractures (experiences failure). Copper has a tensile strength of 360 MPa.

(a) Estimate the maximum strain that copper can experience before it fractures.

(b) A solid copper wire has diameter 0.7 mm and length 2.0 m. By how much might the wire stretch before it breaks?

E.30 A homogeneous rectangular parallelepiped is composed of a substance with shear modulus M_S. The dimensions of the parallelepiped are in proportion $5 : 6 : 10$.

(a) How should one arrange to push on the block (with a transverse shearing force) so as to break it with minimal force?

(b) In what orientation would the block be most resistant to shear fracture?

E.31 A SHEAR PIN is a mechanical linkage which acts in a manner similar to that of a fuse in an electrical circuit. That is, if mechanical stresses become too great, the linkage breaks, thus reducing the likelihood of major mechanical damage. A simple shear pin consists of a homogeneous rectangular parallelepiped with length L, width W, and height H. Here the labels are ordered so that $L > W > H$. There are three orthogonal directions in which this pin can shear, and hence three different critical values of shear force. For the pin to fail for the smallest of these forces, which of the plane surfaces should bear the shear force?

E.32 The densities and elastic moduli of tungsten, molybdenum, aluminium, and platinum (in their solid metallic forms at standard temperature) are given in the table at the beginning of this section. Tungsten is steel-grey, solid elemental molybdenum is silvery-white, aluminium can be silvery to dull grey, and platinum ("white gold"), a precious metal, is greyish-white. The goal in this question is to construct some artful forgeries.

(a) In what ratio should tungsten and molybdenum be combined to produce a mixture with the same density as platinum? Comment.

(b) Propose a simple combination of tungsten and aluminium blocks with total face area A and length L that would have the same response to tensile stress as a sample of platinum with the same dimensions. Comment.

(c) Propose a simple combination of molybdenum and aluminium blocks with total face area A and length L that would have the same response to shear stress as a sample of platinum with the same dimensions. Comment.

(d) How might the response of platinum to bulk compressive stress be mimicked?

E.33
A block of mass $M = 200$ kg is 2 m long. The weight of the block is deemed to act at the location of its CofM, which lies midway between its ends. The block is to be hung by means of a nylon rope and a thin steel wire, as shown in the figure, in such manner that it is perfectly horizontal and 0.75 m below the ceiling. Both the rope and the wire have circular cross-section with diameters $D_{\text{Nylon}} = 1.25$ cm and $D_{\text{Steel}} = 1$ mm, respectively.

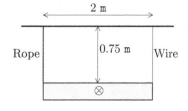

What [initially unstressed] lengths of rope and wire must be employed to achieve the desired configuration?

E.34 It is our intention to assemble two thin rods and a rigid beam (all with negligible masses), some ideal rope, and a block of mass M into the structure illustrated in the nearby figure. The system experiences a uniform gravitational field, $\vec{g} = g\,[\downarrow]$. Our plan is to model the elastic deformation of such a system, were it to be naively fabricated.

The unstressed rods, of length l, have respective cross-sectional areas A_1 and A_2 and Young's moduli Y_1 and Y_2. The rigid beam has length L. The distance from the far end of the beam to the point of attachment of the first (second) rod is d_1 (d_2), and thus the distance between the [centres of the] thin rods is $d_{12} = d_2 - d_1$.

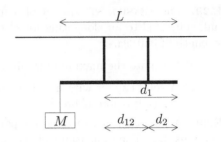

(a) (i) Enumerate the forces acting on the beam, paying particular attention to the precise locations at which the forces act.   (ii) Apply the conditions of static equilibrium to determine the forces exerted by the rods on the beam.

(b) Compute the tensile stresses in the two rods. [Note whether these are compressive or tensile.]

(c) Estimate the strain in each rod and its change in length.

(d) Estimate the angle (with respect to horizontal) at which the rigid beam hangs (when the system is naively fabricated).

(e) Specialise the result in (d) to the case in which the vertical rods are identical in geometry and composition.

F

Fluids Problems

Useful Data

Whenever the value of the acceleration due to gravity at the Earth's surface is needed, take it to be $g = 10\,\mathrm{m/s^2}$.

Standard atmospheric pressure is $101.3\,\mathrm{kPa}$. Use this value unless otherwise specified.

The universal gas constant has the approximate value $R = 8.31447\,\mathrm{J/(mole \cdot K)}$.

Densities $[\,\mathrm{kg/m^3}\,]$ and Bulk Moduli for Select Elements					
HELIUM	ρ_{He}	$=$	0.179	$B_{\mathrm{He}} \simeq 170\,\mathrm{kPa}$	$*$
IRON	ρ_{Fe}	$=$	$7,874$	$B_{\mathrm{Fe}} = 160\,\mathrm{GPa}$	
MERCURY	ρ_{Hg}	$=$	$13,534$	$B_{\mathrm{Hg}} = 28.5\,\mathrm{GPa}$	
$*$ Adiabatic bulk modulus at standard pressure and temperature.					

Densities $[\,\mathrm{kg/m^3}\,]$ and Bulk Moduli for Select Materials					
Air	ρ_{air}	$=$	1.2	$B_{\mathrm{air}} = 142\,\mathrm{kPa}$	$*$
Graphite	$\rho_{\mathrm{C\text{-}gr}}$	$=$	2267	$B_{\mathrm{C\text{-}gr}} = 8\,\mathrm{GPa}$	
Marshmallow	$\rho_{\mathrm{m\text{-}m}}$	$=$	360	$B_{\mathrm{m\text{-}m}} = 80\,\mathrm{kPa}$	
Vegetable Oil	ρ_{oil}	$=$	920		
Vinegar	ρ_{vin}	$=$	$1,005$	$B_{\mathrm{vin}} \simeq 2.2\,\mathrm{GPa}$	
Fresh Water	ρ_{H2O}	$=$	$1,000$	$B_{\mathrm{H2O}} = 2.2\,\mathrm{GPa}$	
Sea Water	ρ_{H2O}	$=$	$1,025$	$B_{\mathrm{H2O}} = 2.3\,\mathrm{GPa}$	
Ice	ρ_{ice}	$=$	917	$B_{\mathrm{ice}} = 8.8\,\mathrm{GPa}$	
Fat Tissue	ρ_f	$\simeq$	900		
Lean Tissue	ρ_l	$\simeq$	1100		
$*$ Adiabatic bulk modulus at standard pressure and temperature.					

F.1 A swimming pool is filled with water to a depth of 2 m. Determine the pressure in the pool at depth   (a) 2 m [the bottom],   (b) 1 m [half-way], and   (c) 0 m [the top surface].

F.2 Roughly how far might one have to descend in a freshwater lake to experience five times atmospheric pressure?

F.3 Estimate the increase in the density of water subjected to additional bulk pressure of 4 atm.

F.4
In many small towns the municipal water system incorporates a prominent water tower located on an elevated section of land. Rank the respective "faucet pressures" in the three houses shown in the figure from highest to lowest.

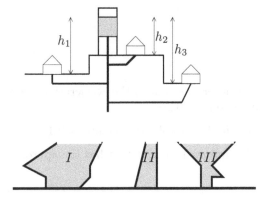

F.5
Three open-topped vessels, $\{I, II, III\}$, are shown in the adjacent figure. Each is filled to common depth with the same type of fluid.

(a) Rank the three containers in order of increasing volume of fluid.

(b) Rank them in order of increasing weight of fluid supported by the base of the container.

(c) Rank the three containers in order of increasing pressure exerted on the bottom of the container.   [This is known colloquially as the HYDROSTATIC PARADOX, although there is nothing paradoxical about it!]

F.6 A swimming pool is filled to a depth of 2 m. A balloon filled with air occupies a volume of 1 litre when it is held just above the surface of the water. What would the volume of the balloon be if it were submerged and brought to the bottom of the pool?

F.7 A balloon filled with helium is roughly spherical, with radius 35 cm under standard atmospheric pressure.

(a) Suppose that the balloon is brought to a high mountaintop where the air pressure is 80% of its standard value. Estimate the radius of the balloon.

(b) Estimate the radius of the balloon when it is submerged to a depth of 1.7 m in fresh water.

(c) At what ocean depth would the balloon have the radius determined in (b)?

F.8 At what ocean depths would samples of   (a) graphite and   (b) air   be reduced to 95 % of their surface volumes?

F.9 Compute the pressure needed to reduce a sample of water to 95 % of its volume under standard conditions and the depth, in water, at which such a pressure would be obtained.

F.10 A pressure gauge consisting of a mercury-filled U-shaped tube, with one end open to the air, is employed to determine the pressure inside a sealed container. If the pressure inside the container is actually 1.75 atm, then what is the height difference between the mercury menisci in the two arms of the U-tube?

F.11 A pressure gauge consisting of a fluid-filled U-shaped tube, with one end open to the air, is employed to determine the pressure inside a sealed container. The pressure inside the container is P_{in}, while the pressure of the surrounding air is P_{atm}. Assuming that the fluid is incompressible, with density ρ_0, determine the height difference between the fluid columns in the pressure gauge.

F.12 A force of 10 N acts directly downward on a piston with face area $0.5\,\text{cm}^2$, connected to a hydraulic reservoir filled with incompressible and frictionless fluid. What upward force is experienced at another piston with face area   (a) $10\,\text{cm}^2$ and   (b) $0.25\,\text{cm}^2$ sharing the same reservoir?

F.13 There is a zone in the Bay of Florida where the flow of fresh water from the Everglades results in 2 m of warm fresh water "floating" on top of [slightly] cooler salt water. A snorkler and his gear have total mass 100 kg, volume $0.0985\,\text{m}^3$, and near infinite bulk modulus. Compute the net force (magnitude and direction) acting on the snorkler when he is swimming at a depth of   (a) 1 m and   (b) 3 m.

F.14 Large marshmallows are cylindrically shaped, with face area $3.0\,\text{cm}^2$ and height 2.5 cm. The density of hot cocoa is $\rho_c = 1.08\,\text{g/cm}^3$ and its bulk modulus is essentially infinite. [In these analyses, we do not consider the buoyancy of the marshmallow in air, or that (over time) the marshmallow dissolves in the cocoa.]

(a) PK puts a marshmallow into a cup of hot cocoa while sitting by the seashore. What fraction of the marshmallow floats above the surface of the cocoa?

(b) PK puts a marshmallow into a cup of hot cocoa while resting in a ski lodge high on a mountain, where the air pressure is reduced to 91.3 kPa. What fraction of the marshmallow floats above the surface of the cocoa?

F.15 The ice shelves found in the Arctic and Antarctic oceans are comprised of nearly pure water because salt is expelled into the surrounding water as freezing occurs. Hence, ice floes found in oceans and in lakes have essentially the same density. Determine the fraction of the volume of an iceberg floating above the surface in   (a) the saltwater ocean and   (b) a freshwater lake.

F.16 A salad dressing consists of two parts vinegar to one part oil by volume. A container with uniform geometry and cross-sectional area $50\,\text{cm}^2$ holds $450\,\text{cm}^3$ of dressing which has completely separated: its oil and vinegar constituents are unmixed.

(a) Determine the pressure as a function of depth within the oil.

(b) Determine the pressure at the bottom of the oil, *i.e.*, at the top of the vinegar.

(c) Determine the pressure as a function of depth within the vinegar.

F.17 A salad dressing consisting of four parts of vegetable oil to one part vinegar is at rest in a cylindrical container. The oil floats on top of the vinegar and the two substances remain unmixed. Describe in words, formulae, and a sketch how the pressure changes with depth below the top surface of the oil.

F.18

One might toast Saint Patrick's Day with a "black and tan" consisting of equal parts Harp Lager and Guinness Stout layered in a single glass. A 20 ounce British "pint" approximately fills a cylindrical glass with radius 4 cm to a depth of 12 cm. Estimates of the specific gravity of Harp Lager and Guinness Stout are 1.065 and 1.035 respectively. Harp Lager occupies the bottom 6 cm of the glass, while Guinness Stout fills the remainder, *i.e.*, to the 12 cm mark. Here we are most concerned with relative pressures, so represent the atmospheric pressure at the top surface of the Guinness by P_0.

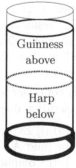

(a) Compute the   (i) volume and   (ii) mass   of Harp Lager in the glass.

(b) Compute the   (i) volume and   (ii) mass   of Guinness Stout in the glass.

(c) (i) Compute the pressure as a function of depth within the Guinness.

(ii) Determine the pressure at the Guinness-Harp boundary.

(d) (i) Compute the pressure as a function of depth below the Harp-Guinness boundary. (ii) Determine the pressure at the bottom of the glass.

(e) Compute the net force acting on the side of the glass within the region occupied by (i) Guinness,   (ii) Harp, and   (iii) the entire beverage.

(f) Compute the net fluid force acting on the bottom surface of the glass.

F.19   Two equal-volume immiscible fluids partially fill a container. The upper fluid has constant density ρ_u, while the lower has ρ_l. A small block of substance with intermediate density ρ_m, where $\rho_u < \rho_m < \rho_l$, falls into the container. Express, in terms of the densities, the fraction of the volume of the block which floats above the fluid boundary.

F.20   Vinegar and olive oil are combined to form a bilayer fluid. A chunk of frozen water with vegetable fragment inclusions, causing it to have a net density of $947 \, \text{kg/m}^3$, falls into the oil–vinegar mix. Neglect melting effects and ascertain the equilibrium position of the chunk of ice. In particular, specify the fractions of the ice's volume immersed in the oil and vinegar.

F.21   Dichloromethane is a liquid often employed as a solvent. It does not mix with water, the most common solvent. Suppose that $90 \, \text{mL}$ of dichloromethane, with specific gravity 1.36, spill into a rectangular pan with length $L = 15 \, \text{cm}$, width $W = 10 \, \text{cm}$, and height $H = 10 \, \text{cm}$ already containing $1.35 \, \text{L}$ of water.

(a) Describe the physical characteristics of the equilibrium mixture of fluids.

(b) Determine a (piecewise) formula for the fluid pressure as a function of depth below the topmost surface of the fluid mixture.

F.22   A spherical plastic bead with radius $R = 3 \, \text{mm}$ and density $\rho = 0.8 \, \text{g/cm}^3$ floats partially submerged in water [density $1.0 \, \text{g/cm}^3$]. Determine the position of the centre of the bead with respect to the waterline. [HINTS: First find the volume fraction protruding above the waterline (into the air). Second, use geometry (the volume of a spherical section minus the volume of its conical part) or calculus (the volume of a solid of revolution generated by a section of circular arc) to determine an expression for the volume of a sphere cap (in terms of its height). Finally, ascertain the cap height corresponding to the protruding volume. WARNING: This step involves solving a cubic equation and can get tricky. If all else fails, resort to a graphical or numerical method. Thence determine the depth of the CofM of the bead below the waterline.]

F.23

Three immiscible and incompressible liquids, labelled "A", "B", and "C," reside in an open container with shape indicated in the adjacent figure.

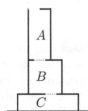

The respective densities and depths of the fluids in the container are

$$\rho_A = 2000 \, \text{kg/m}^3 \qquad h_A = 9 \, \text{m}$$
$$\rho_B = 4000 \, \text{kg/m}^3 \qquad h_B = 6 \, \text{m}$$
$$\rho_C = 8000 \, \text{kg/m}^3 \qquad h_C = 3 \, \text{m}$$

The container is situated on the surface of the moon, where the magnitude of the local acceleration due to gravity is $\frac{10}{6} \, \text{m/s}^2$. Take the lunar "atmospheric pressure" to be ZERO.

(a) (i) Express a formula for the pressure (in terms of depth) within the region occupied by fluid A.   (ii) Determine the fluid pressure at the $A : B$ fluid boundary.
(iii) Compute the average pressure exerted by fluid A on the sides of the container.

(b) (i) Express a formula for the pressure (in terms of depth) within the region occupied by fluid B.   (ii) Determine the fluid pressure at the $B : C$ fluid boundary.
(iii) Compute the average pressure exerted by fluid B on the sides of the container.

(c) (i) Express a formula for the pressure (in terms of depth) within the region occupied

by fluid C.   (ii) Determine the fluid pressure at the bottom of the container.
(iii) Compute the average pressure exerted by fluid C on the sides of the container.

F.24   The specific gravity of tanzanite (a rare mineral) is approximately 3.2. What is the density of tanzanite?

F.25   What is the specific gravity of iron?

F.26   A piece of iron with mass 100.00 g is used to calibrate a [waterproofed] laboratory scale. In parts (b) and (c) below, the scale was zeroed just prior to its receiving the sample.

(a) Compute the volume of this sample.

(b) The scale and sample are completely immersed in pure water. Determine the apparent mass of the sample, *i.e.*, the reading on the scale.

(c) The scale and sample are completely immersed in mercury. Ascertain the apparent mass of the iron sample.

F.27   A certain crown has weight 5.00 ± 0.05 "kg" (sic) in air, and weight 4.50 ± 0.05 "kg" (sic) when immersed in water. Given that the specific gravity of gold is 19.3, what can be concluded about the crown?

F.28   In any kingdom there are always at least two crowns: one is the real one, while any others are for photo-ops. In a land that shall remain nameless, the authentic crown is made of gold, with specific gravity 19.30, while the fake one is made mostly of tinfoil with specific gravity 7.365. Both crowns are bedecked with jewels of inestimable value and relatively low [silicate] density.

It came to pass that after many long years, several moves, and the randomising influences of small children, the royals no longer knew which crown was which. Fortunately, the story of Archimedes was still remembered, albeit somewhat shakily, and the national treasury had a small store of Canadian one-dollar coins ("loonies").

A balance was constructed with a *loonie bin* on the left pan and a special crown sling on the right, as shown in the figure, and measurements were made. Your job is to sift through the data and determine which crown is authentic.

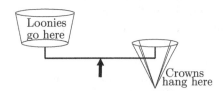

(a) When suspended in air, Crown A was balanced by 169 loonies; when submerged in water, it was balanced by 160 loonies. Ascertain whether this is the real crown or not.

(b) With Crown B suspended in air, the pan held 302 loonies; with Crown B completely submerged in water, it held 260 loonies. Ascertain whether this is the real crown.

F.29

A rectangular bulwark, with length $L = 3\,\mathrm{m}$ and height $H = 1\,\mathrm{m}$, has water completely covering one side and air on the other. The air also lies above the water.

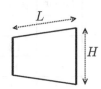

(a) What total force is exerted by the water on the bulwark?

(b) (i) What is the average pressure exerted on the bulwark?
(ii) What is the peak pressure exerted on the bulwark, and where does it occur?

F.30

A triangular bulwark, with top edge $L = 3\,\text{m}$ and
height $H = 1\,\text{m}$, has water completely covering one
side and air on the other. The air also lies above the
water.

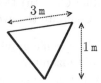

(a) What total force is exerted by the water on the bulwark?

(b) (i) What is the average pressure exerted on the bulwark?   (ii) What is the peak
pressure exerted on the bulwark, and where does it occur?

F.31 A rectangular dam stands $15\,\text{m}$ high and $50\,\text{m}$ wide. Behind the dam is a reservoir
completely filled with slurry. The density of the slurry increases as the square of the depth
from $1100\,\text{kg/m}^3$ at the top to $1500\,\text{kg/m}^3$ at the bottom. Otherwise, the slurry is well-
approximated by an ideal fluid at rest.

(a) Determine a mathematical expression for the density as a function of depth from
the top surface of the slurry. Graph this function.

(b) Determine the [net] pressure on the dam as a function of depth below the top.   [The
constant air pressure acting on the top of the reservoir is also applied on the front of the dam,
and thus cancels.]

(c) Compute the net force acting on the dam.

F.32 The Three Gorges Dam spans the Yangtze River between Chongqing and Wuhan,
and its reservoir volume is almost $4.0 \times 10^7\ \text{m}^3$. The dam may be crudely modelled as a
rectangular wall of concrete $2300\,\text{m}$ long, holding back water with $175\,\text{m}$ depth on one side,
and open to the air on the other. What total force is exerted on the dam by the water in
the reservoir?

F.33 Suppose that a square hole with $10\,\text{cm}$ sides were to open up in the Three Gorges
Dam, discussed in F.32. Assuming that the water is ideal fluid and the flux through the
hole is uniform, determine the speed at which the water emerges from the face of the dam
when the hole is at depth   (a) $50\,\text{m}$,   (b) $100\,\text{m}$, and   (c) $150\,\text{m}$.

F.34 Students pass through a hallway with 10 different doors, labelled $A \to J$. Traffic
into and out of the hallway in a brief interval is described in the table just below.

Door	# In	# Out		Door	# In	# Out
A	10	6		F	5	8
B	12	10		G	11	15
C	8	13		H	4	2
D	11	14		I	21	16
E	7	5		J	?	?

Under the assumption that the number of people in the hallway is the same before and after
the brief interval, determine the net number of people who passed through door **J**.

F.35
Ideal fluid flows in a purely horizontal pipe possessing a constriction. The radius of the pipe in the constricted part is one-half that in the wide portion, *i.e.*, $R_c = R_w/2$. The fluid flows uniformly in the wide part with speed v_w under pressure P_w. Determine the
(a) uniform speed and   (b) pressure
of the fluid in the constricted region.

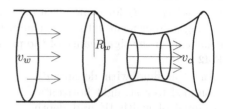

F.36   Horizontal duct work with square cross-section smoothly increases from an initial edge size of 20 cm to 60 cm. The airflow throughout the duct is assumed to be laminar. If the air in the smaller section flows with speed 135 cm/s, then with what speed does the air move in the larger section?

F.37
Water welling up from a narrow basement drain spreads outward along a horizontal floor. At a distance of 3 m from the drain, the edge of the water is observed to be advancing at 0.5 cm/s. Supposing that the water is 3 mm thick, at what rate is water coming from the drain?

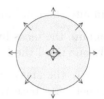

F.38   A crude model treats the human aorta and capillaries as cylindrical tubes of diameter 3.0 cm and 5.0 μm respectively. Assuming that blood typically flows in the aorta at 30 cm/s, and in the capillaries at 0.5 mm/s, estimate the number of capillaries one might expect to find in a person.

F.39

The volume-rate-of-flow [in cubic metres per second] in each of the pipes comprising a network is indicated in the nearby figure. The arrow in each channel specifies whether the flow is inward or outward.

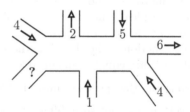

(a) Determine the magnitude and direction of the fluid flux in the channel indicated by the question mark.

(b) Determine the magnitude and direction of the fluid flux in the "?" channel when all of the other fluxes have magnitude 2.

F.40

An ideal fluid with density ρ_0 flows uniformly with speed v_0 along a straight horizontal channel of width W and depth H.

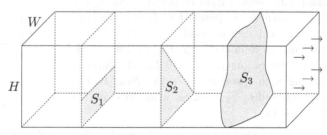

(a) Compute the total mass flux entering and leaving the channel.

(b) Compute the mass flux through the imaginary surface, S_1, spanning the full width but only one-third the height.

(c) Compute the mass flux passing through S_2, a triangular imaginary surface lying in the channel, with base W and height H.

(d) Determine the mass flux through the irregular surface S_3 [spanning the channel].

F.41   Water streaming in a rectangular chute of width W and height H has variable speed $v(h) = 2 v_0 h/H$, for some constant v_0 [with dimensions of speed] and h, the height of the flowing fluid element within the chute. Compute the volume flux of water in the chute, and comment on the significance of the parameter v_0.

F.42

An ideal fluid with density ρ_0, confined to a straight horizontal channel of width W and depth H, flows with variable speed $v(h) = 2 v_0 h/H$, where h represents the height of a particular fluid element from the bottom of the channel.

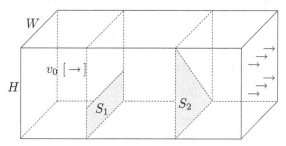

(a) Compute the total mass flux entering and leaving the channel.

(b) Compute the mass flux through the imaginary surface, S_1, spanning the full width but only one-third the height.

(c) Compute the mass flux passing through S_2, a triangular imaginary surface lying in the channel, with base W and height H.

F.43
Three pipes with radii $1.0\,\mathrm{m}$, $0.5\,\mathrm{m}$, and $0.25\,\mathrm{m}$ are fused in a Y-junction. The centres of the pipes lie at the same elevation. [The figure shows a horizontal plane slice through the system.] Dichloromethane, with specific gravity 1.36 and modelled as an ideal fluid, completely fills the system and flows with uniform speed $0.25\,\mathrm{m/s}$ in the large radius pipe toward the junction. The fluxes in the two smaller pipes (both outward) are equal in magnitude.

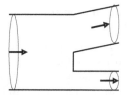

(a) Determine the speeds of uniform flow in the two smaller branches of the Y.

(b) Ascertain the pressure difference between fluid elements located in   (i) the input and the larger-diameter output pipes,   (ii) the input and the smaller-diameter output pipes,   (iii) the two output pipes.

F.44
The pipe shown in the figure is completely filled with perfect fluid of density $1200\,\mathrm{kg/m^3}$. In regions near the points $\{\mathcal{A}, \mathcal{B}, \mathcal{C}\}$, the pipe is cylindrical and, to good approximation, the fluid moves uniformly. In the vicinity of $\mathcal{A}$, where the radius is $r_{\mathcal{A}} = 0.2\,\mathrm{m}$, the fluid is moving with speed $v_{\mathcal{A}} = 3\,\mathrm{m/s}$. Near $\mathcal{B}$, the pipe has radius $r_{\mathcal{B}} = 0.1\,\mathrm{m}$, while at $\mathcal{C}$ its radius is $0.4\,\mathrm{m}$.

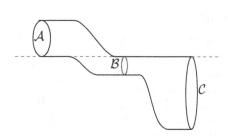

(a) Determine the speed of the fluid   (i) near $\mathcal{B}$ and   (ii) near $\mathcal{C}$.

(b) Suppose that the centres of all three segments of pipe are at the same elevation above ground, *i.e.*, the figure shows a horizontal plane view of the system, and determine the pressure difference between each pair of points in the set $\{\mathcal{A}, \mathcal{B}, \mathcal{C}\}$.

(c) Suppose that the figure shows a vertical slice. Infer the pressure differences among the set of three points, $\{\mathcal{A}, \mathcal{B}, \mathcal{C}\}$.

F.45
The pipe in the adjacent figure is completely
filled with perfect fluid ($\rho_0 = 1280\,\text{kg/m}^3$)
which is moving uniformly in the vicinities
of points $\{A, B, C\}$. At A, the fluid is moving with speed $v_A = 3\,\text{m/s}$ through a circular
pipe with radius $r_A = 0.2\,\text{m}$. Near B, the pipe
has radius $r_B = 0.1\,\text{m}$, while at C the radius is
$0.4\,\text{m}$.

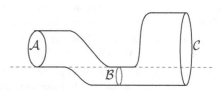

(a) Determine the speed of the fluid   (i) near B and   (ii) near C.

(b) Suppose that the centres of all three segments of pipe are at the same elevation above
ground, *i.e.*, the figure shows a horizontal plane view of the system, and determine the
pressure difference between each pair of points in the set $\{A, B, C\}$.

(c) Suppose that the figure shows a vertical slice. Infer the pressure differences among
the set of three points $\{A, B, C\}$.

F.46
An ideal fluid with density ρ flows in the pipe
shown. The flow through the lower left pipe
is inward and uniform, with speed v_i. The
outward flow through the pipe on the upper
right is assumed to be uniform at an unknown
speed, v_o. The left and right pipes have circular cross-section with radii R_i and R_o, respectively. At the top of the system, a pressure
gauge exposed to the fluid reads P_*.

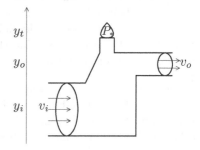

(a) Ascertain the value of Bernoulli's constant, assuming that the fluid near the pressure
gauge is effectively at rest.

(b) Determine the   (i) pressure in the input pipe and   (ii) input flux.

(c) (i) Determine an expression for the pressure in the output pipe.   (ii) Write the
output flux in terms of (the unknown) v_o.   (iii) Apply an equation of continuity
to determine v_o.   (iv) Express the pressure in the output pipe in terms of known
quantities: $\{\rho, g, R_i, R_o, y_i, y_o, y_t, v_i, P_*\}$.

F.47
Fluids gush from a damaged offshore oil well. The depth of the
seawater at the wellhead is $1500\,\text{m}$. The bulk modulus of sea
water is sufficiently large that, even at such depths, it remains
essentially incompressible, and thus its density is effectively
constant. The fluids erupting from the well are presumed to
be ideal in this crude analysis.

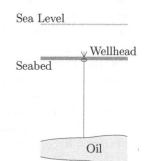

(a) Estimate the water pressure at the wellhead.

(b) The flow rate from the wellhead is estimated to be about
2200 barrels per hour. One barrel is approximately $160\,\text{L}$. The
drill hole is approximately $50\,\text{cm}$ in diameter. Estimate the
speed of the fluid erupting from the wellhead.

(c) Take the seawater at the ocean floor to be effectively at rest and estimate the pressure
which is driving the spilling fluid, given that the density of crude oil is about $850\,\text{kg/m}^3$
and the oil is driven from a depth of about $5000\,\text{m}$ below sea level.

F.48 PK once mistakenly wandered into a spa, believing that he had found a local chapter of the Society of Physicists and Astronomers [SPA]. Once inside, he was weighed twice: the first time on dry land, and the second time submerged in a large tank of water. The two readings were 840 N and 40 N. From this data, we'll compute PK's body mass index [BMI]. The model that we employ assumes that one's body is comprised of adipose [FAT] and non-adipose [LEAN] tissue. These are complementary in both mass and volume: $M_t = M_f + M_l$ and $V_t = V_f + V_l$, where the subscripts signify 'total,' 'fat,' and 'lean,' respectively. The average densities of each type of tissue and the fresh water in the tank, are found in the table at the beginning of this section. Finally, BMI is the percentage measure of the fat-mass fraction, $\text{BMI} = 100 \times \frac{m_f}{m_t}$.

Compute PK's   (a) specific gravity,   (b) average density, and   (c) BMI.

F.49
Ideal fluid with density $\rho = 4000\,\text{kg/m}^3$ flows from left to right in the system of pipes shown in the figure. The pipe on the left has radius 0.6 m, while those on the right both have radius 0.3 m. The flow in all of the cylindrical regions is uniform across the faces of the pipes; on the left and the upper right it is at 0.5 m/s. The difference in height between the centres of the two pipes on the right is 3.0 m. The pressure in the fluid at the midpoint of the pipe on the left is 100 kPa. Determine the speed and the pressure of the fluid at the centre of the lower pipe on the right.

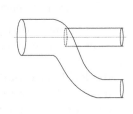

F.50
A small residential water pump is employed to lift water from a lake through a vertical distance of 20 m, as shown in the figure. Assume that the fresh water in the lake is ideal and that it is effectively still. The pump has a circular water channel with average diameter 20 cm. A rotary impeller drives the circulation of water in the channel.

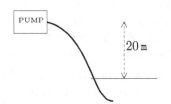

(a) At what minimum speed must the water in the pump's channel circulate so as to draw water uphill from the lake to the residence?

(b) At what angular speed, in revolutions per minute, must the impeller turn to move the water in the channel at the speed determined in (a)?

F.51
Water pours straight down from a cistern (on the roof of a house) through the eaves via a pipe with circular cross-section. Once through the eaves the water falls freely to the ground, a distance H below the open end of the pipe. The radius of the pipe is R and the water leaves it with speed v_r. It is observed that the column of water gets increasingly slender as it gets nearer to the ground.
[Neglect drag and assume that the acceleration due to gravity, g, is constant.]

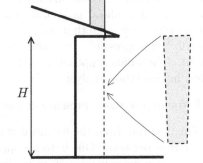

(a) Determine the radius of the stream of water just before it strikes the ground.

(b) Ascertain the difference between the fluid pressure of the water when it leaves the pipe and that of the water just before it strikes the ground. Comment.

F.52   [See also F.51.] Specialise F.51 when $H = 4\,\text{m}$, $R = 1.25\,\text{cm}$, and $v_r = 1\,\text{m/s}$.

F.53 [See also F.52.] Employ the simple model developed in Chapter 9 to estimate the Reynolds number for the falling water in F.52 (a) when it has just left the pipe in the eaves and (b) just before it reaches the ground. (c) Comment.

F.54 A very large cylindrical tank is filled with water to height H. The tank is situated on a horizontal plane surface (the ground).

(a) A tiny hole in the side of the tank appears at depth h below the waterline. Our goal is to determine the distance D from the cylinder that the stream of water hits the ground, under the assumptions that the water exits the hole horizontally, that g is constant, that drag is negligible, and that flux is sufficiently small that H is very slowly changing.

(i) Determine the [horizontal] speed with which the water leaves the hole.

(ii) Ascertain the time that the water takes to fall to the ground.

(iii) Determine the distance D from the side of the tank at which the water strikes the ground.

(b) Repeat the analysis in (a) when the hole in the side of the tank is at a depth of $H - h$ below the waterline.

(c) Comment.

F.55 Verify that the Green's function solution of the diffusion equation (with constant diffusion coefficient), introduced in Chapter 8, does indeed satisfy this equation.

F.56 Verify that the Fourier solutions of the diffusion equation (with constant diffusion coefficient), introduced in Chapter 8, do indeed satisfy this equation.

F.57 [See also F.58.] Write a simple computer program that generates a number of ten-step random walks in 1-d, records the net displacement in each, and produces a [normalised] histogram showing the relative frequency of each valid net displacement (like the ones shown in Chapter 8).

F.58 Determine how the "seed" for the random number generator employed in your program in F.57 is set. Run the code several times, manually resetting the seed to the same initial value each time. Comment on your results.

F.59 A fixed amount of gas in a container is well described by the ideal gas law. How might one keep the pressure constant while the volume is (a) increased and (b) decreased?

F.60 An amount of ideal gas is confined to a sealed box at one-half atmospheric pressure, *i.e.*, 50.65 kPa. With clever control, the volume of the box is reduced to 1/8 of its original value (without any gas leaking in or out), while the temperature is kept constant. Determine the pressure of the gas in the smaller box.

F.61 A sample of ideal gas occupies a volume of one litre at atmospheric pressure and temperature 20 C. What pressure is required to squeeze the gas into one-half litre at -40 C?

F.62 When a car has been parked for a while, its tires look a wee bit low, and when their pressure is measured at 23 C (in the garage) it is found to be 320 kPa. After the car has been driven some distance, the tires look fine and their measured pressure has increased to 360 kPa. What is the temperature of the air inside the tires once they have warmed up?

F.63 The density of carbon dioxide at $T = 0$ C and $P = 101.3$ kPa is 1.97 kg/m^3. Determine the density of CO_2 when the temperature and pressure are $T = 2000$ C and $P = 800$ Pa.

F.64 Compare the number densities (number per unit volume) of atoms in the galaxy and in the air which surrounds us, based on the following considerations. The shape of the Milky Way Galaxy is approximated by a squat cylinder of diameter 10^{21} m and height 10^{19} m. The Milky Way is thought to contain about 3×10^{11} individual stars, each consisting of, on average, about 10^{57} atoms. Reasonable approximate values for the pressure and temperature of the air we are immersed in are: $P_{\text{atm}} \simeq 1.013 \times 10^5$ Pa and $T_{\text{room}} \simeq 300$ K. The molecules comprising the air are, by and large, diatomic.

F.65 A sample of gas initially at temperature 300 K, pressure 120,000 Pa, and volume 0.25 m³ is later observed to have temperature 400 K and volume 2.0 m³. Ascertain the pressure in the gas at the later time.

F.66 A fixed amount of ideal gas had pressure 15 kPa when its temperature was 300 K and its volume was 0.125 m³. Ascertain the pressure of the gas when its temperature is 600 K and its volume is 0.25 m³.

F.67 A fixed amount of gas is held in a cylinder. One of the cylinder's end walls is attached to a movable piston. An external agent applies a force on the piston, compressing the gas into a smaller volume.

(a) Does the external agent do a positive or negative amount of work on the gas?

(b) Assume that the piston is moved very slowly and that the temperature of the gas remains constant. (i) How much work is done by the gas? (ii) Discuss how it is possible for the temperature to stay constant.

(c) Assume that the piston moves quickly. Discuss what happens to the work performed by the external agent.

(d) How might the final pressures in the gasses considered in (b) and (c) compare?

F.68 Consider a gas in which the local fluid pressure varies as: $dP = -\rho g \, dh$. The negative sign arises because we take the height, h, to be increasing upward. Let's suppose that the density of the gas varies [linearly] with pressure according to $\rho(P) = \frac{\rho_0}{P_0} P$. [The proportionality constant is expressed in terms of the reference values of density and pressure.] For the purposes of this analysis, the reference values ρ_0 and P_0 are the density and pressure at sea level, where $h = 0$.

(a) Derive an expression for the air pressure as a function of height above sea level.

(b) How does your formula compare to the prediction based on straightforward application of Pascal's formula, for (i) small and (ii) large deviations from sea level?

F.69 A physics student proposes to rise above all petty concerns with the aid of some (easily produced) hydrogen gas, a big mylar bag, some rope, and a lawn chair. To evaluate the feasibility of this flighty endeavour, we shall predict the amount of H_2 gas required, employing a simple model:

○ The total payload mass is $M \simeq 100$ kg.

○ One mole of molecular hydrogen, H_2, has mass 2 g and occupies a volume of 22.4 L at standard temperature and pressure.

○ It is safe to neglect the small buoyant force produced by the displacement of air by the payload.

(a) Compute the density of molecular hydrogen (in kg/m³).

(b) Compute the net force acting on a cubic metre of molecular hydrogen immersed in air near sea level.

(c) Estimate the number of cubic metres of hydrogen needed to float the payload. The mylar bag must accommodate at least this much gas.

[Please keep your seatbelt fastened securely, and remember that this is a non-smoking flight!]

O

Oscillation Problems

O.1 In Chapter 11, it was realised that, upon dropping PK into a tunnel drilled straight through the centre of the Earth, he underwent SHO with period $T = 2\pi \sqrt{R_\oplus^3/(G M_\oplus)} \simeq$ 84.3 minutes. What would happen should the hole lead directly to a non-antipodal point? [HINT: Apply the same approximations and assumptions as were employed in the chapter. Parameterise PK's position in the tunnel by his (signed) distance from the midway point.]

O.2 [See also O.1.] Employ Kepler's Third Law [discussed in Chapter 50 of VOLUME I] to compute the period of a satellite in a circular orbit about the [centre of the] Earth at radius $R_\oplus$ [so that it lies just at, or slightly above, the surface]. Comment.

O.3 Express the function $\cos(\omega t)$ as a phase-shifted sine.

O.4 Express the function $\sin(\omega t + \pi/6)$ as a phase-shifted cosine.

O.5 Express $\cos(\omega t - \pi/3)$ as a phase-shifted sine.

O.6 Express $\sin(\omega t - \pi/12)$ as a phase-shifted cosine.

O.7 An iceberg in the ocean is composed of nearly pure water ice. The densities of ice and sea water are reported in the preceding section [Fluids Problems].

(a) Determine the fraction of the volume of the iceberg lying above the ocean surface.

(b) A large [and expensive] movie strikes the iceberg, pushing half of the volume computed in (a) beneath the surface. [Thereupon the iceberg is released and moves freely, while the movie founders to the bottom of the video charts.] Assuming that the sea water is incompressible and has zero viscosity, propose a model description for the subsequent motion of the iceberg.

O.8 Six mass–spring systems are presented in the table below.

System	M [kg]	k [N/m]	System	M [kg]	k [N/m]
A	9	3	B	3	9
C	27	3	D	3	27
E	27	9	F	9	27

Which pairs of systems, if any, are dynamically equivalent [i.e., share the same ω_0]?

O.9 An SHO mass–spring system has $M = 2\,\text{kg}$ and $k = 3\,\text{N/m}$.

(a) At $t = 0\,\text{s}$, the block is offset from its equilibrium position and released from rest. Determine the trajectory of the block, $x(t)$, when its initial displacement from equilibrium is   (i) $+5\,\text{m}$ and   (ii) $-5\,\text{m}$.

(b) The block is in motion and passing through its equilibrium position at time $t = 0\,\text{s}$. Determine the trajectory, $x(t)$, of the block when its initial velocity, v_0, is equal to (i) $5\,\text{m/s}$ and   (ii) $-5\,\text{m/s}$.

(c) At $t = 0\,\text{s}$, the block is offset from equilibrium by $5\,\text{m}$ and is in motion with velocity $v_0 = 5\,\text{m/s}$. Determine the trajectory of the block.

O.10 Repeat O.9 with $M = 4\,\text{kg}$ and $k = 9\,\text{N/m}$.

O.11   An SHO trajectory is given by $x(t) = \frac{\pi^4}{90} \sin\left(\frac{7\pi t}{2}\right)$, where t is in seconds and x is in metres. Determine the   (a) equilibrium position,   (b) amplitude,   (c) angular frequency,   (d) cycle frequency, and   (e) period.

O.12

The trajectory of an object undergoing SHO is sketched in the nearby figure. The horizontal axis shows time measured in seconds, while the vertical axis shows position in millimetres.

(a) From the trajectory, determine the   (i) equilibrium position, (ii) amplitude,   (iii) period, (iv) angular frequency, and (v) cycle frequency.

(b) Express the mathematical form of the trajectory illustrated in (a) in terms of a   (i) `sine` and (ii) `cosine`   function.

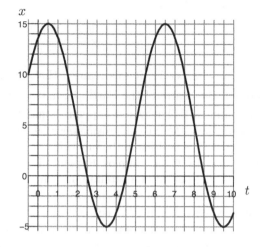

O.13   The two graphs below display the space-time trajectories of six SHO systems: (a–c) and (d–f). Time, in seconds, is on the horizontal axis, while displacement from equilibrium, in metres, is registered on the vertical axis.
(a–f) From each SHO trajectory shown below, ascertain the system's   (i) amplitude, (ii) period,   (iii) cycle frequency, and   (iv) angular frequency.

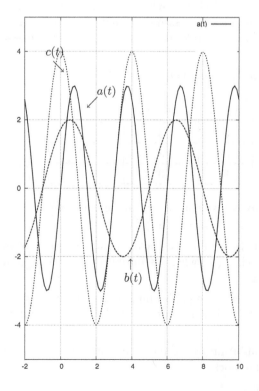

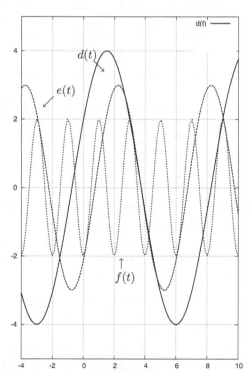

O.14

The trajectory of an oscillating object was carefully measured and is presented on the adjacent graph. The times are in seconds, while the displacement is measured in millimetres.

From the trajectory depicted in the graph, estimate the (a) amplitude, (b) period, and (c) (i) cycle and (ii) angular frequencies of the oscillatory motion. (d) Express the trajectory in terms of a (i) `cosine` and (ii) `sine` function of time, employing suitable phase angles.

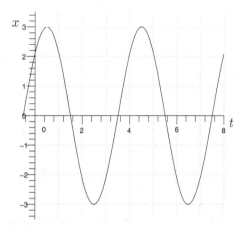

O.15

[See also O.14.] Consider the oscillating object in O.14. Suppose that it consists of a block of mass M riding on a frictionless horizontal plane surface and attached to a fixed anchor by a spring with force constant k.

　　　(a) If $M = 10\,\text{kg}$, what then is k?　　　(b) If $k = 5\,\text{N/m}$, what then is M?

O.16

Careful observations of the trajectory of an oscillating object were used to produce the position (in `cm`) *vs.* time (in `s`) graph plotted nearby.

Estimate the (a) amplitude, (b) period, (c) (i) cycle frequency, and angular frequency associated with the motion depicted in the graph. (d) Write the trajectory in terms of a (i) `cosine` and (ii) `sine` function of time (by employing suitable phase angles).

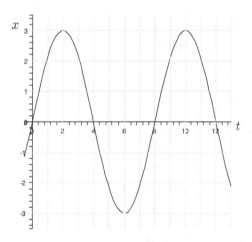

O.17

[See also O.16.] Consider the oscillating object in O.16. Suppose that it consists of a block of mass M riding on a frictionless horizontal plane surface and attached to a fixed anchor by a spring with force constant k.

　　　(a) If $M = 10\,\text{kg}$, what then is k?　　　(b) If $k = 5\,\text{N/m}$, what then is M?

O.18　　Consider the trajectory $x(t) = 5\,\cos(\pi t/4) + \cos(2\pi t)$ [with x in m, t in s].

　　(a) Does $x(t)$ oscillate?

　　(b) If the answer to (a) is yes, determine the period of oscillation.

O.19　　Repeat O.18 for $x(t) = 5\,\cos(\pi t/4) + \cos(\sqrt{2}\,\pi t)$.

O.20　　Repeat O.18 for $x(t) = 5\,\cos(\pi t/4) + \cos(3\pi t)$.

O.21　　Repeat O.18 for $x(t) = 5\,\cos(\pi t/4) + \cos(\pi t/3)$.

O.22　　Repeat O.18 for $\big(x(t),\, y(t)\big) = \big(5\,\cos(\pi t),\, 3\,\sin(3\pi t)\big)$.

O.23　　Repeat O.18 for $\big(x(t),\, y(t)\big) = \big(5\,\cos(\pi t),\, 3\,\sin(3t)\big)$.

O.24 Consider the trajectory $x(t) = 6\cos^2(3\pi t)$ [with x in cm, t in s].

(a) Is this an SHO trajectory?

(b) If the answer to (a) is yes, determine the   (i) equilibrium position,   (ii) amplitude,   (iii) angular frequency,   (iv) cycle frequency, and   (v) period   of oscillation.

O.25 Repeat O.24 for $x(t) = 7\cos^2(5\pi t + 4)$ [with x in mm, t in s].

O.26 Repeat O.24 for $x(t) = 8\sin(4\pi t)\cos(4\pi t)$ [with x in cm, t in s].

O.27 Repeat O.24 for $x(t) = 8\sin(4\pi t)\cos(3\pi t)$ [with x in cm, t in s].

O.28 Repeat O.24 for $x(t) = \cos^2(5\pi t) - \sin^2(5\pi t)$ [with x in m, t in s].

O.29 Repeat O.24 for $x(t) = \cos^2(5\pi t) - \sin^2(3\pi t)$ [with x in m, t in s].

O.30 Two springs, with $k_1 = 5\,\text{N/m}$ and $k_2 = 15\,\text{N/m}$ respectively, are combined and then subjected to an applied tension force[1] of $F_A = 2\,\text{N}$.

(a) Suppose that the two springs are joined in series.   (i) Compute the effective spring constant of this system.   (ii) By how much does the spring combination stretch?   (iii) By what amount does Spring 1 stretch?

(b) Suppose that the springs are arranged in parallel.   (i) Compute the effective spring constant of this system.   (ii) By how much does the spring combination stretch?   (iii) By what amount does Spring 1 stretch?

O.31 Four ideal springs, labelled by $n = 1, 2, 3, 4$, have spring constants (in N/m) given by $k_n = 2n - 1$. The four springs are combined in series.

(a) Compute the effective spring constant.

(b) An applied force, F_A, acts to stretch the combination of springs.   (i) Determine the relative amount of extension of each spring [as a fraction of the total extension of the series combination].   (ii) Determine the fraction of F_A that each spring experiences.

O.32 Repeat O.31 with the four springs combined in parallel.

O.33 Three ideal springs have spring constants $k_1 = 3$, $k_2 = 7$, and $k_3 = 5$ (all in N/m). Compute the effective spring constant when   (a) all three are in series, (123),   (b) spring 3 is in series with the parallel combination of 1 and 2, ([12]3),   (c) spring 2 is in parallel with the series combination of 1 and 3, [(13)2],   (d) all are in parallel, [123].

O.34
Eighteen identical springs, all possessing spring constant k, are arranged in the two nine-spring networks shown in the nearby figures. Compute the effective spring constants $k_{(a)}$ and $k_{(b)}$.

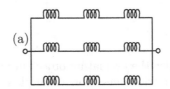
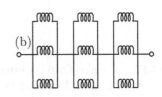

O.35
Various springs are arranged as shown in the figure to the right. The number adjacent to each spring is its spring constant, in N/m.

(a) Compute the effective spring constant of this network of springs.

(b) Compute the amount by which the entire network extends under the application of a 6 N force.

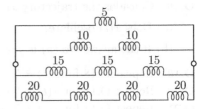

---

[1]For the quibblers, the applied force is increased quasi-statically, from zero to its final value.

(c) By how much does each of the   (i) $10\,\mathrm{N/m}$ and   (ii) $20\,\mathrm{N/m}$ springs stretch under the force applied in (b)?

O.36   Predict the effective spring constant for the combinations of three identical springs.
        (a) $[kkk]$ (b) $[(kk)k]$ (c) $([kk]k)$ (d) (kkk)

O.37   A number of identical ideal springs with force constant k are arranged as indicated in the figure.

(a) Compute the effective force constant for this array of springs.

(b) Suppose that $k = 2\,\mathrm{N/m}$, and that a force of $0.33\,\mathrm{N}$ is applied across the entire network. By how much does the spring indicated by the arrow stretch?

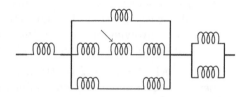

O.38   Ten identical springs, with $k = 12\,\mathrm{N/m}$, are arranged to form the network displayed in the figure to the right.

(a) Compute the effective spring constant of this network of springs.

(b) Suppose that an applied force of $2.4\,\mathrm{N}$ is exerted across the network and determine the amount by which the network stretches in response.

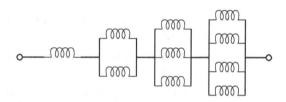

(c) By how much does each of the springs in the parallel   (i) pair,   (ii) triplet,   and (iii) quartet   combinations stretch under the force exerted in (b)?

O.39   A block of mass $M = 150\,\mathrm{g}$ is free to ride on a frictionless horizontal plane. Three identical springs, each with $k = 6\,\mathrm{N/m}$, are used to affix the block to a rigid anchor. The effects of drag are negligible.

(a) Supposing that the springs are configured in parallel, $[kkk]$, determine the period of oscillation, $T_{[kkk]}$, of this block–spring system.

(b) Supposing that the springs are configured in series, (kkk), determine the period of oscillation, $T_{(kkk)}$, of this block–spring system.

(c) There are two mixed-symmetry combinations: $[k(kk)]$ and $(k[kk])$. Determine
(i) $T_{[k(kk)]}$ and   (ii) $T_{(k[kk])}$.

(d) A fourth spring, identical to the other three, becomes available. Enumerate and sketch all of the distinct mixed combinations of these four springs. [Note that some seemingly distinct configurations will give rise to the same effective spring constant, and thus are not operationally distinct.]

O.40   Eight ideal springs, all with spring constant k_0, are arranged as shown in the nearby figure.

(a) Compute the effective spring constant of the network.

(b) Suppose that one end of the network is attached to a fixed anchor and the other end is secured to a block of mass M. Ascertain the natural frequency of this mass–spring system.

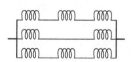

O.41 A torsional pendulum consists of a torsion fibre with constant κ and a load with moment of inertia I. This system experiences an angular drag torque, $\tau_{\text{D}} = -\beta_1\,\omega$, where ω is the angular velocity of the load about the axis of the fibre and β_1 is a constant in which all of the details about the shape of the load and the properties of the fluid in which it resides are subsumed.

(a) Write down the angular equation of motion for the load.

(b) Let's attempt to find a solution to this equation of motion of the form $\theta(t) = \exp(\rho\,t)$, where ρ is a [constant] *Ansatz* parameter. (i) Ascertain that such solutions actually exist and that they fall into the three anticipated classes. (ii) Express the condition involving the torsion fibre constant, κ, the moment of inertia, I, and the angular damping parameter, β_1, which leads to critical damping.

O.42 A block of mass $8\,\text{kg}$ is attached to a fine ideal torsion fibre in such a way that its moment of inertia about the axis defined by the fibre is $5\,\text{kg·m}^2$. The other end of the fibre is firmly fixed. The fibre has torsion constant $\kappa = 3\,\text{N·m/rad}$. Express the angular position of the block as a function of time in each of the following four cases.

(a) At $t = 0\,\text{s}$ the block is released from rest at an angle of 0.15 radians from its equilibrium position.

(b) At $t = 0\,\text{s}$ the block is at $\theta = 0\,\text{rad}$, and moving with angular speed $\omega_0 = 0.5\,\text{rad/s}$ in the direction of increasing θ.

(c) At $t = 0$ the block is at $\theta(0) = 0.10\,\text{rad}$ from its equilibrium position, and moving with angular velocity $\omega(0) = -1/3\,\text{rad/s}$.

(d) At $t = 0$ the block is at $\theta(0) = 0.10\,\text{rad}$ from its equilibrium position, and moving with velocity $\omega(0) = 1/3\,\text{rad/s}$.

O.43 A block of mass $5\,\text{kg}$ riding on a frictionless plane is attached to one end of an ideal spring with force constant $k = 3\,\text{N/m}$. The other end of the spring is fixed. Express the position of the block as a function of time in each of the following four cases.

(a) At time $t = 0\,\text{s}$ the block is released from rest at a distance of 0.15 metres from its equilibrium position.

(b) At $t = 0\,\text{s}$ the block is at $x = 0\,\text{m}$, and moving with speed $v_0 = 0.5\,\text{m/s}$ in the direction of increasing x.

(c) At $t = 0$ the block is at $x(0) = 0.10\,\text{m}$ from its equilibrium position, and moving with velocity $v(0) = -1/3\,\text{m/s}$.

(d) At $t = 0$ the block is at $x(0) = 0.10\,\text{m}$ from its equilibrium position, and moving with velocity $v(0) = 1/3\,\text{m/s}$.

O.44 A system is known to be undergoing simple harmonic oscillation. Careful observation reveals that the acceleration amplitude is $12\,\text{m/s}^2$ and the velocity amplitude is $6\,\text{m/s}$. Determine   (a) the angular frequency of the system and   (b) the amplitude of the SHO trajectory.

O.45 A mass–spring system with $M = 5\,\text{kg}$ and $k = 6\,\text{N/m}$ is known to be undergoing simple harmonic oscillation. The total mechanical energy of the system is $27\,\text{J}$.

(a) Determine the   (i) amplitude and   (ii) maximum speed   of the block.

(b) Determine the   (i) angular frequency and   (ii) period   for this system.

O.46 We model the energetics of a massive (non-ideal) spring in a mass–spring system. [See also O.47, O.48.] A block of mass M, joined to a spring of mass m and force constant k, rides on a horizontal frictionless plane. Prior to $t = 0$, the block was displaced from equilibrium [taken to be $x_0 = 0$] to $x = A$ and held there. At $t = 0$, the block is released.

Our goal is to compute the speed of the block as it passes through its equilibrium position. A very crude model treats all of the mass of the spring as though it were concentrated at its centre. A version of this model (with two ideal springs) is shown in the figure.

For this analysis, we must assume that the motions of the actual block and the synthetic block representing the mass of the spring are exactly correlated. *I.e.*, when M moves with velocity V, m has velocity $v = V/2$. [Do not attempt to assign a potential energy to the particle representing the spring.]

   (a) Ascertain the initial   (i) potential and   (ii) kinetic   energy of the system.

   (b) As the block passes through the equilibrium position, it has [unknown] speed V_0. In terms of V_0, determine the kinetic energies of the   (i) block and   (ii) mass representing the spring   at the instant the system passes through x_0.

   (c) Determine the total kinetic energy of the block–(massive)spring system at the instant that the block passes through its equilibrium position.

   (d) Use conservation of energy to determine V_0 in terms of the initial displacement of the system from equilibrium.

O.47   Generalise the analysis in O.46 to obtain the speed of the system as it passes through any point x, $|x| < A$.

O.48   [See also O.46.] A block of mass M, joined to a spring of mass m and force constant k, rides on a horizontal frictionless plane. Prior to $t = 0$, the block was displaced from equilibrium [taken to be $x_0 = 0$] to $x = A$ and held there. At $t = 0$, the block is released. A better model for the spring with inertia assumes that its mass is uniformly distributed along the entire length, L, from the anchor to the block.

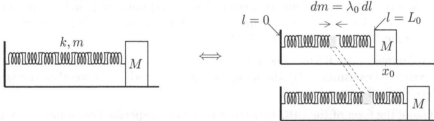

We assume that the velocity of the mass element at distance y from the left end of the unextended spring varies from zero when $y = 0$ [at the left edge] to matching the velocity of the block, V, when $y = L$.

   (a) Express the velocity of a generic mass element of the spring as a linear function of its position y along the unextended spring. [This will be incorporated into a pull-back to a defined parameter space.]

   (b) Compute the differential contribution to the total kinetic energy from the spring mass element located at [unstretched] y. Use your result from (a) to express this in terms of the block's instantaneous velocity.

   (c) Integrate (b) over the [unperturbed] length of the spring.

   (d) Ascertain the total kinetic energy of the system when the block is moving with velocity V.

(e) Infer the speed with which the block passes through the equilibrium position of the spring.

(f) Infer the speed with which the block passes by the point x, where $|x| \leq A$.

O.49   A swing set, not unlike those found in playparks, consists of a thin bench hung from a rigid horizontal bar by ideal ropes of length $L = 2.25\,\text{m}$. We seek to determine the period of [approximate] SHO oscillation of this swing set when it is installed in a variety of locales.

(a) The swing set is on top of a tower extending roughly $300\,\text{km}$ above the Earth's surface. [The mass of the Earth is roughly 6.0×10^{24} **kg**. First compute the local acceleration due to gravity at a distance of 6.67×10^{6} m from the centre.]

(b) The swing set is installed on the surface of the Moon, where the magnitude of the local acceleration due to gravity is $g_m = \frac{5}{3}\,\text{m/s}^2$.

(c) The swing set is transferred to the surface of the Sun. [Don't forget the sunscreen! The mass of the Sun is about 2.0×10^{30} **kg**, while its radius is roughly 7.0×10^{8} m.]

O.50   An oscillating system located on the surface of the Earth consists of a person named Bob, with mass m_B, swinging on a swing set with pendular length L.

(a) When Bob went away to university his mass doubled. How did this affect his swinging?

(b) After university, Bob got a job on the surface of the Moon, where the local acceleration due to gravity is one-sixth that on the Earth. As Bob's employer was willing to pay his moving expenses, Bob brought his swing set with him. How did this relocation affect his swinging?

O.51   A block of mass $M = 2\,\text{kg}$ rides on a horizontal frictionless plane. An ideal spring with $k = 8\,\text{N/m}$ is attached to the block and affixed to an anchor post. The system undergoes simple harmonic oscillation with amplitude $A = 1.5\,\text{m}$.

(a) Determine the   (i) angular and   (ii) cycle   frequency,   (iii) period,   (iv) maximum speed and   (v) maximum acceleration,   (vi) minimum and   (vii) maximum potential energies,   (viii) minimum and   (ix) maximum   kinetic energies, and   (x) total mechanical energy   of this SHO.

(b) A damper with effective linear drag coefficient $b_1 = 2.5\,\text{N·s/m}$ is incorporated into the system. Determine   (i) the damping rate, γ, and   (ii) whether the system is under-/critically-/overdamped.

(c) Quote the form of the DHO trajectory with two unspecified constants of integration [requiring additional data for their determination] for this mass–spring–damper system.

O.52   A block of mass $M = 5\,\text{kg}$, free to move in one dimension on a frictionless horizontal plane, and a spring with force constant $k = 40\,\text{N/m}$, connected to the block and secured to an anchor, comprise an SHO.

(a) Determine the SHO's   (i) angular frequency, ω_0,   (ii) period, $\mathcal{T}$, and   (iii) cycle frequency, ν.

(b) At the instant $t = 0\,\text{s}$ the block was observed to be $+5$ cm [in the forward direction] from its equilibrium position and instantaneously at rest. Determine the trajectory of the block.

(c) An identical system [possessing the same mass, spring constant, and damping parameter] was set into motion and the block was observed to be passing through its equilibrium position, moving backwards with speed $v_0 = 8\sqrt{2}\,\text{cm/s}$, at $t = 0\,\text{s}$. Determine the trajectory of this block.

(d) Plot the trajectories of the blocks in (b) and (c) on common axes.

O.53 A linear drag force acts on the block–spring system considered in O.52. The drag coefficient has magnitude $b_1 = 30$ N·s/m. This block–spring–drag system is a DHO.

(a) Determine the DHO's   (i) angular frequency and   (ii) damping parameter.

(b) Is this system under-/critically/overdamped?

(c) Observations of this system have determined that at $t = 0$ s the block was $+5$ cm from its equilibrium position and instantaneously at rest. Determine the trajectory of the block.

(d) An identical system [possessing the same mass, spring constant, and damping parameter] was set into motion and the block was observed to be passing through its equilibrium position, moving backwards with speed $v_0 = 8\sqrt{2}$ cm/s at $t = 0$ s. Determine the trajectory of this block.

(e) Plot the trajectories of the blocks in (c) and (d) on common axes.

O.54 Repeat parts (a–c) in O.53 when the drag coefficient is reduced to $b_1 = 20$ N·s/m.

(d) An identical system [possessing the same mass, spring constant, and damping parameter] was set into motion and this block was observed to be passing through its equilibrium position, moving backwards with speed $v_0 = 10$ cm/s at $t = 0$ s. Determine the trajectory of this block.

(e) Plot the trajectories of the blocks in (c) and (d) on common axes.

O.55 A block of mass $M = 5$ kg is able to move in 1-d on a frictionless horizontal plane. A spring with force constant $k = 40$ N/m is connected to the block and secured to an anchor. A linear drag force with coefficient $b_1 = 20$ N·s/m acts on the block. An external agent applies an oscillatory force, $F_A(t) = 3\cos(\pi t)$ N, to the block. This [idealised] block–spring–drag system is a driven DHO.

(a) We presume that the steady state response of the system is $x(t) = A\sin(\pi t - \varphi)$ for *Ansatz* parameters $\{A, \varphi\}$. Solve for the response amplitude and phase.

(b) Sketch $F_A(t)$ and $x(t)$ together on a common set of axes.

(c) Compute the average power passing through this system.

(d) Suppose that the driving frequency were to be increased slightly. Ascertain, without doing any additional calculations, whether the power flowing through the system would increase, stay the same, or decrease.

O.56 A damped harmonic oscillator system consists of a block with mass $M = 0.5$ kg, a spring with force constant $k = 32$ N/m, and a damper with linear coefficient $b_1 = 3$ N·s/m. In this system, the damping parameter is easily adjusted.

(a) (i) Determine whether this DHO is over-/critically/underdamped.   (ii) For what value of the damping parameter, $b_{1,c}$, is the system critically damped?

[For the remainder of this question, this $\{M, k, b_{1,c}\}$ system is referred to as the DHO′.]

(b) A sinusoidal driving force with amplitude $F_0 = 5$ N and variable frequency is applied to both the DHO and DHO′ systems.   (i) Compute the power flowing through the DHO when it is driven at $\omega = 10$ rad/s.   (ii) Ascertain the driving frequency at which the power through the DHO is maximised, along with the maximum power.   (iii) Compute the power flowing through the DHO′ when it is driven at $\omega = 10$ rad/s.   (iv) Ascertain the driving frequency at which the power through the DHO′ is maximised, along with the maximum power.

O.57 Three DHO systems have the physical properties listed in the table below.

DHO	Mass [kg]	Drag Coefficient [N·s/m]	Spring Constant [N/m]
A	81	9	1
B	9	1	81
C	1	81	9

(a) Compute the natural frequency, ω_0, and damping rate, γ, for each of the systems.

(b) State whether each system is under-/critically/over-damped.

Determine the steady-state position response, $x(t)$, of each of the systems when subjected to the following driving forces:

    (c) $F_c(t) = 3\,\cos(t/9)$ N   (d) $F_d(t) = 3\,\cos(t)$ N   (e) $F_e(t) = 3\,\cos(3\,t)$ N

O.58 Repeat O.57 (a–b) for the three DHO systems whose physical properties are listed in the table below.

DHO	Mass [kg]	Drag Coefficient [N·s/m]	Spring Constant [N/m]
A	16	4	1
B	4	1	16
C	1	16	4

Determine the steady-state position response, $x(t)$, of each of the systems when subjected to the following driving forces:

    (c) $F_c(t) = 3\,\cos(t/4)$ N   (d) $F_d(t) = 3\,\cos(t)$ N   (e) $F_e(t) = 3\,\cos(2\,t)$ N

O.59 Ten DHO systems have the physical properties listed in the table below.

DHO	Mass [kg]	Drag Coefficient [N·s/m]	Spring Constant [N/m]
A	10	5	1
B	10	1	5
C	5	10	1
D	1	5	10
E	5	1	10
F	20	10	2
G	1	0	1
H	5	0	5
I	10	5	0
J	0	5	1

(a) Determine whether each system is underdamped/critically damped/overdamped or, indeed, not an oscillator at all.

(b) Identify those oscillators which are identical in their dynamical properties.

(c) An external sinusoidal force, $F_A(t) = 4\,\cos(5\,t)$, is applied to every *bona fide* DHO system. For each case, determine the amplitude and relative phase shift of the position-response function describing the steady-state oscillation.

O.60 The equilibrium position of a particular critically damped mass–spring [DHO] system is $x = 2.0\,\text{m}$. Sketch a trajectory representing a possible path of the block if it is released from rest at $x_0 = -1\,\text{m}$.

O.61 Three mass–spring–damper systems have common mass $M = 4\,\text{kg}$ and drag coefficient $b_1 = 16\,\text{N·s/m}$, while possessing differing spring constants: $k_i = 1\,\text{N/m}$, $k_{ii} = 16\,\text{N/m}$, $k_{iii} = 64\,\text{N/m}$.

(a) (i–iii) Determine the characteristic properties of each DHO.

(b) (i–iii) Each system is subjected to the sinusoidal driving force $F_A(t) = 3\cos{(2t)}$. Determine the response of each system.

O.62 We have available various blocks, springs, and dampers possessing the parameters listed below:

$$M \in \{10,\,200,\,1000\}\,[\text{g}]\,,\qquad k \in \{10,\,500,\,10000\}\,[\text{N/m}]\,,\qquad b_1 \in \{0.1,\,10\}\,[\text{N·s/m}]\,.$$

(a) Using components from this set, construct a DHO system that will react vigorously to driving frequencies in a narrow band about $1000\,\text{rad/s}$, and respond poorly to all other frequencies. [This is a narrow-band filter.]

(b) Using components from this list, design a DHO that responds to a very broad range of frequencies centred on $50\,\text{rad/s}$. [This is a broad-band filter.]

O.63 We possess a box full of $20\,\text{N/m}$ springs, a block of mass $M = 0.5\,\text{kg}$, and two dampers with linear drag coefficients $b_1 = 0.05\,\text{N·s/m}$ and $B_1 = 0.5\,\text{N·s/m}$, respectively.

(a) (i) Design a mass–spring–damper system which is strongly responsive to angular frequencies near $\omega = 20\,\text{rad/s}$, and quite unresponsive otherwise. (ii) Design a mass–spring–damper system which is most responsive to a range of angular frequencies $2.5 < \omega < 4\,\text{rad/s}$, centred on $\sqrt{10}\,\text{rad/s}$.

(b) Determine approximate quality factors for the two driven DHO systems whose average power frequency response curves are illustrated below.

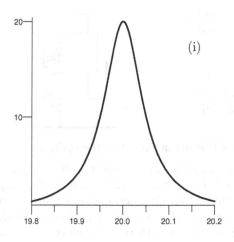

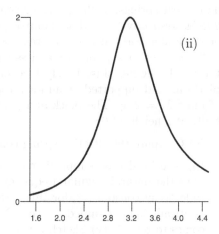

O.64

This figure depicts the power flowing through a DHO with fixed parameters as a function of the driving frequency, ω. Estimate the (a) resonant frequency, (b) full width at half-maximum, and (c) quality factor for this system.

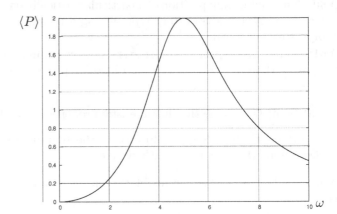

O.65 The expression for the average power flowing through a driven DHO is quoted in Chapters 19 and 20.

(a) Show that for $\omega < \omega_0$ the slope is everywhere positive.

(b) Argue that for $\omega > \omega_0$ the slope is everywhere negative.

(c) The maximum power, $\langle P \rangle_{\text{max}} = F_0^2/(2\,b_1)$, occurs when $\omega = \omega_0$. Determine the upper and lower frequencies at which the power through the oscillator is equal to one-half the maximum power. [HINT: The equation for these frequencies is a quartic. It factorises into a product of quadratic equations. One of the two distinct roots from each quadratic is positive and relevant. The other roots are negative, and hence extraneous.]

(d) Employ the results in (c) to derive an expression for the quality factor.

O.66 A damped mass–spring system is parameterised by: $M = 5\,\text{kg}$, $k = 10\,\text{N/m}$, and $b_1 = 1\,\text{N·s/m}$. The system is subjected to a variety of applied sinusoidal forces, each of amplitude $F_o = 0.5\,\text{N}$ and specified driving frequency. For each of the driving frequencies listed below (in **rad/s**), compute the average power passing through the system.

(a) $\omega = \frac{1}{2}$ (b) $\omega = 1$ (c) $\omega = \sqrt{2}$ (d) $\omega = 2$ (e) $\omega = 4$

(f) From the results (a–e), infer whether the quality factor for the oscillator is less than, equal to, or greater than $\sqrt{2}$.

O.67

Two ideal springs, with spring constant $5\,\text{N/m}$ and $20\,\text{N/m}$, respectively, and a damper [*a.k.a.* dashpot or shock absorber] with linear drag coefficient $b_1 = 12\,\text{N·s/m}$ together support a massless platform, as in the nearby figure. Just barely in contact with the platform, and supported by an ideal rope, is a block of mass $M = 2\,\text{kg}$. The block and platform are initially at a height of $6\,\text{m}$.

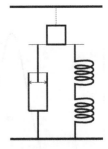

(a) Compute the effective spring constant for this arrangement of the springs.

(b) The ideal rope supporting the block is suddenly cut at time $t = 0$.   (i) Compute the natural angular frequency, ω_0, for this system.   (ii) Compute the amplitude decay constant, γ, for this system.   (iii) Ascertain whether this system is over-/critically/underdamped.   (iv) Determine the height at which the block eventually comes to rest.   (v) Sketch a trajectory illustrating a path that the block might have taken from its initial position at height $6\,\text{m}$ prior to $t = 0\,\text{s}$ to its asymptotic position computed in (iv).

(c) The block is subjected to a sinusoidally-varying external driving force with amplitude 4 N and angular frequency $\omega = 2\sqrt{3}\,\text{rad/s}$. (i) Determine the relative phase of the block's response to the driving force. (ii) Determine the average rate at which power is being absorbed and emitted by the system.

O.68

Repeat O.67 when the springs are combined in parallel (as shown in the figure).

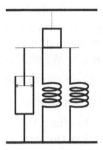

O.69 A block of mass $M = 5\,\text{kg}$ moves in one dimension on a frictionless horizontal plane. A spring with force constant $k = 20\,\text{N/m}$ is connected to the block and its other end is held fixed.

(a) Prior to $t = 0$, the block oscillates about its equilibrium position in a manner given by $x(t) = 7 + 4\cos(2t)$, where x is in mm and t is in s. Determine the (i) equilibrium position, (ii) amplitude, (iii) maximum speed, (iv) maximum kinetic energy, (v) maximum potential energy, and (vi) total mechanical energy of the block.

(b) At $t = 0$, a linear drag force acts on the block and its trajectory becomes $x(t) = 7 + 4e^{-0.345t}\cos(1.97t)$. Determine the (i) damping coefficient, (ii) drag coefficient, (iii) asymptotic position [$x(t)$, as $t \to \infty$], and (iv) fraction of the original energy remaining after the elapse of n half-times.

(c) A system identical to that studied in (b) is observed to be oscillating with trajectory $x(t) = 7 + 2\sin(t)$. Infer the properties of the force which drives the block's motion.

O.70 [See also O.71, O.72, and O.74. Here we begin to develop a model describing a buoy floating in a calm freshwater harbour.] A thin, smooth, cylindrical steel shell has mass 10 kg, radius $\frac{50}{\sqrt{\pi}}$ cm, and height 1.2 m. A thin steel plate of mass 10 kg covers the entire bottom of the cylindrical shell for added strength and stability. The bottom 20 cm of the buoy is packed with sand of density $2000\,\text{kg/m}^3$, while the upper 100 cm is filled with styrofoam of density $200\,\text{kg/m}^3$. The buoy is placed upright in calm water, guided to its equilibrium position, and released. A bright mark is made on the side of the buoy at the waterline.

(a) Determine the total mass of the buoy.

(b) Determine the position of the CofM of the buoy with respect to the thin plate on the bottom of the cylinder.

(c) (i) Determine the position of the waterline on the side of the buoy. (ii) Determine the position of the CofM with respect to the waterline. (iii) Is the buoy stable floating in this position? (iv) Describe the unstable equilibrium position of the buoy.

(d) The buoy is lifted 15 cm straight upward, and released at $t = 0\,\text{s}$. Assume that water is frictionless and verify that the equation of motion describing the buoy's subsequent motion is of SHO form.

(e) (i) From (d), infer the angular frequency of the buoy's motion. (ii) Write down the trajectory of the bright mark on the buoy with respect to the waterline, $y = 0$.

O.71 [See also O.70 and O.73. Here we improve on the previous model by incorporating effects of viscous drag.] A small quantity of biodegradable dye is released into the water near the buoy. Careful observation reveals that the dye 3.0 cm from the edge of the buoy remains undisturbed, while that closer in participates in synchronised up-and-down motion with the oscillating buoy. For our crude model, we take the viscosity of the water to be 1×10^{-3} Pa·s.

(a) As an approximation, we shall neglect the fact that as the buoy rises and falls, the area on its sides in contact with the water is changing. Determine the area on the side of the cylinder which is underwater when the buoy is floating at its equilibrium height.

(b) Reverse-engineer the model developed for viscosity, in Chapter 9, to isolate the shear force needed to maintain motion of the plate on one side of the fluid channel at constant speed v. We will use this as the drag force acting on the moving buoy.

(c) From the model in (b), infer the value of the lineal drag coefficient.

(d) Determine the damping parameter for the motion of the buoy.

(e) Recalling the natural angular frequency of the buoy from O.70 part (e), ascertain whether this system is underdamped, critically damped, or overdamped.

(f) Estimate the half-time for the buoy. [After one half-time interval has elapsed, the mark on the buoy is confined within $\pm \frac{15}{\sqrt{2}}$ of the waterline.]

O.72 [See also O.70.] Repeat O.70 for the buoy floating in a large vat of maple syrup. Take the density of high-grade maple syrup to be $\rho_{\text{ms}} = 1360 \, \text{kg/m}^3$.

O.73 [See also O.72.] Repeat O.71 for maple syrup, with viscosity $\eta_{\text{ms}} = 0.1535$ Pa·s. In this case, the liquid lying within 4 cm of the smooth wall oscillates along with the buoy.

O.74 [See also O.70.] Repeat O.70 for the buoy *jammed* in an enormous "family-sized" jar of peanut butter. Take the density of peanut butter to be $\rho_{\text{pb}} = 1283 \, \text{kg/m}^3$.

O.75 [See also O.74.] Repeat O.71 for peanut butter, with viscosity $\eta_{\text{pb}} = 250$ Pa·s. In this case, the paste lying within 5 cm of the smooth wall oscillates along with the buoy.

W

Wave Problems

W.1 A harmonic wave is represented in the graphs below. The left hand panel shows a snapshot taken at the instant $t = 0$. The right hand one shows the transverse displacement of the wave medium at the position $x = 0$ as a function of time. The units of position, time, and the disturbance are m, s, and cm, respectively. From the graphs, estimate or infer the (a) amplitude, (b) wavelength, (c) period, (d) wavenumber, (e) cycle frequency, (f) angular frequency, (g) speed, and (h) direction of the wave.

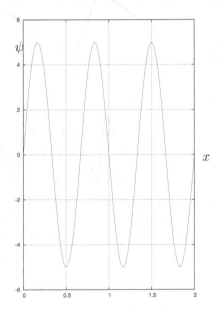 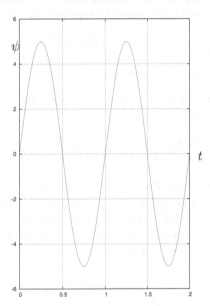

W.2 A particular harmonic wave admits expression as $\psi(t, x) = \frac{17}{3}\cos(51\,x + 3\,t)$, where ψ, t, and x have units of kPa, μs, and cm. Ascertain the (a) amplitude, (b) wavenumber, (c) wavelength, (d) angular frequency, (e) cycle frequency, (f) period, (g) speed, and (h) direction of the wave.

W.3 A harmonic travelling wave has wave function $\psi(t, x) = 8 \sin\left(\frac{\pi}{15}x - \frac{\pi}{3}t + \sqrt{1/8}\,\pi\right)$. The units of time and position, t and x, are seconds and metres, respectively, while the wave disturbance is measured in nanograms per cubic micrometre.

(a) Ascertain the (i) amplitude, (ii) wavenumber, (iii) angular frequency, and (iv) phase of ψ.

(b) Determine (i) the speed with which ψ propagates, and (ii) whether it moves to the right or to the left.

W.4 Repeat W.3 for $\psi(t, x) = 15 \sin\left(\frac{16}{25}x + 6\,t - \frac{\pi}{4}\right)$. In this case, the time and position, t and x, are measured in seconds and centimetres, respectively, while the wave disturbance is quantified in millimetres.

W.5 A harmonic wave is described by $\psi(t,x) = 5.5\cos(3x/2 - \pi t)$. The units for ψ are centimetres, x is expressed in kilometres, and t is in minutes.

(a) Determine the   (i) wavenumber,   (ii) angular frequency,   (iii) amplitude, (iv) wavelength,   (v) frequency, and   (vi) period   of ψ.

(b) Ascertain   (i) the speed of the wave and   (ii) the direction in which it propagates.

W.6 Repeat W.5 for $\psi(t,x) = 5\cos(10\pi x - 120\pi t + 3\pi/2)$. The dimensions of this wave are electron volts, eV, x is in mm, and t is in s.

W.7
Two views of a travelling wave disturbance are shown in the figures on the right. Above is a snapshot taken at $t = 0\,\text{s}$. The lower view tracks the displacement of the medium [from its equilibrium position] at the fixed point in space x_0 as a function of time. The wave disturbance is measured in centimetres.

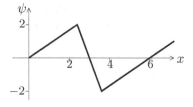

(a) Determine the amplitude of the wave disturbance by inspecting the   (i) upper and (ii) lower   sketch.

(b) Determine the   (i) wavelength   and (ii) wavenumber.

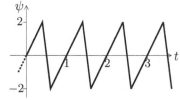

(c) Determine the   (i) period,   (ii) frequency, and   (iii) angular frequency.

(d) Determine the speed of the wave.

W.8 A snapshot of a particular travelling wave is illustrated below on the left. To the right is a view of the time-dependence at fixed position of the same wave. The vertical axes show transverse displacement in mm, while the horizontal axes display position in m on the left and time in s on the right.

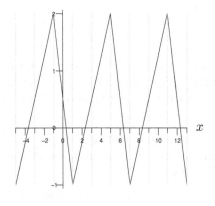
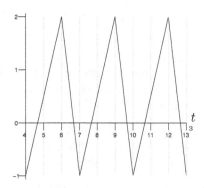

(a) Ascertain the wave's   (i) amplitude,   (ii) wavelength, and   (iii) wavenumber [from the sketch on the left].

(b) Ascertain the wave's   (i) amplitude,   (ii) period,   (iii) cycle frequency, and (iv) angular frequency [from the sketch on the right].

(c) Infer the speed of the wave.

W.9 Two images of a harmonic wave are found below. The first is a snapshot taken at the instant $t = 0\,\text{s}$, while the second shows the trajectory of a chunk of medium located at $x = 2\,\text{m}$. The disturbance is measured in centimetres. From these curves, determine the (a) amplitude, (b) (i) wavelength, (ii) wavenumber, (c) (i) period, (ii) cycle frequency, (iii) angular frequency, (d) (i) speed and (ii) direction of the wave.

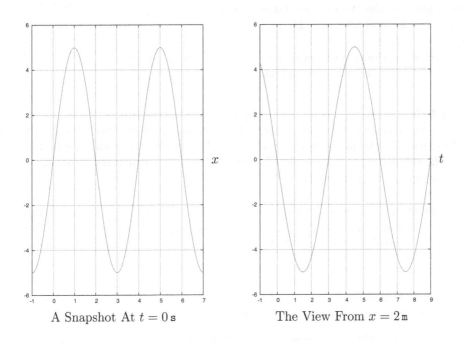

A Snapshot At $t = 0\,\text{s}$ The View From $x = 2\,\text{m}$

W.10 A harmonic wave is described by $\psi(t, x) = 8\,\cos(5\,x - t/3)$, where x is in metres, t is in seconds, and the disturbance, ψ, is measured in millimetres. Ascertain the (a) amplitude, (b) (i) wavelength and (ii) wavenumber, (c) (i) period, (ii) cycle frequency, and (iii) angular frequency, (d) (i) speed and (ii) direction of the wave.

W.11 A left-moving harmonic wave is observed to have frequency $15\,\text{Hz}$, wavelength $5\,\text{m}$, and amplitude $8\,\text{cm}$.

(a) From the description of the wave, infer its (i) angular frequency, (ii) wavenumber, (iii) period, and (iv) speed.

(b) At $t = 0$, the wave disturbance has value $4\,\text{cm}$ at the origin, $x = 0$. Also at that instant, the disturbance is greater going forward, *i.e.*, $\psi(0, x) > \psi(0, 0)$ for $x > 0$, and lesser going backward, *i.e.*, $\psi(0, x) < \psi(0, 0)$ for $x < 0$, in small spatial neighbourhoods of the origin. [The disturbance is said to be "increasing in the forward direction."] With this added information, write down a mathematical description of the wave, *i.e.*, its wave function, $\psi_{(b)}(t, x)$.

(c) Suppose that, contrary to what was claimed in (b), the wave disturbance was $4\,\text{cm}$ and increasing in the forward direction at the position $x = 0$ at the instant $t = \frac{1}{4}\,\text{s}$. Determine the wave function, $\psi_{(c)}$, in this case.

W.12
Three seismic monitoring stations, $\{A, B, C\}$, are positioned on a broad, flat plain [plane] as illustrated in the nearby figure. [The curvature of the Earth's surface is neglected.] North is toward the top, east is to the right. The distances between the stations are known to be 500 km, 1200 km, and 1300 km to high precision.

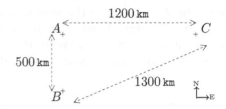

The Earth's crust in this region is nearly homogeneous. Pressure waves travel at 5 km/s and shear waves move at 3 km/s through the crust.

(a) A shallow-focus earthquake (originating high in the crust near the surface) occurred one day, and all three stations detected the arrival of the first pressure and shear waves. At station A, the first pressure wave arrived precisely 40 s before the first shear wave. At B, the delay was $53\frac{1}{3}$ s. At C it was $130\frac{1}{4}$ s. Determine the approximate location of the epicentre of the earthquake [*i.e.*, the source of the waves].

(b) Twelve years previously, another earthquake was detected by the same three stations. The time delays at stations A and B were the same as measured above. The delay at station C on this occasion was 193.5 s. Was the epicentre the same for both earthquakes? Comment.

W.13 Show that for $\psi(z)$, a twice differentiable function of a single variable, $\psi(k\,x - \omega\,t)$ satisfies the wave equation with wave speed $c = \omega/k$.

W.14 Two harmonic waves, differing only in phase, are
$$\psi_1(t, x) = 3\sin\left(\tfrac{\pi}{4}x - \tfrac{\pi}{3}t\right) \qquad \text{and} \qquad \psi_2(t, x) = 3\sin\left(\tfrac{\pi}{4}x - \tfrac{\pi}{3}t - \tfrac{\pi}{2}\right),$$
where x is in m, t is in s, and $\psi_{1,2}$ are measured in mm. Determine the wave which results from the superposition of these two waves.

W.15 Repeat W.14 for $\psi_1(t, x) = 5\sin(5\,x - 7\,t + \pi/6)$ and $\psi_2(t, x) = 5\sin(5\,x - 7\,t - \pi/3)$.

W.16 The snapshot [taken at a particular instant] of a wave on the spatial region $-4 \le x \le 4$ is described by $\psi(x) = 5\sin\left(\tfrac{7\pi x}{4} + 0.643501109\right)$. Determine the Fourier components of $\psi(x)$ on the symmetric interval $-4 \le x \le 4$. [HINTS: The Fourier basis functions are orthonormal on any fixed interval (*cf.* Chapter 22), and the **sine** of a sum of angles may be expressed in terms of factors of the **sine** and **cosine** of the summands.]

W.17 Two waves on the common interval $x \in [-6, 6]$ at the instant $t = 0$ are shown in the nearby figure.

(a) Sketch the wave which results from pointwise superposition of the two waves.

(b) Suppose that one were to perform a Fourier analysis of the superposition. Without performing any computations, what might be inferred about the Fourier coefficients of the **sine** and **cosine** terms?

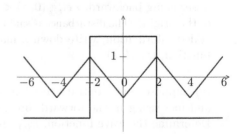

W.18

A saw-tooth-like wave pulse is described and
exhibited in the formulae and figure.

$$\psi = \begin{cases} 0 \;, \; y < -\dfrac{L}{2} \;\; \text{and} \;\; y > \dfrac{L}{2} \\[2mm] \dfrac{4\,y}{L} \;, \; -\dfrac{L}{2} \le y \le \dfrac{L}{2} \end{cases}$$

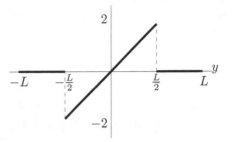

(a) Work out in detail the Fourier series approximation to this wave function.

(b) Comment.

W.19
The **Dirac delta-function** is a curious beastie.
[Strictly speaking, Dirac's delta is not a function at all.
It is more aptly considered a *distribution* and kept safely
within the ambit of an integral.]

$$\delta(x) = \begin{cases} 0, \; \forall \, x \ne 0, \\ \infty, \;\;\; x = 0. \end{cases}$$

Our goal is to determine the Fourier components of the delta function centred on the
origin in a domain of width $2\,L$. An essential property is that, for all functions $f(x)$,
$\int_{-L}^{L} f(x)\,\delta(x)\,dx = f(0)$. [That is, integrating f along with the Dirac delta "picks out" the value
of the function evaluated at the zero of the delta's argument.]

(a) Evaluate the projection of $\delta(x)$ onto the constant function basis vector.

(b) Project $\delta(x)$ onto the nth $\quad$ (i) **sine** and $\quad$ (ii) **cosine** $\quad$ basis function.

(c) Comment on your results.

(d) Set $L = 4$ and sketch the Fourier series approximation to the Dirac delta function
developed just above on the extended range $-12 < x < 12$. [Include a fairly large number
of Fourier components.]

[HINT: One might want to write a computer program to plot these approximations for small
and intermediate values of n, and then extrapolate these results to produce the sketch.]

W.20 $\quad$ Repeat W.19 for the "double-delta function" identified and illustrated below.

$$f(x) = \frac{1}{2}\Big[\,\delta(x - L/2) - \delta(x + L/2)\,\Big]$$

$$= \begin{cases} 0, \; \forall \, x < -L/2, \\ -\infty, \;\;\; x = -L/2, \\ 0, \; \forall \; -L/2 < x < L/2, \\ +\infty, \;\;\; x = L/2, \\ 0, \; \forall \, x > L/2. \end{cases}$$

Note that, for all functions $g(x)$, $\displaystyle\int_{-L}^{L} g(x)\,\delta(x - a)\,dx = g(a)$, provided that $-L < a < L$.

W.21 A double triangular wave on the interval $x \in [-4, 4]$ is expressed and sketched below. Set up, but do not evaluate, the projections of this function onto the Fourier basis functions.

$$f(x) = \begin{cases} -4 - x\,, & -4 \leq x < -2 \\ x\,, & -2 \leq x \leq 2 \\ 4 - x\,, & 2 \leq x \leq 4 \end{cases}$$

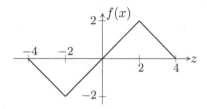

W.22 Fourier coefficients often exhibit regularities. Plot curves which result from taking the first ten or so [non-zero] terms in the following. [$c_0 = 0$ in all cases.]

(a) $c_n = 0$, $s_n = \frac{1}{n}$ (b) $c_n = \frac{1}{n}$, $s_n = 0$

(c) $c_n = 0$, $s_n = \frac{1}{n^2}$ (d) $c_n = \frac{1}{n^2}$, $s_n = 0$

(e) $c_n = 0$, for n odd, $c_n = \frac{(-1)^{n/2}}{n}$, for n even, $s_n = 0$

(f) $c_n = 0$, $s_n = 0$ for n even, $s_n = \frac{(-1)^{\frac{n-1}{2}}}{n}$, for n odd

W.23 Reconsider the Three-Phase Miracle example in Chapter 24 using the waves: $A\sin(k\,x - \omega\,t + \pi/6)$, $A\sin(k\,x - \omega\,t + 5\,\pi/6)$, and $A\sin(k\,x - \omega\,t + 3\,\pi/2)$. Comment.

W.24 [See also W.25 and W.70.] Recollect the Three-Phase Miracle example in Chapter 24, and construct an analogous four-phase miracle.

W.25 [See also W.24.] Recollect the Three-Phase Miracle example in Chapter 24, and construct an analogous five-phase miracle.

W.26 Two musicians simultaneously play what each hopes is high C (C5's frequency is approximately 523 Hz) and both hear beating at a frequency of 3 Hz. Assuming that one of the instruments is precisely in tune, ascertain the possible frequencies of the other.

W.27 A piano tuner listens carefully to a tuning fork calibrated to vibrate at 264.13 Hz and to the piano string corresponding to Middle C. The combined tone beats at 2.50 Hz. Determine the two possible frequencies at which the string might be vibrating.

W.28 Light steel wire with mass per unit length 0.003 kg/m is fixed at both ends and held at a tension of 120 N. Compute the speed with which transverse waves move on this wire.

W.29 A homogeneous harp string has mass $M = 1.5\,\mathrm{g}$ and length 1.5 m. The string is fixed at both ends and is held under tension $T = 10\,\mathrm{N}$. Compute   (a) the speed with which transverse waves move and   (b) the frequency of the third harmonic   on this string.

W.30 Repeat W.29 with tension 3.6 N.

W.31 A uniform pipe filled with water is 5 m long. Water has density 1000 kg/m^3 and bulk modulus $B = 2.25\,\mathrm{GPa}$. Determine the frequency of the fifth harmonic in the pipe when the pipe has   (a) one open and one closed end and   (b) both ends closed.

W.32 Repeat W.31 for the third harmonic.

W.33 A pipe open at one end and capped at the other has overtones at 440 Hz (A4) and 1320 Hz (approx. E6). Propose several possibilities for the number of overtones which lie between these notes.

W.34 Repeat W.33 when the pipe is open at both ends.

W.35 Sound is observed to travel at 2250 m/s through a sample of polycarbonate material with density 1.2 g/cm^3. Estimate the bulk modulus of this material.

W.36 A neutron star has density somewhat greater than that of an atomic nucleus, or roughly 5×10^{17} kg/m³. The bulk modulus of nuclear matter is approximately 5×10^{31} Pa. Estimate the speed at which sound might travel in a neutron star. Comment.

W.37 Gaseous helium with density 0.1788 kg/m³ completely fills a pipe of radius $R = 5$ cm and length $L = 12.5$ m. The bulk modulus of helium is approximately 170 kPa. The pipe is capped at both ends. Determine the fundamental frequency of this system.

W.38 Very cold air with density 1.3 kg/m³ fills a pipe of radius $R = 5$ cm and length $L = 1.100$ m which is open at both ends. The bulk modulus of this air is 141.6 kPa. Determine the fundamental frequency of this system.

W.39 A string of length $L = 125$ cm and constant mass per unit length $\mu = \frac{1}{10}$ g/cm is fixed at both ends and maintained at tension $T = 225$ N.

(a) Determine the speed of transverse waves on such a string.

(b) Ascertain the (i) wavelength and (ii) frequency of the fifth harmonic.

W.40 An air-filled pipe of length $L = 1548$ cm is closed at one end. The density and bulk modulus of air are 1.2 kg/m³ and 1.42×10^5 Pa.

(a) Determine the speed of longitudinal waves in such a pipe.

(b) Ascertain the (i) wavelength and (ii) frequency of the fifth harmonic.

W.41 A rope of length L and constant mass per unit length μ hangs vertically downward from an anchor point. Suspended from the lower end of the rope is a block of mass M. [Assume that the rope does not stretch, and that the acceleration due to gravity, g, is constant.]

(a) Compute the tension in the rope as a function of position, measured positive downward, below the anchor point.

(b) Determine the speed of waves propagating in the rope as a function of position below the anchor point.

(c) Suppose that the block is suddenly jerked a horizontal distance δx and returned to its equilibrium position directly below the anchor. This produces a pulse of height δx in the rope. As the pulse propagates up the rope, how do (i) its vertical extent and (ii) its height change?

W.42 A transverse harmonic wave with amplitude 0.8 cm and frequency 440 Hz propagates in a very long wire. The wire has mass per unit length $\mu = 1.5$ g/m and experiences tension of 3.75 N. Compute the average power borne by the wave as it moves along the wire.

W.43 Repeat W.42 for a string with mass per unit length 25 g/m and tension 10 N, bearing a wave with frequency 1.5 Hz and amplitude 15 cm.

W.44 Three transverse waves with the properties listed below propagate in a single medium. Rank these waves in terms of the average power each conveys [from greatest to least].
$$\psi_A = 5 \sin(3\pi x - 4\pi t) \quad \psi_B = 3 \sin(4\pi x - 5\pi t) \quad \psi_C = 4 \sin(5\pi x - 3\pi t)$$

W.45 A longitudinal harmonic sound wave with pressure amplitude 4.0×10^{-4} Pa and angular frequency 1643.87 rad/s [Middle C] propagates at 344 m/s through an air-filled channel of cross-sectional area 10 cm². Compute the average power borne by the wave through the channel.

W.46 Repeat W.45 for $P_0 = 6 \times 10^{-2}$ Pa, $\omega = 3217$ rad/s, and $\mathcal{A} = 2.53$ m².

W.47 Three longitudinal waves with the properties listed below propagate in a waveguide.
$$\{\psi_A : P_{0A} = 3 \times 10^{-5} \text{ Pa}, \nu_A = 400 \text{ Hz}\} \quad \{\psi_B : P_{0B} = 5 \times 10^{-4} \text{ Pa}, \nu_B = 300 \text{ Hz}\}$$
$$\{\psi_C : P_{0C} = 4 \times 10^{-3} \text{ Pa}, \nu_C = 500 \text{ Hz}\}$$
Rank these waves in terms of the average power each conveys [from greatest to least].

W.48   An incident harmonic wave, $\psi_a(t,x)\big|_{x<0}$, with angular frequency ω_a and wavenumber k_a, moves with speed c_a toward $x = 0$ from the left. There the medium changes abruptly. The transmitted wave, $\psi_b(t,x)\big|_{x>0}$, has angular frequency ω_b, wavenumber k_b, and speed c_b, where $c_b = \frac{4}{3}\,c_a$. The amplitude of the incident wave is 1 [in well-chosen units]. Determine the   (i) angular frequency and   (ii) amplitude   of the   (a) transmitted and   (b) reflected   waves.

W.49   Repeat W.48 when $c_b = \frac{5}{8}\,c_a$.

W.50   Two wave media are fused at $x = 0$. In Medium a, where $x < 0$, an incident rightmoving wave is given by $\psi_i(t,x) = A_i \cos(k_a\,x - \omega_a\,t)$. A rightmoving transmitted wave, $\psi_t(t,x) = A_t \cos(k_b\,x - \omega_b\,t)$, propagates in Medium b, where $x > 0$. A reflected wave, $\psi_r(t,x) = A_r \cos(k_a\,x + \omega_a\,t)$, propagates to the left in Medium a.

(a) Derive the relation among the amplitudes of the three harmonic waves arising from the constraint that the wave be continuous at the medium boundary, *i.e.*, at $x = 0$.

(b) Derive the relation among the amplitudes stemming from the demand that the wave be differentiable at the boundary.

(c) Using the relations from (a–b), specify the   (i) transmitted and   (ii) reflected amplitude [in terms of the incident amplitude].

W.51   Determine the wavelength and frequency of pulse-trains received by   (a) leeward and   (b) windward   observers [remaining at rest with respect to the medium] when the emitter is moving [toward leeward, away from windward] at speed $c/3$. Use temporal and spatial units of periods and wave-periods in your analysis.

W.52   Determine the wavelength and frequency of pulse-trains received from a stationary emitter by   (a) advanced and   (b) retarded   observers moving at speed $c/3$. [The advanced observer approaches the emitter, while the retarded observer recedes.] Use temporal and spatial units of periods and wave-periods in your analysis.

W.53   Repeat W.51 when the emitter is moving at speed $c/4$.

W.54   Repeat W.52 when the receivers are moving at speed $c/4$.

W.55   Repeat W.51 when the emitter is moving at speed $c/5$.

W.56   Repeat W.52 when the receivers are moving at speed $c/5$.

W.57   Ascertain the wavelength and frequency of pulse-trains received by   (a) leeward and   (b) windward   observers [remaining at rest with respect to the medium] when the emitter is moving [toward leeward, away from windward] at speed c/n, where $n > 1$. Use temporal and spatial units of periods and wave-periods in your analysis.

W.58   Ascertain the wavelength and frequency of pulse-trains received from a stationary emitter by   (a) advanced and   (b) retarded   observers moving at speed c/n, where $n > 1$. [The advanced observer approaches the emitter, while the retarded observer recedes.] Use temporal and spatial units of periods and wave-periods in your analysis.

W.59   Ascertain the Doppler-shifted frequency and wavelength when the emitter is moving with speed $c/2$ and the observer is   (a) receding and   (b) approaching   at speed $c/2$.

W.60   Ascertain the Doppler-shifted frequency and wavelength when the emitter is moving with speed $c/3$ and the observer is   (a) receding and   (b) approaching   at speed $c/4$.

W.61   Ascertain the Doppler-shifted frequency and wavelength when the emitter is moving with speed $c/4$ and the observer is   (a) receding and   (b) approaching   at speed $c/3$.

W.62   Ascertain the Doppler-shifted frequency and wavelength when the emitter is moving with speed $c/4$ and the observer is   (a) receding and   (b) approaching   at speed $c/4$.

W.63   A racing ambulance passes by your vehicle [which is stopped and pulled to the side of the road]. While the ambulance is approaching, the apparent frequency of the siren is 864 Hz. As the ambulance recedes, the siren sounds at 672 Hz. On this particular day the speed of sound in air is 344 m/s. Determine the frequency of the siren and the speed of the ambulance.

W.64 PK rowed a boat up and down a deep lake at a constant speed of 3 m/s with respect to the water. Water waves (assumed to be effectively transverse) with amplitude 8 cm and approximate wavelength 63 cm were present on the lake. Estimate the frequency at which the waves lapped at the bow when the boat was moving   (a) with and   (b) against   the waves.

W.65 Repeat W.64 when PK is waterskiing up and down the lake at 15 m/s.

W.66 Laser light with wavelength 400 nm is incident upon two narrow slits separated by 150 μm. How widely spaced are the maxima in the interference pattern produced on a flat screen perpendicular to the laser beam and 3.0 m from the slits?

W.67 Repeat W.66 for laser light with wavelength 600 nm.

W.68 Each of the slits in W.66 is 12 μm wide. One of the slits is covered. How far apart on the screen are neighbouring diffraction minima?

W.69 Repeat W.68 when the laser has wavelength 600 nm, as in W.67.

W.70 [See also W.24.] Reinterpret the four-phase miracle, W.24, in terms of phasors.

A

Acoustics and Optics Problems

A.1 The range of human hearing spans twelve orders of magnitude in energy intensity. For comparision, let's imagine that our sense of touch were equally acute over a corresponding range of sizes.

(a) Suppose that 1 mm corresponds to the largest ridge that we can feel with a stationary fingertip. [Detecting larger ridges would require that we move our fingertip.] What sorts of objects might we encounter if our fingertips were able to sense ridges of size 10^{-12} mm $= 10^{-15}$ m?

(b) Suppose that 1 mm corresponds to the smallest ridge that we can feel with a moving elbow. [Detecting smaller ridges would require that we use our fingers.] What sorts of objects are of size 10^{12} mm $= 10^9$ m $= 10^6$ km?

A.2 A reference sound has energy intensity 3.0×10^{-9} W/m^2. Determine the energy intensity of a sound which is (a) $+6$ dB [louder], (b) $+17$ dB [louder], (c) -23 dB [softer].

A.3 A source of sound with known energy intensity is determined to have loudness 65 dB. Determine the new loudness if the energy intensity is (a) diminished and (b) increased by a factor of (i) 10, (ii) 5, and (iii) 2.

A.4 An unmuffled machine produces 90 dB sound loudness in the vicinity of its operator. Prolonged exposure to noise of this intensity is considered in some jurisdictions to be an occupational hazard requiring mitigation. How much of the energy intensity must be absorbed by a sound-muffling system in order to reduce the loudness perceived by the operator to 70 dB?

A.5 Employ the crude—logarithmically equidistant—model to estimate the frequencies of the notes { do, re, me, fa, so, la, ti, do }, spanning the octave from middle to high C, given that la is 440 Hz.

A.6 A simple logarithmic model for frequency sensitivity [the counterpart to the decibel description of loudness] would imply that a measure of perceived frequency difference is

$$\Delta v_{12} = 100 \log \left[\frac{v_2}{v_{\text{ref.}}} \right] - 100 \log \left[\frac{v_1}{v_{\text{ref.}}} \right] = 100 \log \left[\frac{v_2}{v_1} \right] .$$

(a) Middle C [approximately 261.6 Hz] and A440 differ in frequency by 178.4 Hz. Compute Δv for this pair of notes.

(b) Consider a note lying 178.4 Hz below the highest C on the piano keyboard [approx. 4186 Hz]. Compute Δv for this pair of notes.

(c) Comment.

A.7 Adult men's vocal tracts produce sounds whose fundamental frequencies typically fall within the range from 85 Hz to 180 Hz. For adult women (and larger children) the range is shifted higher: $165 \rightarrow 255$ Hz. Suppose that a man with a fundamental frequency of 120 Hz and a woman with fundamental frequency of 200 Hz are carrying on a conversation.

(a) Assuming that they are speaking directly to one another (under good listening conditions) each can hear the fundamental frequency and overtones up to about 15 kHz. (i) How many tones is the man able to hear in the woman's speech? (ii) How many tones is the woman able to hear in the man's speech?

(b) Suppose that they are conversing on a telephone which only transmits sounds with frequencies between 300 and 3000 Hz.   (i) Are the fundamental frequencies transmitted?   (ii) How many tones is the man now able to hear in the woman's speech?   (iii) How many tones is the woman now able to hear in the man's speech?

[In telephone speech large swathes of the signal are omitted. However, we are able to reconstruct much of the missing information (and even recognise voices!) by inference.]

A.8
A cafe has sixteen small round tables crammed into a 6 m by 6 m space. The centres of nearest-neighbour tables are 1.5 m apart. The centres of those along the periphery are 0.75 m from the wall. The cafe is very popular and every table is occupied by a couple deeply engaged in quiet heartfelt conversation:

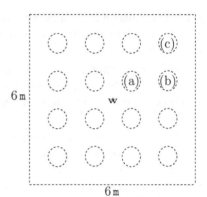

> ... and then The Doctor and Clara ...
> ... Leafs failed to make it to the playoffs ...
> ... Do you know how strongly I am attracted to ...
> ... calculation has me thoroughly stymied ...

A waiter [w in the figure] stands at the cafe's centre.

(a) Two patrons at one of the inner tables, seated 1 m apart, are conversing. Each speaks softly with an apparent loudness of 40 dB [as heard by his interlocutor].   (i) How loud does this conversation seem to the waiter who is the same distance from the speaker as is the listener? [Assume that the sound intensity is attenuated as the inverse-square of the distance to its source. We neglect the effects of reflections of sound from the walls, floors, and other hard surfaces and absorption of sound by softer materials (and people, too) present.]   (ii) There are four such conversations going on. How loud do all of these together seem to the waiter?

(b) A pair of patrons is seated at one of the tables along the wall, at approximate distance $\sqrt{5}$ m from the waiter.   (i) Under the same assumptions as in part (a), ascertain the loudness of this conversation at the waiter's position.   (ii) Ascertain the loudness (determined at the centre of the room) of the eight conversations taking place at these tables.

(c) A pair of patrons is seated at one of the tables in the corner of the cafe, at an approximate distance of $\sqrt{8}$ m from the waiter.   (i) Under the same assumptions as above, ascertain the loudness of this conversation at the waiter's position.   (ii) Ascertain the loudness (determined at the centre of the room) of the four conversations taking place at these tables.

(d) Estimate the total sound intensity in the centre of the cafe. Comment.

A.9   The average solar intensity [measured above the Earth's atmosphere] is approximately 1360 W/m^2. Given that the Earth's orbit is approximately circular, with mean radius 1.5×10^{11} m, estimate the power output by the Sun.

A.10   The average solar intensity [measured at the Earth's surface] is approximately 700 W/m^2. How much solar energy falls upon an area of 1/2 square metre in the course of a 6 hour exposure on a sunny day?

A.11   The work function for caesium is $W_{Cs} \simeq 2.1$ eV. The effective area presented by a caesium atom is roughly $a_0 \simeq 30 \times 10^{-20}$ m^2. [This is huge! Caesium is a very large atom.] Two monochromatic light sources each have power 0.1 mW uniformly distributed over a beam of cross-sectional area 5 mm^2. The light waves in one beam have wavelength 250 nm, the other

500 nm. The beams shine directly (at normal incidence) on a flat surface of liquid caesium (at room temperature).

(a) (i) Determine the rate at which the light energy from either beam impinges upon a caesium atom on the top surface. (ii) Assuming that all of this energy is absorbed and transferred to a single electron, determine the time which must elapse for an amount of energy equal to the work function to be collected.

(b) Determine whether or not either of these beams is capable of liberating electrons from caesium according to the Einsteinian analysis of the photoelectric effect.

A.12 The work function for platinum is $W_{\mathrm{Pt}} \simeq 6.35\,\mathrm{eV}$. The effective area presented by a platinum atom is roughly $a_0 \simeq 10 \times 10^{-20}\,\mathrm{m}^2$. Analyse the effects of the same two light sources as in A.11 shining directly on a platinum surface.

A.13
A concave mirror has constant radius of curvature $R = 5\,\mathrm{m}$. An object is placed before the mirror, along the axis, at a point which is 8 m from the centre of the mirror. The object extends 50 cm outward from the axis.
Determine the optical properties of the image of the object by employing the master equations, and by sketching the canonical rays on a copy of the figure.

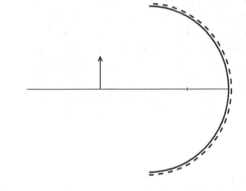

A.14
An object lies $p = 50\,\mathrm{cm}$ along the optical axis, directly in front of a concave mirror with radius of curvature $R = 1.5\,\mathrm{m}$.
(a) Employ the master equations to determine the location and properties of the image of the object formed by the mirror. (b) Confirm your results with a sketch.

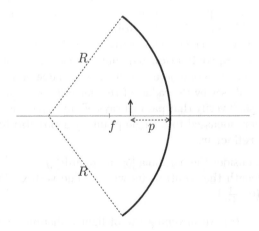

A.15 Repeat A.14 when the object lies at $p = 1.0\,\mathrm{m}$ along the optical axis of a concave mirror with radius of curvature $R = 1.5\,\mathrm{m}$.

A.16 Repeat A.14 when the object lies at $p = 250\,\mathrm{cm}$ along the optical axis of a concave mirror with radius of curvature $R = 1.5\,\mathrm{m}$.

A.17 Repeat A.14 when the object lies at $p = 0.25\,\mathrm{m}$ along the optical axis of a concave mirror with radius of curvature $R = 1.5\,\mathrm{m}$.

A.18
An object lies along the optical axis of a convex mirror. The mirror has radius of curvature $R = 1.5\,\mathrm{m}$. The distance from the object to the mirror is $p = 50\,\mathrm{cm}$.   (a) Employ the master equations to determine the location and properties of the image of the object formed by the mirror.   (b) Confirm your results with a sketch.

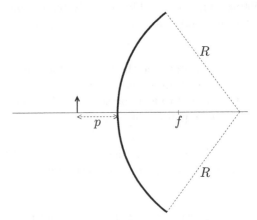

A.19   Repeat A.18 when the object lies at $p = 250\,\mathrm{cm}$ along the optical axis of a convex mirror with radius of curvature $R = 1.5\,\mathrm{m}$.

A.20
Pictured nearby is a convex mirror with uniform radius of curvature $R = 3\,\mathrm{m}$. An object $3\,\mathrm{m}$ from the mirror extends $0.3\,\mathrm{m}$ from the optical axis.   (a) Determine the location and optical properties of the image by drawing the three canonical rays on a copy of the figure. (b) Confirm your results by calculation using the optical master equations.

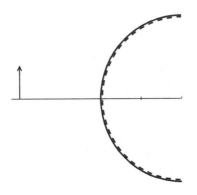

A.21
Spherical mirrors do not focus all widely separated paraxial light rays to a single point. This effect is called **spherical aberration**. [In Chapter 34 we implicitly adopted the **paraxial approximation** and limited ourselves to consideration of objects which were not large on the scale set by the radius of the mirror.] Here, we shall verify that parallel rays of any separation are focussed to a single point by a **parabolic reflector**.

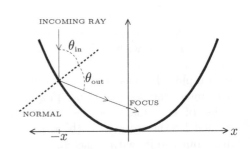

Consider the parabola [open upwards] $y = \alpha\,x^2$, where α is a parameter with units of inverse length that controls its width. The vertex of this parabola is at the origin; its focus is at $\left(0\,,\frac{1}{4\alpha}\right)$.

   (a) An incoming ray of light a distance x from the midline of the parabola is directed straight down, *i.e.*, with unit vector $(0\,,-1)$. Determine   (i) the point of intersection of the ray with the parabola and   (ii) the slope of the parabola at this point.

   (b) Characterise the ray directed perpendicular to the parabola at the point of intersection. This is the normal direction.

   (c) Use the dot product to obtain the **cosine** of the angle lying between the normal direction and the incoming ray. [WLOG fix the sign to be positive.]

(d) Determine the vector from the point of intersection of the incoming ray and the parabola to the parabola's focal point.

(e) Use the dot product to obtain the **cosine** of the angle lying between the normal direction and the ray in (d).

(f) Compare your results for (c) and (e). Comment.

A.22 The Floating Coin Trick Explained[1]

Take two identical glasses and place a coin in each. Fill one of the glasses with an amount of water sufficient to completely cover the coin. When viewed from above, the wet coin appears closer than the dry coin, *i.e.*, it appears to be floating in the water. From the side, both coins are seen to lie on the bottom.

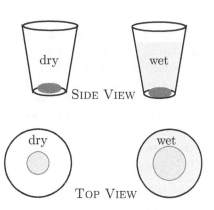

We denote the actual depth of the coin (beneath the surface of the water) by d_A and its apparent depth by d_a. Binocular rays showing the apparent depth are drawn in the adjacent figure. The angle at which these rays diverge from the normal in air are labelled θ_1, while the angle at which light from the coin meets the surface is θ_2. That the refractive index of water is greater than that of air implies $\theta_1 > \theta_2$. The distance from each emerging ray to the point on the surface directly above both object and image is dubbed l.

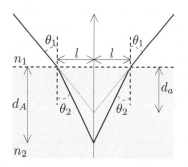

Justify each of the following statements.

(a) The small angle approximation to SNELL'S LAW is $n_1 \theta_1 \simeq n_2 \theta_2$.

(b) For small angles, $\theta_2 \simeq \frac{l}{d_A}$.

(c) For small angles, $\theta_1 \simeq \frac{l}{d_a}$.

(d) Hence, the apparent distance is $d_a = \frac{n_1}{n_2} d_A$. Comment.

---

[1] Canadian coins are ideally suited for this trick. The dollar coin—the loonie—bears the image of a loon (an aquatic bird). The nickel, the quarter, and the two-dollar coin sport a beaver, a moose, and a polar bear (all are aquatic mammals) on their reverse sides. The Bluenose (a sailing ship) graces the dime. [Perhaps this one is ill-advised.] The penny has been removed from circulation.

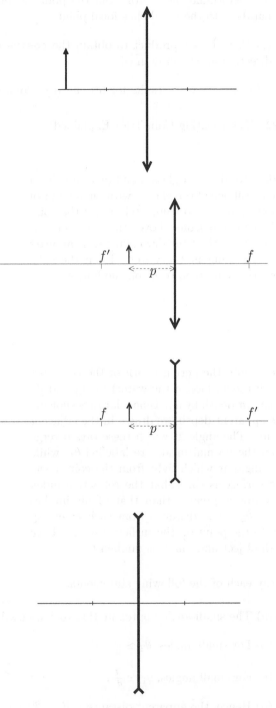

A.23
A large-diameter, thin, symmetric double-convex lens has focal length $|f| = 4\,\mathrm{m}$. An object is placed $8\,\mathrm{m}$ from the midpoint of the lens. The object extends $50\,\mathrm{cm}$ outward from the axis. Determine the optical properties of the image of the object   (a) by employing the master equations, and   (b) by sketching the canonical rays on a copy of the figure.

A.24
An object lies at $p = 50\,\mathrm{cm}$ along the optical axis of a symmetric double-convex lens with focal length $f = 0.75\,\mathrm{m}$.   (a) Employ the master equations to determine the location and properties of the image of the object formed by the lens.   (b) Confirm your results with a sketch.

A.25
An object lies at $p = 50\,\mathrm{cm}$ along the optical axis of a symmetric double-concave lens with focal length $f = 0.75\,\mathrm{m}$.   (a) Employ the master equations to determine the location and properties of the image of the object formed by the lens.   (b) Confirm your results with a sketch.

A.26

The magnitude of the focal length of a symmetric double-concave lens is $|f| = 0.5\,\mathrm{m}$. An object extending $30\,\mathrm{cm}$ vertically upward from the optical axis lies at a point $1\,\mathrm{m}$ to the right of the lens.
(a) Sketch a figure showing the object, and ascertain the position of the image by means of the three canonical rays.   (b) Confirm your results via the master equations.

A.27   Two thin double-convex lenses are placed in series along a common optical axis. The first lens has $f_1 = 20\,\mathrm{cm}$; the second has $f_2 = 15\,\mathrm{cm}$. The lenses are separated by $25\,\mathrm{cm}$. An object lies $60\,\mathrm{cm}$ in front of the first lens. Ascertain the location and properties of its image.

A.28   Repeat A.27 when the two thin lenses are separated by $5\,\mathrm{cm}$.

A.29   A thin double-convex lens and a thin double-concave lens are arranged in series and share a common optical axis. The first lens has $f_1 = 20\,\mathrm{cm}$; the second has $f_2 = -15\,\mathrm{cm}$.

The lenses are separated by 25 cm. An object lies 60 cm in front of the first lens. Ascertain the location and properties of its image.

A.30 Repeat A.29 when these thin lenses are separated by 5 cm.

A.31 Bertal and his classmates would like to measure the speed of light using a laser pointer, a plane mirror, a [deluxe] disco ball with 128 small plane mirrors glued along its equator, and a viewing screen. With the ball at rest, the apparatus is arranged so that a steady spot of laser light appears on the screen, as shown in the figure.

The distance from the ball to the plane mirror is denoted by L. We shall assume that L is sufficiently large [equivalently, that the incidence and reflection angles at the plane mirror are sufficiently small] that the distance through with the light travels is effectively equal to $2\,L$.

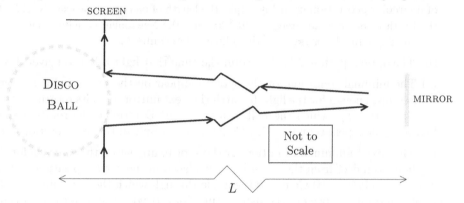

The idea behind the experiment is to rotate the ball at just the right angular speed that light which is reflected toward the plane mirror returns at the instant that a mirror on the ball is in position to redirect it to the proper spot on the screen.

○ This requires tuning the rotation frequency of the disco ball or adjusting the distance L.

○ An infinite number of frequency and path-length combinations redirect the beam to the original point on the screen. We seek the minimal solution.

○ **Q:** Isn't the light going to flicker back and forth [or at least switch on and off]

A: Yes. However, the phenomenon of visual persistence[2] smooths out the flickering. [The intensity will diminish once the ball begins rotating, since not all of the light from the laser pointer will reach the screen.]

(a) Assuming that the ball's equator is a regular 128-gon, through what minimum angle must the ball rotate to position a mirror to catch the returning light and redirect it to the point on the screen?

(b) The disco ball normally rotates at 15 RPM. (i) Convert 15 RPM to radians per second. (ii) At this rate, how long does it take the ball to rotate through the angle determined in (a)? (iii) Through what distance does light, moving at $c = 3 \times 10^8$ m/s, travel in this time interval? (iv) From (iii), determine the necessary distance to the mirror. Comment.

(c) By reconfiguring the gears that drive the disco ball, Bertal and his friends get it to rotate at 90000 RPM! [Don't try this at home.] (i) Convert 90000 RPM to radians per second. (ii) At this rate, how long does it take the ball to rotate through the angle determined in (a)? (iii) Through what distance does light, moving at $c = 3 \times 10^8$ m/s,

---

[2]Our eyes retain a visual signal for a brief period.

travel in this time interval?   (iv) From (iii), determine the necessary distance to the mirror. Comment.

(d) How might adding another [plane] mirror help?

A.32 [See also A.31.] Realising that the disco ball employed in A.31 cannot withstand the internal stresses that arise when it is rotated at 90000 RPM, Bertal casts his eyes heavenward for inspiration just in time to see a satellite pass overhead. Aha! Bertal and his classmates move the disco ball, screen, and laser to a point on the equator directly beneath a geostationary satellite. A shiny plate on the satellite will serve admirably as a mirror.

(a) (i) Using Kepler's Third Law [or a centripetal force argument], determine the radius of a circular geostationary orbit. [Take the length of one day to be exactly 24 h, the mass of the Earth to be approximately 6×10^{24} kg, and the Newtonian Gravitational constant to be roughly $6\frac{2}{3} \times 10^{-11}$ N·m²/kg².]   (ii) Thence determine L.

(b) The optical path is $2L$. Ascertain the time that light takes to travel this distance.

(c) The minimal condition for the spot to appear on the screen is for the ball to have rotated just enough for the light to catch the next mirror upon its return.   (i) Compute the angle through which the ball rotates.   (ii) Determine the rotation speed of the ball in radians per second.   (iii) Convert your result in (ii) to revolutions per minute.

(d) The next-to-minimal condition for the spot to appear on the screen is for the returning light to reflect from the second mirror placed in the correct position by the rotation of the disco ball.   (i) Compute the angle through which the ball rotates.   (ii) Determine the rotation speed of the ball.   (iii) Convert your result in (ii) to revolutions per minute.

(e) The nominal rotation speed for the disco ball [in normal operation] is around 15 RPM. If the rotation speed can only be changed by a maximum of 10% from its operational value, how might Bertal and his friends still perform this experiment?

T

Thermodynamics Problems

Useful Data

Thermal Expansion Coefficients for Select Elements				
ALUMINIUM	α_{Al} =	2.3×10^{-5} c^{-1}	β_{Al} =	6.9×10^{-5} c^{-1}
COPPER	α_{Cu} =	1.7×10^{-5} c^{-1}	β_{Cu} =	5.1×10^{-5} c^{-1}
GOLD	α_{Au} =	1.4×10^{-5} c^{-1}	β_{Au} =	4.2×10^{-5} c^{-1}
MERCURY	α_{Hg} =	6.1×10^{-5} c^{-1}	β_{Hg} =	1.82×10^{-4} c^{-1}

Thermal Expansion Coefficients for Select Materials				
Diamond	$\alpha_{C\text{-}d}$ =	1×10^{-6} c^{-1}	$\beta_{C\text{-}d}$ =	3×10^{-6} c^{-1}
Ethanol	α_{alc} =	2.5×10^{-4} c^{-1}	β_{alc} =	7.5×10^{-4} c^{-1}
Glass	α_{glass} =	8×10^{-6} c^{-1}	β_{glass} =	2.5×10^{-5} c^{-1}
Steel	α_{steel} =	1.2×10^{-5} c^{-1}	β_{steel} =	3.6×10^{-5} c^{-1}

Specific Heat Capacities [J/(kg·C)] for Select Elements				
ALUMINIUM	C_{Al} =	897	COPPER	C_{Cu} = 385
IRON	C_{Fe} =	450	LEAD	C_{Pb} = 129

Specific Heat Capacities [J/(kg·C)] for Select Materials				
Air	C_{air} =	1000	(Red) Cedar	C_{cedar} = 1500
Ethanol	C_{alc} =	2400	Glass	C_{glass} = 800
Granite	C_{gran} =	800	Pasta	C_{pasta} = 1800
Water (ice)	$C_{H2O,i}$ =	2100	Water (liquid)	C_{H2O} = 4186
Water (steam)	$C_{H2O,s}$ =	2080		

Latent Heats [kJ/kg] for Water	
Water	$L_{F,\text{H2O}} = 333$ $\quad$ $L_{V,\text{H2O}} = 2256$

Thermal Conductivities [W/(m·C)] for Select Elements			
ALUMINIUM	$k_{\text{Al}} = 220$	COPPER	$k_{\text{Cu}} = 390$
IRON	$k_{\text{Fe}} = 75$	LEAD	$k_{\text{Pb}} = 35$

Thermal Conductivities [W/(m·C)] for Select Materials			
Air (dry)	$k_{\text{air}} = 0.025$	Concrete	$k_{\text{conc}} = 0.8$
Glass	$k_{\text{glass}} = 1$	Plywood	$k_{\text{plywd}} = 0.08$
Polyurethane Foam	$k_{\text{polyu}} = 0.02$	Polystyrene	$k_{\text{polys}} = 0.03$

Pertinent Constants	
Boltzmann Constant	$k_B = 1.38 \times 10^{-23} \; \frac{\text{J}}{\text{molecule} \cdot \text{K}}$
Universal Gas Constant	$R = 8.31 \; \frac{\text{J}}{\text{mole} \cdot \text{K}}$
Stefan–Boltzmann Constant	$\sigma = 5\frac{2}{3} \times 10^{-8} \; \frac{\text{W}}{\text{m}^2 \cdot \text{K}^4}$

T.1 PK's recipe for pumpkin pie specifies that the pie be baked at 425 °F for fifteen minutes, and then at 350 °F for 35–40 minutes. PK's oven is calibrated in 5 C increments. To what temperatures should he set his oven to best approximate the recipe's instructions?

T.2 A seldom employed temperature unit is the **Rankine**, °R. Zero on the Rankine scale corresponds to absolute zero, and each increment or Rankine is equal to 1 °F. How does the numerical value of a temperature expressed in Rankine compare with its kelvin equivalent?

T.3 The three temperature scales C, K, and °F exhibit coincidental agreements.

(a) Find the temperature whose numerical values on the Celsius and Fahrenheit scales coincide. [Set 32 °F equal to 0 C.]

(b) Find the temperature whose numerical values on the kelvin and Fahrenheit scales coincide.

(c) Argue that no such coincidence occurs for temperatures measured in kelvin and Celsius.

T.4 Peter the Great (1672–1725), Tsar and Emperor of Russia, possessing a restlessly inquisitive disposition and striving to advance his state's interests, travelled extensively. During a Siberian sojourn in his early 20s, Peter secretly uncovered the basic principles of thermometry. Many years later, during his Grand Embassy to Western Europe, Tsar Peter met with Gabriel Fahrenheit (then 11 years old), informed him of his findings, and inspired him to study thermodynamics. Fahrenheit's diary/lab-book entry on that date reads as follows.

Jan. 14, 1698   Met with Tsar Peter - he is very tall - who recounted his system of measuring temperatures in "Retep."

    0 Retep   Water, poured from a glass, freezes before it reaches the ground. (Yakutsk, January 29, 1693)

100 Retep   Mosquitoes fall from the air owing to extreme heat prostration. (Irkutsk, July 07, 1693)

A later entry shows how Fahrenheit re-interpreted Tsar Peter's data.

Mar. 14, 1725   News of Tsar Peter's death arrived this week, spurring me to recollect and review my notes from our conversation on 1698-01-14. Replicating Peter's experiments using yet-to-be-discovered refrigeration techniques for the Yakutsk data, and a sauna for the Irkutsk data, reveals the approximate correspondence: 0 Retep $\iff$ $-67\,°F$ 100 Retep $\iff$ $113\,°F$!

From Fahrenheit's re-interpretation of Tsar Peter's system, determine conversion formulae from Retep to   (i) degrees Fahrenheit,   (ii) Celsius, and   (iii) kelvin.

T.5   An electrical wire consists of a copper core wrapped with a flexible and stretchable insulating material. Suppose that the wire has length $L_0 = 20\,m$ at temperature $20\,c$. Determine the wire's length when its temperature is: (i) $50\,c$ and   (ii) $-10\,c$.

T.6   [See also T.20.] What change in temperature is needed to change the length of a slab of copper by a factor of 0.0051 (*i.e.*, slightly more than one-half of one percent)? By what approximate factor does the volume of the slab change?

T.7   Electrical service to homes is often provided through a length of copper wire extending from a local step-down transformer to a connection point high up on an exterior wall. On a fine spring day, when the temperature was $5\,c$, one such wire was measured to be exactly $20\,m$ long.

(a) What length does the wire have on a scorchingly hot summer's day when the outside temperature is $40\,c$?

(b) What is the length of the wire on a bitterly cold winter's day when it is $-35\,c$?

T.8   A square table with edge length $1.3\,m$ has legs which are exactly $0.5\,m$ long at $20\,c$. The legs on one end of the table are made of diamond, while those on the other are made of gold. By what angle and in which direction is the table tilted at   (a) $40\,c$ and   (b) $0\,c$?

T.9   An amount of mercury [Hg] held in a glass cylinder with cross-sectional area equal to $0.01\,cm^2$ occupies two cubic centimetres at $T = 20\,c$. A line is drawn to note the topmost extent of the mercury column in the glass cylinder.

(a) Determine the increase in the volume of the mercury should the temperature increase to $37\,c$.

(b) Determine the increase in the interior volume of the glass cylinder should its temperature increase to $37\,c$.

(c) What excess volume of mercury must then lie above the line on the cylinder?

(d) At what height above the mark is the top of the mercury when the temperature of the system is $37\,c$? [Assume that the cross-sectional area of the segment of glass tubing into which the mercury rises remains effectively constant.]

T.10   Re-analyse the Thermal Expansion of Volume example appearing in Chapter 37 to take into account the increase in the cross-sectional area of the segment of tube into which the excess mercury rises. [Note that this particular type of glass has greater response to temperature shifts than does the glass in T.9.]

(a) Let us approximate the areal thermal expansion coefficient of the glass by the geometric mean of the lineal and volume coefficients, *i.e.*, $\tilde{\alpha} \simeq \sqrt{\alpha_{\text{glass}} \beta_{\text{glass}}}$, using $\beta_{\text{glass}} = 2.7 \times 10^{-5}$ c^{-1}, as quoted in Chapter 37, and $\alpha_{\text{glass}} = 0.833 \times 10^{-5}$ c^{-1}. Compute the value of $\tilde{\alpha}$.

(b) Compute the cross-sectional area of the tube at $T = 100\,c$.

(c) Recompute the excess volume of mercury thrust above the line marking $0\,c$.

(d) Comment.

T.11   Apply the correction described in T.10 to the system analysed in T.9.

T.12   A glass vessel of mass 0.5 kg is filled to the brim with 1.5 L of liquid ethanol at initial [ambient] temperature $T_0 = 20\,c$. The density of ethanol is approximately $800\,\text{kg/m}^3$.

(a) Suppose that the ambient temperature is slowly raised to $T = 40\,c$.   (i)   Does the vessel overflow?   (ii)   If it overflows, determine the amount of ethanol spilled; otherwise determine how much additional ethanol must be supplied to bring the liquid level back to the brim.   (iii)   Determine the amount of heat that was needed to raise the temperature of the vessel and its contents from $T = 20\,c$ to $T = 40\,c$.

(b) Repeat (a) supposing instead that the temperature is reduced from $T = 20\,c$ to $0\,c$.

T.13   A solid cube of steel, with edge length equal to 1 cm at 20 c, just barely fits into an aluminium frame. Describe in words how the cube fits at   (a)   cooler $(T < 20\,c)$ and (b)   warmer $(T > 20\,c)$ temperatures.

T.14   Two hundred grams of 70 c water are added to 300 g of water at 40 c. Assuming no heat flows into or out of the water, ascertain the temperature of the mixture.

T.15   Compute the amount of heat needed to raise the temperature of 150 g of ice from $-20\,c$ to $-10\,c$.

T.16   Compute the amount of heat needed to raise the temperature of 150 g of liquid water from 10 c to 20 c.

T.17   Compute the amount of heat needed to convert 150 g of ice (solid water), initially at a temperature of $-20\,c$, to liquid at 20 c.

T.18   Compute the amount of heat needed to convert 150 g of ice, initially at $-20\,c$, to steam at 120 c.

T.19   Determine the amount of thermal energy needed to convert 100 g of liquid water at 25 c to 100 g of steam at 110 c.

T.20   [See also T.6.] The slab of copper considered in problem T.6 has mass 3 kg. What quantity of heat must be added or subtracted to effect the 0.51% change in the slab's length?

T.21   Preparing dinner one day, PK heated 1.5 L [1.5 kg] of water to a boil, *i.e.*, 100 c. Before a significant amount of water had evaporated, PK tossed in 400 g of room-temperature, *i.e.*, 25 c, pasta.

(a) Assuming that the pasta and water quickly equilibrate [allowing us to neglect the effects of heat input from the stove burner and heat loss to the surrounding air], determine the equilibrium temperature of the water-pasta system.

(b) The average net heat flow into the pot is roughly 2000 W. Estimate the time that will elapse before the water again begins to boil.

T.22   A sample of pure aluminium with mass $m = 20\,\text{g}$ is held at temperature $T = 50\,c$. Determine the amount of thermal energy that must be added to, or removed from, the sample to bring its temperature to: (a)   100 c,   (b)   200 c,   (c)   0 c,   (d) $-100\,c$.

T.23   Ascertain the factor which converts specific heat capacities expressed in $c/(\text{kg·K})$ to $c/(\text{g·c})$.

T.24 An isolated mixture of two substances is at temperature T_i. A quantity of heat, ΔQ, is added to the system, which then equilibrates at final temperature T_f.

(a) Is this process best described as SERIES or PARALLEL?

(b) Identify the (i) SUM and (ii) SAME constraints.

T.25 Materials A and B have specific heat capacities $C_A = 2000\,\text{J/kg·C}$ and $C_B = 4000\,\text{J/kg·C}$. Samples of A with mass $M_A = 500\,\text{g}$, and of B with $M_B = 1500\text{g}$, are both initially at $T = 20\,\text{C}$.

(a) What would the temperature of sample A become were 6000 J of heat added to it?

(b) What would the temperature of sample B become were 6000 J of heat added to it?

(c) Suppose that the two samples in (a) and (b) are combined and thermally isolated. (i) At what temperature does the mixture equilibrate? (ii) What is the effective specific heat capacity of the mixture? (iii) What would the temperature of the mixture be should 12000 J of heat be removed? Comment.

T.26 *Double, double toil and trouble; Fire burn and cauldron bubble* Macbeth, Act 4, scene 1

(a) In a stage production of *That Scottish Play*, the director originally proposed using a witches' cauldron made of 9 kg of cast iron, containing 6 kg of water (with colouring and thickening agents), in the cavern scene which begins Act IV. Eerie wisps of steam form when the water in the cauldron is about to boil. The "fire" beneath the cauldron was to consist of twenty-five 100 W light bulbs.

(i) How much heat energy is required to warm the cauldron from 20 c to 100 c?

(ii) Suppose that the foul contents of the cauldron have a net specific heat capacity of $C_{\text{Foul}} = 4200\,\text{J/(kg·C)}$. How much heat energy is required to warm the cauldron's contents from 20 c to 100 c? [Don't forget the *eye of newt and toe of frog*.]

(iii) Determine the total amount of energy needed to warm the cauldron and contents under the assumptions made above.

(iv) The light bulbs have a combined power of 2500 W. About 1/5 of the power is absorbed by the cauldron, while the rest goes elsewhere. How long will it take for the temperature of the cauldron and its contents to increase from 20 c to 100 c, assuming that all of the heat absorbed remains in the system and no evaporation occurs before the cauldron and its contents reach 100 c? [It is doubtful that any audience will sit still for this long.]

(b) Searching for an alternative way to stage the scene, the director opted to use 1.5 kg of dry ice (solid CO_2), at an initial temperature of -60 c. The dry ice sublimates, absorbing heat, and thus cools the surrounding air, causing nearby bits of water vapour to condense, creating a sinister ground-hugging fog. The latent heat of sublimation of dry ice is 571 kJ/kg under the conditions prevailing in the theatrical cauldron.

(i) Compute the energy required to sublimate the 1.5 kg of dry ice, neglecting any warming of the solid or gaseous CO_2.

(ii) The specific heat capacity of gaseous CO_2 is roughly 840 J/(kg·c) throughout the range of conditions [pressures and temperatures] found in the cauldron and the theatre. Compute the amount of heat needed to bring the sublimated CO_2 from its initial temperature of -60 c to room temperature, $+20$ c.

(iii) Determine the total amount of heat energy required to sublimate the dry ice and bring it to room temperature.

(iv) Before the show can go on, the lawyers inform the producer that there are insurance concerns about the use of the dry ice in the enclosed theatrical space. [Someone might suffer a chill, catch a cold, and sue!] We shall develop a rough model to assuage these lawyerly concerns.

Suppose that the theatrical space in which the play is to be performed has an open air volume of about $3000\,\mathrm{m}^3$. [Say 750 for the stage area and 2250 for the house.] Take the density of the air to be $\rho_{\mathrm{air}} = 1.2\,\mathrm{kg/m}^3$. Under the assumption that all of the CO_2 sublimates and warms to $+20\,C$ by drawing heat (instantaneously) from the air filling the venue, ascertain the change in the temperature of the air.

T.27 An ice cube of mass $20\,\mathrm{g}$, initially at $0\,C$, is dropped into an insulated container containing $100\,\mathrm{g}$ of liquid water at $20\,C$. If there is more than enough ice to reduce the temperature of the water to $0\,C$, compute the total amount of water available (including the melted ice) to slake your thirst; otherwise compute the final temperature of the water.

T.28 A restaurant patron put $0.2\,\mathrm{kg}$ of crushed water ice, at a temperature of $-5\,C$, into an insulating cup from a soft drink dispenser. [The ice was slightly colder before it left the machine, but its warming to $-5\,C$ cooled the inside surface of the cup.] Then, he quickly added $0.4\,\mathrm{kg}$ of soda pop at a temperature of $27\,C$. [This is warmer than room temperature because the refrigeration unit for the ice reservoir emits heat into its surroundings.] The pop is composed almost entirely of liquid water, and so we take its specific heat capacity to be [approximately] $4200\,\mathrm{J/(kg\cdot C)}$.

(a) Assuming that no heat is exchanged with the environment, determine   (i)   the heat required to warm the ice to $0\,C$ and   (ii)   the temperature of the pop in the cup once the ice reaches $0\,C$.

(b) Compute the amount of heat that must be absorbed by the ice to melt all of it.

(c) Determine the heat that would be emitted by the pop, were it to cool to $0\,C$.

(d) Comment on the results for (b) and (c).

(e) Do this problem again, this time beginning with $0.02\,\mathrm{kg}$ of ice in the cup.

T.29 Compute the rate at which heat is conducted through a (medium density) concrete wall under the following conditions. The wall area is $160\,\mathrm{m}^2$. Its thickness is $0.25\,\mathrm{m}$. The temperatures on the two sides are $20\,C$ and $35\,C$. The thermal conductivity of this grade of concrete is $k = 0.5\,\mathrm{W/(m\cdot C)}$.

T.30 The ski-lift attendants who stand by to shut off the lift in the event that a skier encounters difficulty usually have the option of standing on a concrete slab or a piece of wood (on top of the concrete). Which surface would you choose to stand on, and why?

T.31 Take the result obtained for the intermediate temperature in the series composition of two materials example appearing in Chapter 39 and compute the steady state rate at which heat flows through   (a) Medium 1,   (b) Medium 2, and   (c) their series composite.

T.32 A fireplace poker consists of a wrought iron rod, $50\,\mathrm{cm}$ long and $5\,\mathrm{cm}^2$ in cross-section, fused to a piece of cork $10\,\mathrm{cm}$ long (with the same cross-section). One end of the poker is in contact with embers at a temperature of $700\,C$, while the other end is on the hearth at $30\,C$. Estimate the temperature at the iron-cork boundary and answer the *burning* question: Is it safe to grasp the handle? [The thermal conductivities of (this grade of) iron and cork are 60 and $0.06\,\mathrm{W/(m\cdot C)}$, respectively.]

T.33 A child's toy oven employs a $60\,\mathrm{W}$ light bulb shining in a confined area to bake a small cake. Suppose that the cake has effective emissivity $\varepsilon = 1$ and total surface area $A = 200\,\mathrm{cm}^2$. Let's also assume that all of the energy released from the bulb is absorbed by the cake.

(a) Imagine that, while the cake is being irradiated by the bulb, it somehow also exchanges heat radiatively with the surrounding room, which is at effective temperature $300\,\mathrm{K}$. Compute the steady state temperature of the cake under these conditions.

(b) Instead of a 60 W bulb, let's use a 25 W bulb and line the walls of the cavity with a reflective coating which returns 80 % of the energy incident upon it. Compute the steady state temperature of the cake under these circumstances. [HINT: The coating reflects heat radiation from both the cake and the room. *I.e.*, only 20 % is lost from the cake, and similarly only 20 % of the room's radiation reaches the cake. The net effect of the reflecting walls is to reduce the emissivity of the cake to 1/5 of its original value.]

T.34 A puddle of water at temperature $T_p = 2$ C, with roughly equal top and bottom surface areas $A = 2\,\text{m}^2$, loses heat by conduction to the ground below and the air above. The effective temperature of the ground is $T_g = -3$ C and that of the air is $T_a = -13$ C. The top surface of the puddle also radiates thermal energy to the night sky, at an effective temperature of $T_s = -23$ C.

(a) The thermal resistance for heat transfer to the ground is about $0.2\,\text{m}^2\!\cdot\!\text{C/W}$. Compute the rate at which heat flows from the water in the puddle into the ground.

(b) The resistance for heat flow into the air is approximately $1\,\text{m}^2\!\cdot\!\text{C/W}$. Compute the rate at which heat flows from the water in the puddle into the air.

(c) Compute the rate at which the top surface of the puddle radiates heat. The surface of the puddle appears black, so take its emissivity to be $e \simeq 0.9$.

(d) Compute the rate at which the top surface of the puddle absorbs heat from the night sky.

(e) Compute the net rate of radiative heat flow from the puddle to the night sky.

(f) Comment.

T.35 Three slabs of material are arranged as in the figures below. The heavy line in Figure (a) represents a very thin layer of a material, *e.g.*, gold, with such high thermal conductivity that all points within the substance have essentially the same temperature at every instant. [The interface lying between A and B in (b) may also be thought to consist of such material, although here it is not so necessary.] The dashed lines represent thin fully-insulating layers through which no heat passes. Materials A and B have the same cross-sectional area, A, and length, L, and different thermal conductivities: $k_A = k$ and $k_B = 3\,k$. Material C has thermal conductivity $k_C = 2\,k$, area $2\,A$, and length $2\,L$. Compute the effective thermal conductivity of the slabs as arranged in Figure (a) and (b).

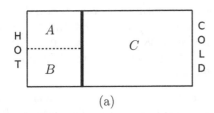

(a)

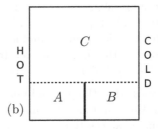

(b)

T.36 A large living room picture window with an area of $3.0\,\text{m}^2$ consists of a single pane of glass 5 mm thick. Suppose that the temperature inside the room is a comfortable 20 C, while outside it is a bracing -10 C.

(a) Compute the rate of conductive heat flow through the single pane of glass from inside the room to outside.

(b) Perhaps the temperature within the room is maintained at 20 C by means of an electric heater at a cost of 0.10 per kilowatt-hour [kWh]. To put the magnitude of the rate of heat loss determined in (a) into context, compute (i) the number of kWh of and

(ii) the dollar cost of replacing the heat which flows out the window in the course of one week during which the outside temperature averages $-10\,$C.

(c) A not-so-good response to the above is to cover the inside of the glass window with polyurethane foam to a thickness of 25 mm. (i) What is the temperature at the glass–foam interface when the inside and outside temperatures are $+20\,$C and $-10\,$C, respectively? (ii) What is the rate of conductive heat flow through the [now opaque] window? [This works to reduce heat flow and cost, but one's view suffers!]

(d) A better response is to replace the single-pane window with a double-pane system consisting of two 5 mm glass panes separated by 25 mm of dry air. Compute the rate of conductive heat flow through the window system (inner pane, air gap, and outer pane) from inside the room to outside. Comment.

T.37 A section of wall alongside the window in the previous problem has area $12.0\,\mathrm{m}^2$. The wall has effective thermal conductivity $k_w = 0.05\,\mathrm{W/(m{\cdot}K)}$, and is 15 cm thick. The temperature in the room is $+20\,$C and it is $-10\,$C outside.

(a) What is the rate of conductive heat flow from the room to outside through the wall?

(b) How much does it cost (at $0.10 per kilowatt-hour) to replace the heat which is lost through the wall in a week during which the outside temperature averages $-10\,$C?

T.38 A large living room picture window has an area of $3.0\,\mathrm{m}^2$. Suppose that the temperature inside the room is a comfortable 20 C, while outside [and all around] it is a bracing $-10\,$C. The window allows radiant energy to flow through it without impediment, so it may be assigned an emissivity of $e \equiv 1$.

(a) Compute the rate at which the room radiates heat out through the window.

(b) Compute the rate at which the outside radiates heat in through the window.

(c) Compute the net rate of heat loss from inside to outside.

(d) At $0.10 per kilowatt-hour, compute the energy-loss cost over the course of a week.

T.39 Many young healthy active people are most comfortable when maintaining a skin temperature of 33 ± 1 C and a body core temperature of 38 C.

(a) Assume that the change in temperature from the core value to that on the external surface of the skin occurs through 2 cm of tissue with an average thermal conductivity of $0.40\,\mathrm{W/(m{\cdot}K)}$. Take the person's total surface (skin) area to be $2\,\mathrm{m}^2$. (i) Compute the rate at which heat is conducted from core at 38 C to skin at 33 C. (ii) The heat conducted to the skin is emitted from the body. Estimate the total amount of thermal energy loss, in Calories, over a period of 24 hours.

(b) Suppose that the person in part (a) wears a fleece track suit [with hood and feet] that is 0.5 cm thick and has thermal conductivity $k_t = 0.05\,\mathrm{W/(m{\cdot}K)}$. The outer surface of the track suit is at the temperature of the surrounding air, T_{room}, which is assumed to be constant. Assume that all surfaces have area $2\,\mathrm{m}^2$ [neglect the increase in area associated with the thickness of the tracksuit] and compute the person's skin temperature when: (i) $T_{\mathrm{room}} = 20\,$C and (ii) $T_{\mathrm{room}} = -10\,$C.

(c) The person in part (b) dons a 3 cm thick full-body snowsuit with $k_s = 0.03\,\mathrm{W/(m{\cdot}K)}$. [The snowsuit is worn over the track suit.] The temperature of the outer surface of the snowsuit is fixed at T_{room}. Assume that all surfaces have area $2\,\mathrm{m}^2$ [again ignoring any increase in area] and compute the temperature of the person's skin under the following conditions: (i) $T_{\mathrm{room}} = 15\,$C and (ii) $T_{\mathrm{room}} = -54\,$C.

T.40 Twenty kilograms of granite constitute the "hot rock" part of a sauna. Granite (with quartz inclusions) has specific heat capacity $C_{\text{granite}} \approx 800\,\text{J}/(\text{kg·C})$ and emissivity $e_{\text{granite}} \approx 0.75$. The walls of the sauna are lined with red cedar. Red cedar has specific heat capacity $C_{\text{cedar}} \approx 1500\,\text{J}/(\text{kg·C})$, and emissivity $e_{\text{cedar}} \approx 0.33$. The effective surface area of the rocks is $0.5\,\text{m}^2$, while that of the interior of the sauna is $24\,\text{m}^2$. Outside, it is wintertime. [The granite is kept from contact with the walls so as to not set them on fire.]

(a) At initial time t_i, the rock is at an initial temperature of $T_{g,i} = 227\,\text{C}$, while the walls are at $T_w = 27\,\text{C}$. The wall temperature remains effectively constant over the time scale of this question. Compute the rates at which the rocks are   (i) radiating heat, (ii) absorbing heat from the walls, and   (iii) gaining or losing heat.

(b) Roughly how long does it take for the temperature of the granite rocks to reach 225 C? [HINT: Feel free to make reasonable approximations.]

(c) At the instant that the temperature of the granite reaches 225 C, 200 g of water at an initial temperature of 30 C are splashed upon the rocks. The water is very quickly [instantaneously] heated to 100 C, vaporises, and spreads throughout the sauna as steam at 100 C.

(i)   How much energy is absorbed by the water as it undergoes its transformation to steam?

(ii)   What is the temperature of the rocks immediately after the water boils away? [Neglect the effects of thermal radiation on this short time scale.]

T.41 Let us suppose that the temperatures inside your refrigerator and freezer are +5 C and −20 C respectively, while the kitchen is at +20 C. One day, because of a space crunch in the freezer, you shift the open container of baking soda in the freezer into the fridge. This inspires you to replace the open container of baking soda in the fridge with a new one from the cupboard.

(a) Each box of baking soda is at a temperature different from that of its immediate environment. Which one will experience the more rapid rate of temperature change?

(b) Suppose instead that the ambient temperature in the kitchen is 35 C. Does your answer to (a) change under this change of circumstance?

T.42
A solar-powered water heater consists of a light-gathering mirror and a light-absorbing tube full of cool water runnning along the mirror's focal line.   The mirror is made from a 1/6 section of a cylinder of radius 1.5 m. Sunlight strikes the mirror with intensity $700\,\text{W}/\text{m}^2$.

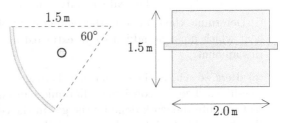

(a) Compute the energy flux (power) impinging upon the water-heater system.

(b) Suppose that the tube holds 20 kg of water, and that all of the power in (a) is redirected to it. [In operation, water would circulate through the tube.]   (i) Crudely estimate the rate at which the temperature of the water in the tube is increasing. [HINT: Assume that the water is originally at, or near, the ambient temperature, and neglect mechanisms of energy loss.]   (ii) Express this rate in C/min.

(c) What effect(s) might act to reduce the rate of increase of the water temperature?

T.43 Explain why quickly stretching a thick rubber band causes it to warm up, and why allowing a previously stretched band to snap back can lead it to feel cool.

T.44 Compute the net changes in the internal energy of a thermodynamic system under the following circumstances.

(a) 50 J of heat is added to the system and 25 J of work is done on the system.

(b) 50 J of heat is added to the system and 25 J of work is done by it.

(c) 50 J of heat is removed from the system and 25 J of work is done on it.

(d) 50 J of heat is removed from the system and 25 J of work is done by it.

T.45 Compute the change in the internal energy of a thermodynamic system under the following scenarios:

(a) 50 J of heat are added, while 40 J work are extracted,

(b) 50 J of heat are added, while 75 J work are extracted,

(c) 50 J of heat are removed and 40 J work extracted,

(d) 50 J of heat are removed and 75 J work input.

T.46 Compute the change in the internal energy of 0.25 kg of water when

(a) a liquid sample is heated from 10 c to 15 c

(b) a solid sample is warmed from -10 c to -5 c

(c) a solid sample at 0 c melts into liquid at 0 c

(d) a solid sample at -5 c is transformed into liquid at 5 c

[HINT: Obtain the necessary heat flows using the specific heat capacity approach realising that negligible amounts of work are done in these transformations.]

T.47 A fixed amount of gaseous nitrogen with $n R = 5$ J/K and mass 0.016846 kg is confined in a fancy box. At the temperatures and pressures considered here, the gas may be considered ideal. Initially, the gas occupies $\frac{1}{6}$ m^3 at a pressure of 9000 Pa. The gas is isobarically heated until it reaches a final volume of $\frac{1}{5}$ m^3.

(a) Determine the   (i) initial and   (ii) final   temperature of the gas.

(b) [Here we adopt a phenomenological approach, using specific heat capacity, to analyse this transition.] The specific heat capacity of nitrogen at constant pressure is $C_P \simeq 1039 \frac{\text{J}}{\text{kg·C}}$. (i) Determine the change in the temperature of the gas.   (ii) Compute the amount of heat which must be added to [or extracted from] the gas to change its temperature by this amount.

(c) [Here we subject the system to a thoroughgoing thermodynamical analysis of its isobaric transition.] The gas consists of diatomic molecules, thus $\gamma \simeq 5$, and hence $U = \frac{5}{2} n R T$. (i) Compute the work done by the gas in the course of this transition.   (ii) Determine the change in the internal energy of the gas.   (iii) Infer the quantity of heat added to [or removed from] the system.   (iv) Ascertain whether the entropy of the system increased or decreased.   (v) Ascertain the direction of change (increase or decrease) in the entropy of the box's environment.

(d) Comment on the results for (b) and (c).

T.48 A copy of the box of gas in T.47 maintains a constant volume of $\frac{1}{6}$ m^3. The initial pressure of the gas is again, 9000 Pa. Heat is added to the gas until it reaches a final temperature of 360 K.

(a) Determine the initial temperature of the gas in the box.

(b) [Here we adopt a phenomenological approach, using specific heat capacity, to analyse this transition.] The specific heat capacity of nitrogen at constant volume is $C_V \simeq 743.1 \, \frac{J}{kg \cdot C}$.
(i) Determine the change in the temperature of the gas.   (ii) Compute the amount of heat which must be added to [or extracted from] the gas to change its temperature by this amount.

(c) [Here we subject the system to a thoroughgoing thermodynamical analysis of its isochoric transition.] The gas consists of diatomic molecules, thus $\gamma \simeq 5$, and hence $U = \frac{5}{2} n R T$.
(i) Compute the final pressure of the gas.   (ii) Compute the work done by the gas.
(iii) Determine the change in the internal energy of the gas.   (iv) Infer the quantity of heat added to [or removed from] the system.   (iv) Ascertain whether the entropy of the system increased or decreased.   (v) Ascertain the direction of change (increase or decrease) in the entropy of the box's environment.

(d) Comment on the results for (b) and (c).

T.49 The box of gas in T.47 is adjusted so as to have initial pressure 8000 Pa and volume $1/4 \, m^3$, and is isobarically compressed to a final volume of $1/5 \, m^3$. Repeat T.47 in this case.

T.50 The box of gas in T.48 is adjusted so as to have initial pressure 8000 Pa and volume $1/4 \, m^3$, and is isochorically cooled to a final temperature of 320 K. Repeat T.48 in this case.

T.51 A particular fixed amount of gas is confined in a thermally isolated box in which the pressure, volume, and temperature are monitored and controlled. Compute the net amount of work done by the gas in the course of the following transitions.

(a) The pressure remains at $500 \, N/m^2$, while the volume increases from $2.0 \, m^3$ to $2.5 \, m^3$.

(b) The pressure is reduced from $500 \, N/m^2$ to $300 \, N/m^2$, while the volume remains unchanged at $2.5 \, m^3$.

(c) The pressure remains at $300 \, N/m^2$, while the volume decreases from $2.5 \, m^3$ to $0.5 \, m^3$.

T.52 A fixed amount of gas with $n R = 2 \, J/K$ is confined to a thermally isolated box in which the pressure, volume, and temperature are monitored and controlled. The gas is then subjected to a series of actions which together comprise a thermodynamic box cycle. The four state-corners of the box are at

$$\begin{Bmatrix} P_1 = 800 \\ V_1 = 1.25 \\ T_1 = 500 \end{Bmatrix}, \quad \begin{Bmatrix} P_2 = 800 \\ V_2 = 2.5 \\ T_2 = 1000 \end{Bmatrix}, \quad \begin{Bmatrix} P_3 = 400 \\ V_3 = 2.5 \\ T_3 = 500 \end{Bmatrix}, \quad \begin{Bmatrix} P_4 = 400 \\ V_4 = 1.25 \\ T_4 = 250 \end{Bmatrix},$$

where the pressures are quoted in Pa, the volumes in m^3, and temperatures in K. The cycle consists of two isobaric (denoted by $\longrightarrow$) and two isochoric ($\Longrightarrow$) transitions:

$$(1) \longrightarrow (2) \Longrightarrow (3) \longrightarrow (4) \Longrightarrow (1).$$

(a) Compute the work output from, and the heat input to, the system along each of the four steps in the cycle.

(b) Determine the net work and the net input of heat through an entire cycle.

(c) Compute the efficiency of this heat engine.

T.53 A fixed amount of ideal monatomic gas with $n R = 20 \, J/K$ is confined in a fancy box. The system undergoes the series of transitions displayed in the figure and chart below. All pressures, volumes, and temperatures are in Pa, m^3, and K, respectively.

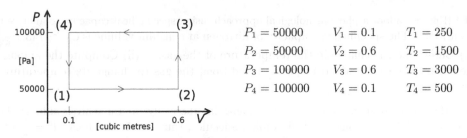

$P_1 = 50000$ $V_1 = 0.1$ $T_1 = 250$
$P_2 = 50000$ $V_2 = 0.6$ $T_2 = 1500$
$P_3 = 100000$ $V_3 = 0.6$ $T_3 = 3000$
$P_4 = 100000$ $V_4 = 0.1$ $T_4 = 500$

(a) Do the thermodynamic properties of the cycle depend on which of the states is considered the initial/final state?

(b) Do the thermodynamic properties of the cycle depend on whether the cycle is traversed in clockwise or anti-clockwise manner?

(c) Compute the mechanical work done by the gas as it undergoes the transitions:
(i) $(1) \to (2)$,   (ii) $(2) \to (3)$,   (iii) $(3) \to (4)$,   (iv) $(4) \to (1)$,
(v) $(1) \to (2) \to (3) \to (4) \to (1)$.

(d) By invoking the ideal gas law, the appropriate internal energy equation of state, and the First Law of Thermodynamics, ascertain the flows of heat into and out of the system during the transitions: (i) $(1) \to (2)$,   (ii) $(2) \to (3)$,   (iii) $(3) \to (4)$,   (iv) $(4) \to (1)$,   (v) $(1) \to (2) \to (3) \to (4) \to (1)$.

(e) Ascertain whether this thermodynamic cycle has an efficiency or a coefficient of performance, and compute its value.

T.54

A thermodynamic system consists of a fixed amount of monatomic gas confined in a box with initial volume V_0 and pressure P_0. The system undergoes two different 4-step cycles, which combine to form an 8-step grand cycle.

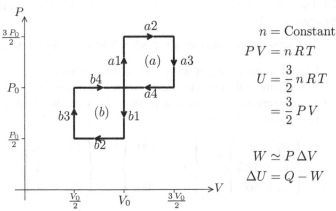

$n = $ Constant
$PV = nRT$
$U = \dfrac{3}{2} nRT$
$ = \dfrac{3}{2} PV$

$W \simeq P \Delta V$
$\Delta U = Q - W$

(a) Determine the flows of heat and work which occur throughout: (i) $a1$,   (ii) $a2$,   (iii) $a3$,   (iv) $a4$, and   (v) the entire a-cycle.   (vi) Determine whether this is a heat engine or pump and compute the efficiency or CofP accordingly.

(b) Determine the flows of heat and work which occur throughout: (i) $b1$,   (ii) $b2$,   (iii) $b3$,   (iv) $b4$, and   (v) the entire b-cycle.   (vi) Determine whether this is a heat engine or pump and compute the efficiency or CofP accordingly.

(c) (i) Ascertain the net flows of heat and work throughout the combined ab cycle.
(ii) Determine the overall efficiency or CofP, if possible.

T.55   Three six-sided dice are tossed. What is the likelihood that their sum is 4?
T.56   At a particular instant, a mug holding $350\,\mathrm{g}$ of coffee at temperature $77\,\mathrm{C}$ is in a room with ambient temperature $27\,\mathrm{C}$. During a brief subsequent interval, $0.5\,\mathrm{J}$ of heat flows from the coffee into the surrounding air.

(a) Verify that the temperature of the coffee is barely affected by this loss of heat. Take the specific heat capacity of coffee to be equal to that of water.

(b) Verify that the temperature of the air is barely affected by its thermal energy gain. Take the volume of the room to be $24\,\text{m}^3$ and the density of air to be $1.2\,\text{kg/m}^3$.

(c) Under the assumption that the temperature of the coffee remained (for all intents and purposes) constant, compute the change in entropy of the coffee.

(d) Under the assumption that the temperature of the air remained constant, compute the change in entropy of the air.

(e) Compute the net change in entropy of the system. Comment.

T.57   A Carnot heat pump draws a small amount of heat, Q, from a COLD reservoir at fixed temperature T_C and dumps it, along with the work required to operate the pump, into a HOT reservoir at [constant] temperature T_H.   [HINT: Remember that Carnot engines and heat pumps are reversible.]

(a) State the CofP of this Carnot heat pump.

(b) Determine the work required to pump Q from COLD to HOT.

(c) Determine the total amount of heat dumped into the HOT reservoir.

(d) Determine the change in entropy of the   (i) COLD and   (ii) HOT   reservoir.

(e) Determine the total, or net, change in entropy. Comment.

T.58   Discuss this paraphrase of the Laws of Thermodynamics. [It is often attributed to Richard Feynman (1918–1988, Nobel Prize 1965).]

RULE 0: You must play the game.

RULE 1: You cannot win the game; the best you can hope for is to break even.

RULE 2: You cannot break even unless you get out of the game.

RULE 3: You cannot leave the game.

T.59   Verify that the probabilities computed for the two-level system studied in Chapter 49 sum to 1. Note that this must occur irrespective of the energy split [intrinsic to the system] and the [reservoir's] temperature.

T.60   A two-level system with level-splitting ϵ is in contact with a thermal reservoir at temperature T. At what temperature is the system $1/e$ times as likely to be in its higher-energy state than its ground-state?

T.61   Consider a three-level system [akin to that studied in Chapter 49] in which the energies are: $E_n = E_0 + n^2\,\epsilon$, $n = \{0, 1, 2\}$.

(a) Express the partition function for this system.

(b) Compute the probability of the system's being in each of its three possible states.

(c) Compute the ensemble average energy by constructing the probability-weighted average of the level energies.

(d) Take $-\frac{\partial \ln[Z]}{\partial \beta}$. Comment.

T.62   [See also T.63.] Given the form of the Maxwell–Boltzmann speed distribution,

$$\mathcal{P}(v) = 4\,\pi \left(\frac{\beta\,m}{2\,\pi} \right)^{3/2} v^2\, e^{-\frac{\beta\,m}{2}\,v^2}\,,$$

verify that the modal speed is $v_{\text{modal}} = \left(\frac{2}{\beta\,m} \right)^{1/2}$.

[HINT: The modal speed corresponds to the peak in the distribution.]

T.63 [See also T.62.] Given the Maxwell–Boltzmann distribution for speeds derived in Chapter 50, verify by explicit calculation that average speed and RMS speed are $\frac{2}{\sqrt{\pi}}$ and $\sqrt{\frac{3}{2}}$ times the modal speed, respectively.

List of Symbols

(ij...k) Series composition of elastic materials or springs labelled i, j, ..., k

[ij...k] Parallel composition of elastic materials or springs labelled i, j, ..., k

α, $\tilde{\alpha}$, β Thermal expansion coefficients

β Reciprocal temperature in units of (energy)$^{-1}$

$\cos(\varphi)$ Power factor for a driven DHO system

$\Delta\omega$ Full width at half-maximum for a DHO; beat frequency

ΔE Increment in mechanical energy

ΔK Increment in kinetic energy

ΔL Change in length

ΔP_{av} A contribution to the average pressure in a fluid medium

ΔQ Incremental quantity of heat

ΔS Change in or increment of entropy

ΔT Change in temperature; temperature increment

Δt A time interval

ΔU Increment in potential or internal energy

ΔV Change in volume

η Viscosity

γ Damping parameter

$\hat{\imath}$, $\hat{\jmath}$, $\hat{k}$ Unit vectors in the x, y, and z Cartesian basis directions

κ Torsion fibre constant

λ Wavelength

λ_n, ν_n Wavelength and frequency associated with the nth harmonic in a standing wave

λ_{MAX}, ν_{MAX} Wavelength and frequency corresponding to peak blackbody radiative intensity

$\langle \mathcal{Q} \rangle$ Average value of $\mathcal{Q}$

μ Chemical potential

μ Lineal mass density, especially for a 1-d medium bearing transverse waves

ν Cycle frequency; wave frequency

ν Kinematic viscosity

Ω Number of accessible microstates; degeneracy

ω Angular speed; angular frequency

ω_0 Natural, or resonant, frequency

$\mathcal{T}$ Oscillation period; wave period

$\Phi_{m,\mathrm{S}}$ Mass flux through specified surface S

$\Phi_{V,\mathrm{S}}$ Volume flux through specified surface S

ψ Wave function; amplitude of the wave disturbance as a function of time and position

$\psi_R,\ \psi_L$ Rightmoving wave and leftmoving wave

$\rho;\ \rho_0$ Local mass density; constant or initial density

ρ_{av} Average mass density

ρ_{thermal} Thermal resistivity

σ Stefan–Boltzmann constant

τ Torque

θ An angle

$\theta(t)$ Time-dependent angle; Heaviside step function

θ_i Incident angle

θ_r Reflected angle; refracted angle

φ Phase angle

$\vec{F_{\mathrm{A}}};\ F_{\mathrm{A}}$ Applied force; its magnitude

$\vec{F_{\mathrm{B}}}$ Buoyant force

$\vec{F_{\mathrm{D}}}$ Linear drag force

$\vec{F_{\mathrm{S}}},\ F_{\mathrm{S}}$ Spring force; its magnitude

$\vec{J}$ Current density

$\vec{N}$ Normal force

$\vec{\nabla}$ Gradient operator

$\vec{W}$ Weight force

$\vec{a}\cdot\vec{b}$ The DOT PRODUCT, or scalar product, of vectors $\vec{a}$ and $\vec{b}$ [in n dimensions]

$\vec{a}\times\vec{b}$ The CROSS PRODUCT, or vector product, of three-dimensional vectors $\vec{a}$ and $\vec{b}$

$\vec{a};\ a$ Instantaneous acceleration; its magnitude

$\vec{g}$ The local gravitational field; acceleration due to gravity

$\vec{v};\ v$ Instantaneous velocity; its magnitude

$A;\ A_0$ Amplitude; constant or initial amplitude

$A;\ \mathcal{A}$ Area; cross-sectional or face area

B Bulk modulus

b_1 Linear drag coefficient

C Specific heat capacity

c Wave speed

$c(t,\vec{r})$ Local concentration of a substance [*i.e.*, number density]

C_{molar} Molar specific heat

E	Total mechanical energy
e	Efficiency of a heat engine
e	Emissivity
f	Focal distance
g	$\|\vec{g}\|$, magnitude of the acceleration due to gravity
h	A height; a vertical distance
h	Planck's constant
H, A, G	Thermodynamic potentials: Enthalpy, and Helmholtz and Gibbs free energies
I	Intensity
i	Image distance
K	Kinetic energy: some authors use "T"
k	Hookian spring constant
k	Thermal conductivity
k	Wave number
k_B	Boltzmann's constant
$L; L_0$	Length; original length
L_F, L_V	Latent heat of fusion, of vaporisation
M	Magnification
m, M	Mass
M_S	Shear modulus
n, N	Number of moles or molecules of a substance
N_A	Avogadro's number
p	Object distance
$P; \langle P \rangle$	Mechanical power; its average value throughout one oscillation period [equivalently, a long time]
$P; P_0$	Local fluid pressure; a specific or reference pressure
P_{av}	Average fluid pressure
P_{atm}	Standard atmospheric pressure
Q	Quality factor for a DHO
Q	Quantity of heat present in, or added to, a thermodynamic system
R	Universal gas constant
R, r	Radius; radial parameter; moment arm
$R_{thermal}, R$	Thermal resistance
Re	Reynolds number
S	Entropy
t	Time parameter
$T; T_0$	Temperature; initial, reference or fixed temperature

U Potential energy associated with a conservative force; internal energy of a thermo-dynamic system

V; V_0 Volume; original volume

v_0, V_0 Initial or reference velocity components

W Amount of mechanical work, work performed by a thermodynamic system

$W_{if}\left[\vec{F}\right]_\gamma$ Mechanical work: general form

x, y, z 1-d position; Cartesian component of position

x_0, X_0 Initial or reference position components

Y Young's modulus

Z Impedance

Z Partition function

$\mathcal{D}(c,\vec{r})$ Diffusion coefficient

$\mathcal{P}$ Probability of a given macrostate

$\mathcal{P}(X)$ Probability density dependent upon X

$\mathcal{P}_s$ Probability (relative probability) of finding a system in state s among an ensemble of equivalent systems

a   Constant (instantaneous) acceleration

DHO   Damped harmonic oscillation; a damped harmonic oscillator system

FBD   Free body diagram

LHS, RHS Left and right hand sides of a mathematical expression

RMS   Root-Mean-Square

WRT   With respect to

SHO   Simple harmonic oscillation; a simple harmonic oscillator system

sys, res System and reservoir distinguishers

WLOG Without loss of generality

S.G.   Specific gravity

CofP   Coefficient of performance for a heat pump

Index